Chemistry of Graphene - Synthesis, Reactivity, Applications and Toxicities

Edited by Enos W. Wambu

Published in London, United Kingdom

Chemistry of Graphene - Synthesis, Reactivity, Applications and Toxicities
http://dx.doi.org/10.5772/intechopen.111179
Edited by Enos W. Wambu

Contributors
Aafreen, Ajit D. Kelkar, Carmen Ioana Fort, Chuanlei Qi, Enos W. Wambu, Faycal Znidi, Haris Saeed, Jiaran Wang, Liviu Cosmin Cotet, Lucian Baia, Lucian Cristian Pop, Luhai Wang, Marzieh Azizi, Md. Aliahsan Bappy, Md. Nizam Uddin, Md Fazle Rab, Meena Bhandari, Mohamed Morsy, Mohammad Mahdi Sepahi, Monica Baia, Neeraj Kumari, Priyanka Verma, Shengping Li, Vishwas Jadhav, Yindong Liu, Yuting Cao

First published in London, United Kingdom, 2024 by IntechOpen
IntechOpen is the global imprint of INTECHOPEN LIMITED, registered in England and Wales, registration number: 11086078, 167-169 Great Portland Street, London, W1W 5PF, United Kingdom

British Library Cataloguing-in-Publication Data
A catalogue record for this book is available from the British Library

Additional hard and PDF copies can be obtained from orders@intechopen.com

Chemistry of Graphene - Synthesis, Reactivity, Applications and Toxicities
Edited by Enos W. Wambu
p. cm.

This title is part of the Nanotechnology and Nanomaterials Book Series, Volume 4
Topic: Nanomaterials and Nanostructures
Series Editor: Jung Huang
Topic Editor: Alberto Jiménez-Suárez
Associate Topic Editor: Gonzalo Seisdedos

Print ISBN 978-1-83769-282-8
Online ISBN 978-1-83769-283-5
eBook (PDF) ISBN 978-1-83769-284-2
ISSN 3029-0538

For EU product safety concerns:
IN TECH d.o.o., Prolaz Marije Krucifikse Kozulić 3, 51000 Rijeka, Croatia,
info@intechopen.com or visit our website at intechopen.com.

IntechOpen Book Series
Nanotechnology and Nanomaterials
Volume 4

Aims and Scope of the Series

Humans face growing challenges in environmental sustainability. We need better materials, more effective devices, and higher computing power to conquer these challenges. Researchers have long explored the realm of science at the nanometer scale for developing novel problem-solving technologies. In the past decades, breakthroughs in multiscale computer simulation, novel hierarchical material and device structures, nanometrology, and new functionalities from interfacial and quantum size effects have enhanced our ability to utilize nanotechnology to solve real-world problems. This book series addresses important advancements in nanotechnology and nanomaterials. It showcases how nanotechnology is being continually developed and implemented in a variety of domains of science and technology. The two main topics this series covers are Nanotechnology and Nanodevices, and Nanomaterials and Nanostructures.

Meet the Series Editor

Jung Y. Huang, a university educator and researcher, has been working to unravel the structures and functional properties of materials and cellular events in living cells with varying optical methodologies. He has co-authored hundreds of journal papers and five book chapters. He also holds tens of patents in laser techniques and single-molecule/hyperspectral imaging and has developed architectural photonics based on hierarchically structured materials. Currently, his research focuses on the integration of artificial intelligence methodology with optics to automatically discover meaningful information from optical sensing/imaging data cubes. He has been an editorial board member and reviewer for several scientific journals. As a member of the global scientific community, he sincerely supports and endeavors to promote the spread of scientific knowledge.

Meet the Volume Editor

Dr. Enos Wamalwa Wambu currently heads the Department of Chemistry and Biochemistry at the University of Eldoret in Kenya, where he has taught physical and applied chemistry since 2014. He holds a Ph.D. in chemistry from Kenyatta University, Nairobi, Kenya. He has previously taught chemistry at Jaramogi Oginga Odinga University of Science and Technology (2011–2014). Until then, he had been a long-serving and highly decorated teacher of high school mathematics and chemistry serving in some of the most prestigious educational institutions in Kenya. With more than 18 years of experience in research, Dr. Wambu has distinguished himself as a researcher and an extensively published author in the area of physical applied chemistry and materials science.

Contents

Preface

The existence of "single hexagonal layered" graphene was first conjectured by scientists during the late 1940s [1] However, it was not until the early 1960s that the material was formally reported [2]. Even so, it remained essentially unknown until it was successfully isolated and investigated at the University of Manchester in 2004 [3, 4]. The successful preparation of graphene then triggered massive interest across the world, and soon scientists from around the globe were reporting dozens of variant improved methods for its production. The material gained traction, 4 and by 2014, just a decade after it was isolated, it was reported that over 14,000 scientific papers had been published on this subject. Later in 2022, the global graphene annual revenue was estimated at US$380 million with projections to reach US$1.5 billion in the year 2027 [5].

As graphene continues to attract interest, a growing number of researchers have been inspired to write a rising number of review papers to improve the understanding of its science. The current book was designed to contribute to this knowledge development by consolidating and curating existing knowledge on this important subject. The topics selected and included in *Chemistry of Graphene – Synthesis, Reactivity, Applications, and Toxicities* were aimed to provide readers with a unique resource for promoting our understanding of the new trends, concepts, and phenomena in the field of science that have captivated material scientists for a period of three decades since 2004 like no other. In compiling the literature, attempts have been made toward conducting comprehensive searches for present readers with complete syntheses of the research status and the emerging trends in this field. The goal was to submit a succinct overview of the research status and to point out the key research knowledge gaps and opportunities that are yet to be fulfilled.

The book discusses implications for both academia and practice in a multidimensional manner. The first is to serve as a reference on multiple topics allowing for seamless integration of new ideas into the existing knowledge base of the field while providing new insights into the rapidly changing research areas. Second, the longitudinal approach adopted permits for the materials dealt with in the book to transcend the important phases in the development of graphene science right from the laboratory and theoretical research to technology innovations, applications, and commercialization in a manner that stimulates contribution to lifelong learning.

The individual portions of the book ought to be viewed as representative cases of important concepts in graphene science. We endeavored to offer a deeper exploration of ideas and concepts laying firm foundations for creating new research ideas. We sought a simple approach to the materials to facilitate easier readership and digestion by students and early career scholars to provide definitive foundations for research careers across the representative fields while also hoping to stimulate new research topics, ideas, debates, and critical insights among the readership that is more experienced in research and related practice in this area.

Enos W. Wambu
Department of Chemistry and Biochemistry,
University of Eldoret,
Eldoret, Kenya

References

[1] Wallace PR. The band theory of graphite. Physical Review. 1947;**71**(9):622-634

[2] Boehh VH, Clauss A, Fischer G. Das Adsorptionsverhalten—sehr dunner Ko hlenstoff Folien. Zeitschrift für Anorg und Allg Chemie. 1962;**316**(3-4):117-250

[3] Novoselov KS, Geim AK, Morozov SV, Jiang D, Zhang Y, Dubonos SV, et al. Electric field effect in atomically thin carbon films science. Science (80–). 2004;**306**(5696):666-669

[4] Page S, Receives A, Led P, Grant NSF, Byer R, Ke WR. APS news. APS News. 2009;**18**(9):2

[5] Graphene Flagship. Article. Global graphene market will continue to grow, researchers find. 2024. Available from: https://graphene-flagship.eu/materials/news/global-graphene-market-will-continue-to-grow-researchers-find/#:~:text=The global graphene market has, 1.5 billion US%24 in 2027

Section 1

Current Trends in Synthesis

Chapter 1

Recent Advances in the Synthesis of Graphene and Its Derivative Materials

Aafreen, Priyanka Verma and Haris Saeed

Abstract

Graphene (G) is a 2D material of sp^2 hybridized carbon atoms, discovered by Geim and Novoselov in 2004. The material presents a wide range of peculiar electronics and thermal, optical, mechanical, structural, and surface properties, which have attracted considerable interest from researchers and technologists. The conevntional techniques for graphenization have presented some drawbacks including low yield, costliness, high chances of contamination, and their time-consuming nature. These limitations have led to proliferation of research, which has led to the discovery of more advanced techniques for G synthesis over the years. At the moment, trending approaches to G production include chemical vapor disposition (CVD), epitaxial growth on silicon carbide (SiC), G oxide reduction, chemical synthesis, electrochemical synthesis, and laser-induced graphenization. There is a growing demand to produce G in large quantities and good quality. Nonetheless, because the conventional techniques have presented significant difficulties and imperfections in large-scale G production, various investigations have been conducted to identify new techniques for manufacturing cost-effective, large-scale, and high-quality G with novel applications such as energy storage, sensors, drug delivery, and biomedical devices. Each technique can be used for certain applications and has its own advantages. This chapter deals with the two approaches (top-down and bottom-up) for the synthesis of G and their procedure, limitations, and applications.

Keywords: biosensors, carbon nanotubes, exfoliation, graphene, graphene oxide, graphene synthesis

1. Introduction

In 1960, researchers Ubbelohde and Lewis successfully isolated a monolayer of graphite and determined that graphite is composed of layers, consisting of interconnected hexagonal carbon atom rings [1]. However, as stated by Mouras et al. in 1987, the term "graphene" was initially coined to refer to a solitary sheet of graphite during that time. Graphite serves as the foundational element in graphitic materials such as graphite itself, fullerene, and carbon nanotubes [2]. Although G was discovered to be in existance by Benjamin Collins Brodie in 1859 [3], since its first discovery in 1962 [4] which was observed in an electron microscope, Wallace has spent many years studying it theoretically [5, 6]. However, the properties of the graphenic material were only attained in 2004 when Novoselov and Geim successfully isolated and studied a single-atom-thick crystallite (G) from bulk graphite and transferred them onto thin silicon dioxide on a silicon wafer using the well-known scotch tape technique. Notably, the pioneering method of producing the first G involved micromechanical cleavage of graphite [7]. Konstantin Novoselov and Andre Geim shared the 2010 Nobel Prize in Physics for their ground-breaking work on this two-dimensional material as a result of this feat [8, 9]. The popular nanomaterial G is currently taking the place of silicon in a variety of scientific disciplines. This is because of their nanoscale mechanical, chemical, thermal, and physical characteristics, and due to its potential, G is attracting sponsors and significant donations [10].

The exploration of various commercial applications for these materials is currently underway with a particular emphasis on a range of fields, which include sensors, energy generation, and energy storage devices, which represent some of the fastest growing domains of technology [11–13]. Additionally, the realm of 2D-layered materials exhibits an extensive array of crystalline structures, corresponding to a diverse spectrum of physical properties [1, 14].

The atoms of these bulk layered materials are weakly connected by inter-layer van der Waals forces, but they are strongly bound to one another within the same plane [14]. The creation of 2D-layered nanoscale materials from these bound sheets through chemical or physical interactions opens up interesting possibilities for novel devices that are distinct from those made from conventional bulk materials [15, 16]. Plus, due to the wide range of shapes and sizes that a single graphitic layer can take, G is sometimes referred to as the "mother" of all graphitic-based nanostructures. It can be folded into one-dimensional (1D) carbon nanotubes (CNTs), wrapped into a zero-dimensional (0D) "buckyball" structure, or layered into multiple-layer 3D G sheets [17–22].

2. Graphene synthesis

There are numerous ways by which G is produced. The conventional techniques include (1) mechanical exfoliation, (2) chemical vapor deposition, and (3) epitaxial growth. At the moment, the trending G production techniques are electrochemical exfoliation, laser-induced graphenization, hydrothermal synthesis, and microwave-assisted synthesis. This chapter provides a summary of conventional and trending techniques that are used for the synthesis of G. Nonotheless, G needs to be produced at a price that is comparable to or lower than that of existing materials in order to be successfully applied in industry. The development of G synthesis processes with the qualities of high product yield, high product quality, cost-effectiveness, and scalable production is, however, difficult. As a result, these properties are the main subject of the discussion in the subsequent sections but G's full production and characterization mechanisms have been documented elsewhere in the literature [22].

G must be produced at rates that are compatible with existing materials in order to be utilized by various industries [23, 24]. The development of the production processes that deliver high yields and high-quality products while being cost-effective, reliable, and scalable remains to be a big challenge [25]. As a result, these attributes are the main focus of the following discussion on techniques of G synthesis [26]. Depending on the desired outcome and purity, different methods have been developed to create layers and thin films of G. The schematic representation of the bottom-up (construction) and top-down (destruction) mechanism is depicted in **Figure 1**.

DOI: http://dx.doi.org/10.5772/intechopen.114280

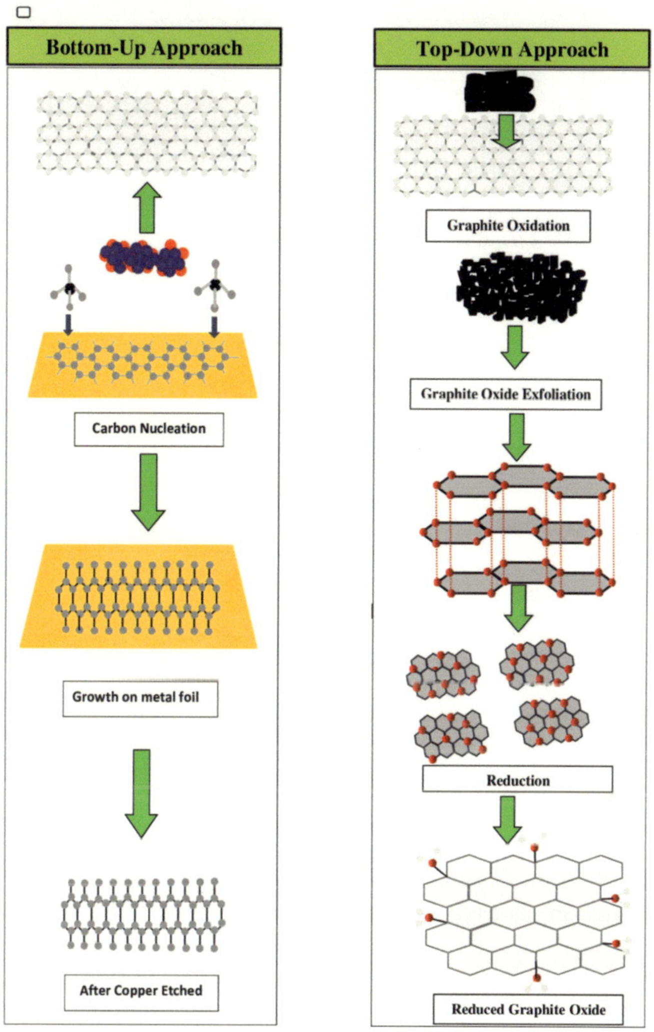

Figure 1.
Synthesis of G from top-down and bottom-up approaches.

Thus, a bottom-up process is the one in which the initial raw materials are found in a smaller entity and are then transferred to a larger entity or sustrate. Chemical vapor deposition (CVD) is an example of a bottom-up strategy for synthesizing G. Because it is more likely to result in fewer flaws, a more homogeneous chemical composition, and better short- and long-range ordering of G, the bottom-up strategy is preferable to the top-down approach. On the other hand, a top-down strategy begins with large or small-scale structures and reduces the size using a variety of techniques. Furthermore, it is believed that the planes or layers (the building blocks) necessary to create the desired nanostructure are taken out of the substrate during the synthesis process. For better control of G synthesis, the top-down method is typically used. The graphite exfoliation in the top-down techniques can be achieved by both mechanically and chemically in the top-down approach [27, 28].

The conventional top-down and bottom-up techniques used for G synthesis are displayed in **Figure 2**.

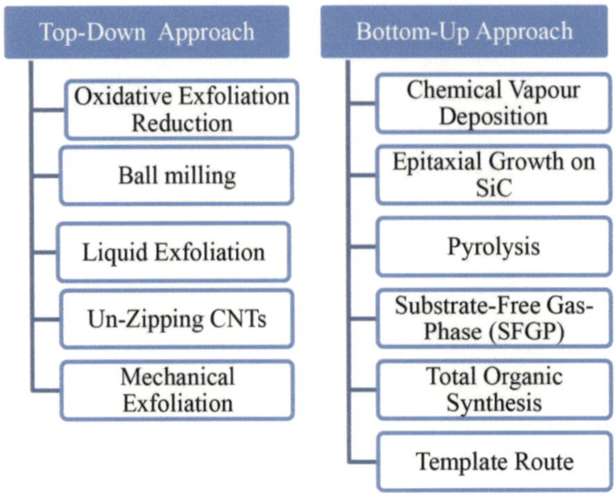

Figure 2.
Conventional G synthesis approaches.

2.1 Top-down approaches

The top-down approach is defined as a tactic that concentrates on the attack of powdered raw graphite. Eventually, the attack will split its layers and create G sheets. Chemical synthesis and mechanical or chemical exfoliation are some of the commonly used top-down approaches for G synthesis [29] and are discussed below.

2.1.1 Mechanical exfoliation

Multiple single-atomic layers of G make up the structure of graphite. Weak van der Waals forces accumulate and hold everything together. The interlayer spacing between each layer is 3.3 Å (0.33 nm), and the interbond energy is 2 eV/nm^2 ($eVnm^{-2}$). Mechanical exfoliation is the rarest and earliest recognized technique for extracting G flakes from a graphitic substrat. This is a top-down approach in nanotechnology in which longitudinal or transverse stress is applied to the graphite surface using a cheap adhesive tape (scotch tape) or AFM (atomic force microscopy) tip. Mechanically cleaving graphitic materials such as highly oriented pyrolytic graphite (HOPG), single-crystal graphite, or natural graphite, can produce single-layer graphene (SLG) to few-layer graphene (FLG) by slicing down the layers [30, 31]. A single G layer may be separated from graphite using this technique by applying an external strain of approximately 300 $nN/\mu m^2$ [32]. A variety of methods such as scotch tape, use of an electric field [33], and ultrasonication [34] or even by transfer printing techniques [35, 36] can be used to exfoliate G from graphite. To increase the yield of SLG and FLG flakes, the HOPG has occasionally been bonded to the substrate using a common adhesives like epoxy resin [34, 37]. A recent study has demonstrated how gold films can be used to transfer print macroscopic G designs from patterned HOPG [38]. It is undoubtedly the least expensive way to make high-quality G. G is exfoliated from a graphite crystal by adhesive tape in this micromechanical process. Multiple-layer graphene that is still on the tape after being removed from the graphite is separated into numerous flakes of few-layer G through repeated peeling. In order to detach the tape, it is then bonded to the acetone substrate, and the last peeling is done with an unused tape. The resulting flakes have varied sizes and thickness, with diameters of a single-layer G based on wafer ranging from nanometers to many tens of micrometers. Due to interference effects, single-layer G on SiO_2/Si can still be seen under a light microscope despite having an absorption rate of 2% [39]. Optical microscopy, Raman spectroscopy, and AFM are typically used to characterize G flakes produced by mechanical exfoliation techniques.

Despite the absence of sustainable flakes, it is actually difficult to produce high quantity of G with this exfoliation technique. Although this process has very little difficulty, it requires a lot of labor to locate the G flakes on the substrate surface. The prepared G has nearly no defects and a very high quality, which is suited for the manufacture of FET (field effect transistor) devices. However, for large-scale, defect-free, high-purity G for mass manufacturing in the field of nanotechnology, the mechanical exfoliation method still needs to be improved.

2.1.2 Chemical exfoliation

One of the finest methods for producing G is chemical synthesis. The first chemical production of G oxide occurred in 1860 using the Brodie method [40]. This was followed by the development of the Hummers [41] and Staudenmaier [42] processes. G is modified from graphite and a graphite intercalation compound using a chemical processs that produces colloidal suspension. Chemical exfoliation involves two steps; it increases the interlayer separation by first reducing the van der Waals forces between the graphite layers by creating G intercalated compounds (GICs) [43, 44]. Then, fast heating or sonication are used to exfoliate single to few layers of G. Ultrasonification is used to produce single-layer G oxide (SGO) [2, 42, 45–48], while density gradient ultracentrifugation is used to create different layer thicknesses [49, 50]. By using the Hummers process, which involves oxidizing graphite using potent oxidizing agents like $KMnO_4$ and $NaNO_3$ in H_2SO_4/H_3PO_4, it is simple to produce G oxide (GO) [42, 51]. SLG was produced via ultrasonication in a DMF/water (9:1) combination. Interlayer spacing therefore increases from 3.7 to 9.5 A. High density of functional groups make oxidation necessary, and reduction is required to get G-like characteristics. Chemical reduction with hydrazine monohydrate is used to scatter single-layer G sheets [2, 47].

Elsewhere, G has been produced using polycyclic aromatic hydrocarbons (PAHs) [51–53] by utilizing a precursor to the dendrict that has undergone cyclodehydrogenation and planarization [54] to create tiny G domains. According to these authors, small domains of G can be produced by using this approach, but larger flakes are produced by the precursor of poly-dispersed hyper-branched polyphenylene. The first one was synthesized via oxidative cyclodehydrogenation with $FeCl_3$. Orthodichlorobenzene [55], perfluorinated aromatic solvents [56], and even low-boiling solvents like chloroform and isopropanol [57, 58] are employed to disperse G. G on SiO_2/Si substrates exhibits electrostatic interaction between HOPG and the Si substrate [59]. HOPG has also been subjected to pulsed neodymium-doped yttrium aluminum garnet (Nd:YAG) laser exfoliation in order to create FG [60, 61]. Good-quality G, also known as reduced G oxide (rGO), is produced via reduction of graphite oxide (GO). Fabrication of energy storage materials [62], polymer composites [63], transparent conductive electrodes [64], and several types of paper [40], among others, have already employed the chemical technique to produce G.

2.1.3 Oxidative exfoliation reduction

The majority of GO is created through the reduction of reduced G oxide (rGO) or from G sheets after oxidative graphite exfoliation. Staudenmaier, Brodie, Hummers, and Hofmann are the most common methods for producing GO [27].

2.1.3.1 Brodie's method

The first person to use $KClO_3$ in severely fuming HNO_3 to produce a novel compound containing carbon, oxygen, and hydrogen was Benjamin Collins Brodie, at the University of Oxford, London. The quality of flake G has improved as a result of this innovative approach. The batch was washed to remove any salts that had formed throughout the process, dehydrated at 100°C, and then put back into the oxidizing atmosphere. A substance with a "light yellow color" that did not change with additional oxidation treatment was produced after three successive attempts of that method. He provided the final molecular composition of the oxidized graphite as $C_{11}H_4O_5$, in accordance with the elemental analysis of his product. In addition, he discovered that the chemical precipitated in acidic surroundings but dispersed in basic or pure water, after which he came up with the name "graphic acid" for the newly synthesized substance. The shift in the material's C:H:O composition to 80.13:0.58:19.29 was brought about by the reduction in carbonic acid and carbonic oxide after heating at 220°C. However, despite the fact that this method can oxidize graphite, its application is constrained by the lengthy response time and dangerous toxic gas emissions [41, 65].

2.1.3.2 Staudenmaier's method

Staudenmaier from Freising university, München, Germany, used enormous quantities of sulfuric acid and an excess of oxidizing chemicals to enhance Brodie's method. Numerous aliquots of concentrated sulfuric acid (H_2SO_4) were added, and throughout the procedure, potassium chlorate solution was added to the reaction mixture in order to improve Brodie's $KClO_3$-fuming HNO_3 formulation's ability to oxidize. These modifications made it possible to produce a highly oxidized GO product in a single reaction vessel, greatly simplifying the GO synthesis process. Although Staudenmaeir's approach was improved, Brodie's method's shortcomings remained, making the oxidation process time-consuming and dangerous. When potassium chlorate is introduced, it can last for more than a week, and when chlorine dioxide is removed from the inert gas, deadly gas explosions can occur continuously [43, 65].

2.1.3.3 Hofmann's method

To improve the work of both Brodie and Staudenmaier, Hofmann et al. used concentrated sulfuric acid and concentrated nitric acid and $KClO_3$ to synthesize GO in place of fuming nitric acid because fuming HNO_3 is exceedingly deadly and dangerous. In addition to acting as an in situ source of dioxygen in acid solutions, $KClO_3$ demonstrated a significant degree of oxidation capacity as the principal oxidant. So far, numerous research teams have used this technique to effectively manufacture GO [65].

2.1.3.4 Hummer's and modified Hummer's method

So, a different method for producing GO was developed by chemists Hummers and Offeman at the Mellon Institution of Industrial Research around 60 years after Staudenmaier's approach. A concentrated sulfuric acid, sodium nitrate, and potassium permanganate solution was prepared and held at temperatures under 45°C to oxidize graphite. First, 100 g of graphite powder and 50 g of sodium nitrate were added to 2.3 L of sulfuric acid to chill it to 0°C in an ice bath. After then, the suspension received 300 g of potassium permanganate gradually. The suspension's internal temperature was raised to 35° when the ice bath was removed, and it was maintained there for 30 min. The combination turned brownish-gray and pasty, and barely any gas had emerged after 20 minutes. After 30 minutes, 4.6 L of water was gradually added to the paste, causing it to fizz vigorously and increase in temperature to 98°C. When this reaction was allowed to continue for 15 minutes at this temperature, the diluted sample turned brown. In order to convert the leftover permanganate and manganese dioxide into colorless soluble manganese sulfate, 3% H_2O_2 was added after the operation. To remove the soluble salt of mellitic acid, the diluted solution was filtered and constantly washed with warm water. Centrifugation was used initially to obtain the dry form of GO and then dehydration at 40°C over phosphorous pentoxide under vacuum.

2.1.3.5 Recent advances in G and G oxide synthesis

Figure 3 illustrates the response patterns of these low-temperature operation G production approaches; normally conducted at lowest possible reaction temperatures to keep production costs down. However, most of these methods except Hummers method result in the production of toxic gases such as nitrogen dioxide (NO_2) and dinitrogen tetroxide (N_2O_4) [29] raising public health and environmental safety concerns. The Hummers technique is presently extensively employed for the synthesis of GO due to its notable efficiency and safety. Additionally, it does not produce dangerous gases like ClO_2 (chlorine dioxide) or acidic fog as it uses sodium nitrate ($NaNO_3$)

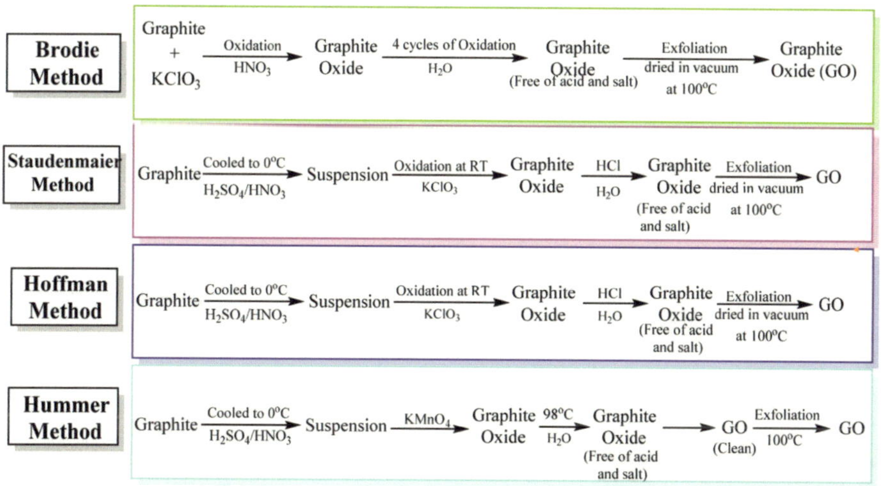

Figure 3.
Various routes of G oxidation.

and potassium permanganate ($KMnO_4$) rather than nitric acid (HNO_3) and potassium perchlorate ($KClO_4$). Overtime, improvements to the Hummers method have led to a more environment-friendly method of producing GO.

However, the G research "gold-rush" started in 2004, and GO also became a prominent material. Numerous articles on its preparation, reduction, and structure have been written. In 2010, Marcano et al. revealed an improved method for making GO. They reduced the amount of $NaNO_3$ and raised the amount of $KMnO_4$ in this new procedure. They also introduced H_3PO_4 to the reaction container in the place of the original acid. This improved method prevents a significant exotherm and produces no toxic gas. GO produced using this approach has a higher yield and degree of oxidation than GO produced using Hummers' process. They also found that the new method affects the graphite's basal plane less than Hummers' approach [42, 65].

The addition of graphite intercalation components in the oxidation process of graphite leads to a notable enhancement in the separation between graphite layers. Consequently, this enhancement may facilitate the formation of well-dispersed single-, bi-, and few-layer G oxide (GO) structures in suitable solvents. Tetrahydrofuran (THF), water, and N-methyl-2 pyrrolidone (NMP) are some of the fluids in which GO can disperse due to oxygen-containing functional groups including carboxylic, hydroxyl, and carbonyl groups. Electrochemical, thermal, and chemical reduction is then employed to eliminate the functional groups and reconstruct the honeycomb lattice of GO as its sp^2 bonding is broken. In the chemical reduction techniques, hydrazine (N_2H_4) is typically used to lower the oxygen content of GO. However, N_2H_4 use is restricted due to its high cost and dangerous nature. As a result, a safer alternative to the dangerous N_2H_4 used in GO reduction was developed. Among the reducing agents used were proteins, microorganisms, plant extracts, amino acids, metal-alkaline, metal-acid, reagents (nitrogen, sulfur, oxygen), hydrohalic acid, aluminum hydrides, borohydrides, and hormones.

Nonetheless, even though numerous GO reduction techniques are mentioned in the literature [66] including electrochemical reducation, plasma therapy, sonochemical, photocatalytic, laser, photothermal, and microwave technology, the electrochemical GO reduction has gained more prominence because of its economic viability, quick reduction, and simplicity of implementation. More importantly, it uses safer reductants than chemical reduction does.

In general, a reasonable cost and good yield are associated with the production of GO through the oxidative exfoliation of graphite and subsequent reduction to rGO. However, it is found that the product from this approach has a small surface area, lower solubility, and weaker electrical conductivity because of van der Waals attraction, and it is liable to irreversible sheet restacking [67, 68].

Furthermore, it is still not possible to fully reduce GO to produce pure G [69]. Nonetheless, although it has some imperfections and different sizes from pure G, the resulting rGO is pretty comparable to it. Greater rGO quality is produced via high deoxygenation, which is caused by a high C:O ratio.

2.1.4 Liquid phase exfoliation

Liquid phase exfoliation or LPE has traditionally involved two basic methods for exfoliating graphite: (1) cavitation in sonication, and (2) shear forces in high-shear mixers [68]. Recently, it has been discovered that a microfluidizer may effectively exfoliate graphite in appropriate aqueous solutions at high shear rates [70]. LPE is an easy-to-use high-shear mixing or sonication instrument that is widely accessible.

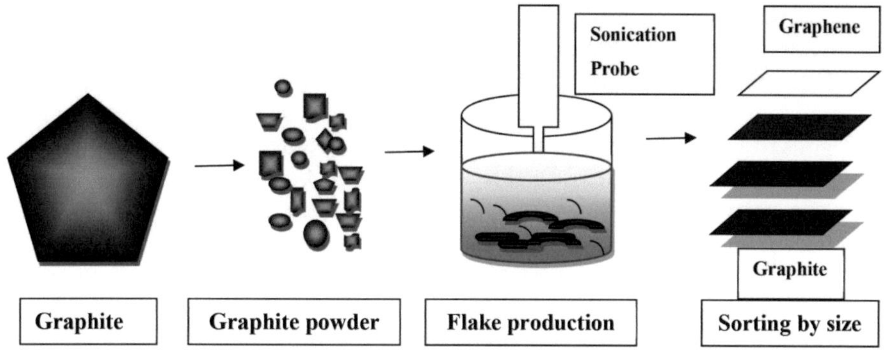

Figure 4.
Liquid exfoliation.

Additionally, LPE does not need a vacuum or high-temperature systems to operate because its working conditions are mild. However, the low amount of G and the significant energy consumption during the fabrication process has limited the widespread use of sonication-assisted LPE as shown in **Figure 4**. The microfluidizer or high-shear mixing is a developing LPE technique, which successfully exfoliates graphite using fluid dynamics powered by a high-shear mixer [71, 72].

2.1.4.1 Sonication

At high concentrations, sonication is an effective exfoliation method that has the potential to produce monolayer- or few-layer G. By creating cavitation bubbles, sonication is frequently used to effect physical or chemical changes in a variety of systems [73]. Compressions and rarefactions impose high and low pressure, pushing and pulling molecules, as ultrasonic waves move across the medium. With each cycle of rarefaction, microbubbles get larger until they reach an unstable state and burst, producing enormous shockwaves [74].

Sonication is classified into two types, bath sonication (BS) and tip sonication (TS), which have been used together or separately to generate monolayer- or few-layer G sheets by exfoliating G [75]. Sonication-assisted LPE typically consists of three phases [76]: (1) production of graphite dispersion in a particular solvent; (2) exfoliation of graphite dispersion using sonication; and (3) G purification. The cavitation-induced pressure pulsations are responsible for the formation and collapse of microbubbles in liquids during sonication. The cavitation action produces high-speed microjets and shockwaves, which create normal and shear stresses on graphite [77], which are important in the exfoliation of graphite to produce G [75]. The exfoliation impact is generally affected by the sonication power, the liquid medium, and the centrifugation rate utilized to disperse the G nanosheets [78]. The appropriate liquid media are chosen to establish an environment that allows for stable G dispersions during sonication, and centrifugation is used in removing big and unevenly distributed graphite particles or aggregates.

2.1.4.2 High-shear mixing

Until recently, the exfoliation of graphite by high-shear mixing has not been well researched; in particular, few investigations have focused on high-shear mixing in aqueous systems utilizing ionic- or non-ionic surfactants. Nonetheless, it has been demonstrated that G may be exfoliated from graphite using shear force in a suitable liquid [79]. Furthermore, strong shear pressures applied by high-shear mixers are considered to be scalable techniques of graphite exfoliation [80]. Shear exfoliation is analogous to sonication exfoliation in that aqueous liquids may be utilized to aid graphite exfoliation and generate stable G dispersions, hence removing the need for use of toxic organic liquids. It is critical to develop industrially scalable ways for producing high-quality G using novel exfoliation processes such as sonication. Coleman and co-workers achieved substantial success in G production by shear exfoliation in 2014 [80], which fueled the great growth of the shear exfoliation process. They proved that high-shear mixing of graphite in suitable solvents might produce high-concentrated G nanosheet dispersions. The resulting G flakes had not been oxidized and had no basal-plane flaws. Notably, when the local shear rate surpassed 104 s^{-1}, graphite was exfoliated in both laminar and turbulent zones. When comparing shear exfoliation to sonication, shear exfoliation is more efficient, and when a liquid volume of 10 m^3 is reached, their scaling rule allowed for a rate of up to 100 gh^{-1}. Recently, the modified Hummers' approach was used to generate G oxide using high-shear mixing [81]. **Table 1** displays the concentration of G dispersion produced by high-shear mixing with various solvents, with 10 $mgmL^{-1}$ being the greatest concentration.

2.1.4.3 Microfluidization

High pressure is applied to the fluid during the high-pressure homogenization process known as microfluidization, which drives the fluid through a microchannel with a diameter of $d < 100$ μm [83, 84]. It produces mild exfoliation conditions, which can assist to reduce the creation of flaws. In general, microfluidization is used in nanoemulsification [85], the food sector [86], cell disintegration [87], carbon nanotube dispersion [85], and fluidizing active medicinal components [88]. Recently, G quantum dots [89] and G-based conductive inks were produced using a microfluidizer. The fundamental advantage of mircrofluidization over sonication and high-shear mixing is that a high shear rate (less than 10^6 cm – ^{1}) exists throughout the fluid area.

Exfoliation medium	Exfoliation time (h)	Concentration (mg/ml)	Reference
NMP	0.5	0.01	[82]
SC/water	2	1.1	[65]
PVP/water	2	0.7	[65]
Black liquor/water	10	10	[82]
Water/black tea	0.5	0.032	[82]
NMP	1	0.65	[78]
40% vol IPA water	1	0.27	[78]
NMP	1	1	[82]
NMP	6	0.251	[48]
OCBD	1.5	0.03	[49]

Table 1.
G concentrations obtained via high-shear mixing in different solvents.

2.1.5 Unzipping carbon nanotubes (CNTs)

In this process, cylindrical carbon nanotubes (CNT) are sliced into flat G sheets with one, two, or a few layers in either the axial or longitudinal orientation. Starting materials can be single-wall carbon nanotubes (SWCNT) or multi-wall carbon nanotubes (MWCNT) [90] The unzipping of CNT can be accomplished using a number of techniques, including plasma etching [91], chemical unzipping [92], intercalation and exfoliation [93], and metal catalyzed cutting [94]. The generation of nanoribbons is caused by the longitudinal unzipping of CNT.

In chemical unzipping, H_2SO_4 treatment and oxidation with $KMnO_4$ results in longitudinal cutting of CNTs. After that, chemical reduction of the oxidized G will take place utilizing NH_4OH and hydrazine monohydrate ($N_2H_4H_2O$) solution [92]. Because the precursor is destroyed during oxidation and G loses some of its electrical properties due to the presence of oxygen defect sites, this approach is regarded as being of low significance [95]. When MWCNT is intercalated in oxalic acid, as a chemical medium, before the chemical unzipping, G yield is increased. The suitable size of oxalic acid, which intercalates nicely between the interlayers of the MWCNT [96], resulting in an increase in G yield is 0.34 nm. Polymer films like poly-methyl methacrylate are utilized in plasma etching. This technique involves embedding CNT into the film, followed by the separation of the film and CNT combination in a KOH solution. After that, CNT is exposed to plasma made of argon, and the CNT's longitudinal C−C link is dismantled to provide G with clean edges [91]. Another way to unzip CNT is through intercalation and exfoliation, where CNTs are exposed to a reaction between lithium and ammonia. The separation of the G layers occurs as a result of a significant amount of stress being applied between the CNT layers during this procedure [93]. Researchers looked at other metal nanoparticles like nickel, cobalt, and copper and discovered that these nanoparticles broke the C−C bonds and hydrogen bonds in MWCNT [97], but the techniques employed expensive and dangerous chemicals. Using an electric field from a tungsten electrode, MWCNT can be unzipped [98] producing highly pure, defect-free G nanoribbons. Therefore, due to its accuracy, viability, and environmental friendliness, the electric field approach is preferred.

2.1.6 Ball milling

An innovative technique for producing high-quality G by dissolving layered graphite into G is ball-milling method, which is said to have begun some 150 years ago, when it was used to produce talc powder, communicate ore size, and for a number of other purposes, but now, the method has recently been suggested for the producing nanoparticles at room temperature [99–101]. It is an easy and highly successful solid-state approach for reducing a variety of materials to fine powders, synthesizing nanocomposites, and foxides, making it a potential method for producing G in large quantities at a reasonable cost [102]. Large graphite sheets undergo shear stresses during milling, whereas normal force is used to crush graphite flakes into nano-sized materials and expose cracks in the basal plane. Both wet and dry conditions can be used to perform the milling procedure [103]. Lv et al. employed Na_2SO_4 salt to make G nanosheets with ripple-like corrugations in hundreds of square nanometer range [103, 104]. Elsewhere, ball milling was carried out in a planetary ball-mill machine with graphite, dry ice, and stainless-steel balls as shown in **Figure 5** [105]. It is found that the type of media utilized affects the size and quality of the materials produced. By using wet ball milling to exfoliate graphite into G flakes in a

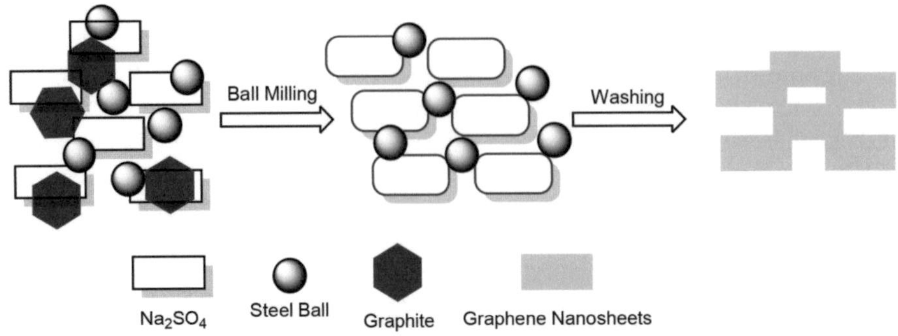

Figure 5.
Ball milling method.

liquid media, Zhao et al. provided a fresh perspective on the ball milling process. The measured thickness was found to be between 0.8 and 1.8 nm, which is equivalent to discrete monolayer and few-layer G (up to three layers). Because of this, there has been a lot of recent research activities in ball milling [106]. Additionally, Caicedo et al. reported on the oxidation of graphite using the ball-milling process in order to exfoliate G from graphite using $KClO_4$ and filtered water. It was shown that, as the milling time was increased, the degree of oxidation also did. Furthermore, the effects of oxidation were investigated using different ball milling time intervals (6, 12, 18, 24, and 30 h) in this manner, and the results were compared to those of the samples that were obtained form the Hummers methods. Thus, after 18 h of milling, the sample displayed improved dispersion and a darker hue as a result of the elimination of functional groups including carboxyl, hydroxyl, and epoxy [99, 107]. Here, the sample obtained after 16 h of milling was the best sample in terms of the level of oxidation, length, and energy utilization parameters assessed. The advantage of the ball milling approach is its ability to produce high-quality and low-cost G. It is a useful technique for exfoliating and functionalizing G. However, the prolonged processing times have resulted in a significant decrease in synthesis of G [99].

2.2 Bottom-up approaches

In the bottom-up techniques of synthesis, G sheets are generated directly from organic precursors such as methane and other hydrocarbon sources as illustrated in **Figure 2**. These methods transform molecular source materials into G-based substances [108]. Various examples of these approaches include chemical vapor deposition (CVD) [109], epitaxial growth [110], substrate-free gas-phase synthesis (SFGP) [111], template routes [112], and comprehensive organic synthesis [113]. While bottom-up strategies yield G products characterized by large surface areas and minimal defects, they often entail elevated production expenses.

2.2.1 Chemical vapor disposition

Methane (CH_4), acetylene (C_2H_2), ethylene (C_2H_4), and hexane (C_6H_{14}) are a few of the hydrocarbon gases that are broken down during CVD in order to develop G sheets on metallic catalysts (such as Cu and Ni films) at high temperatures (650–1000°C) [27]. The carbon precursor separates into free carbon and hydrogen atoms when it comes into contact with a metal catalyst's heated surface. Once the carbon atom attains the carbon solubility threshold, it undergoes diffusion throughout the surface and the bulk of the metal catalyst, ultimately resulting in the formation of a G layer on the metal substrate. Substrates such as glass, quartz, silicon, silicon oxide, boron nitride, and sapphire have been utilized for G production in an effort to optimize CVD techniques described in the work of Chen et al. [114].

Chemical vapor deposition (CVD) enables the production of high-quality G characterized by, a densely interconnected structure, substantial surface area, and minimal structural defects. Nevertheless, it presents drawbacks such as elevated production expenses, modest output, the necessity for additional purification to eliminate catalyst remnants, and the challenge of transferring G to alternative substrates [115]. Furthermore, CVD alone falls short of meeting commercialization needs due to the demand for substantial improvements in manufacturing cost and yield. To surmount these challenges without compromising G quality, researchers have directed their efforts toward refining synthesis conditions, targeting lower temperatures and ambient pressure [116]. For example, by employing surface wave plasma-enhanced CVD (PECVD), Kalita et al. achieved G synthesis at 450°C. This innovation considerably enhanced the overall process by significantly reducing growth temperature and deposition time (5 min) [113, 117]. The conventional CVD setup can be adapted to accommodate PECVD. To be practical on a commercial scale, the development of CVD must continue in conjunction with other technologies like thermal-based and plasma-based CVD.

2.2.2 Epitaxial growth on silicon carbide

G can also be synthesized by thermally decomposing silicon carbide (SiC) hexagonal substrate at temperatures between 120o and 1600°C in an inert or vacuum environment, as shown in **Figure 6**. Because silicon (Si) melts at a high temperature (1100°C), too many C atoms remain behind and create a sp^2 hybridized network, which promotes the formation of G [117]. The epitaxial growth of G on SiC is the term used to describe this process [118]. The G created using this process, however, is not homogeneous [119]. It has been documented that SiC and polytetrafluoroethylene (PTFE) undergo a unique exothermic reaction [120]. Due to the energy-intensiveness of the technique and constrained size of commercial SiC substrate, epitaxial development of G would be expensive under current synthesis conditions. Additionally, the

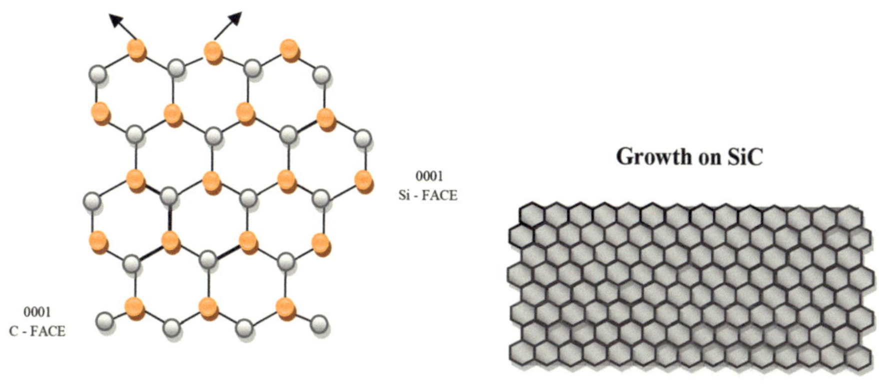

Figure 6.
Epitaxial growth on SiC showing Si (yellow spheres) and C (gray spheres) atoms (under elevated temperatures, Si atoms evaporate (arrow), creating a carbon-rich surface that gives rise to the formation of G sheets).

epitaxial development might result in polar faces like Si-face or C-face that degrade the quality of the finished G product as shown in **Figure 6** [113, 121]. This technique makes it simple to modify the amount of G layers that are dependent on the heating temperature [122].

The most noticeable benefit of G on the Si-face is that we can easily regulate the G thickness at the wafer size on the semi-insulating SiC substrate. By adjusting the growth temperature optimally, this control is accomplished. Images, captured by a high-resolution transmission electron microscope (HRTEM), of the G produced is monolayer, bilayer, trilayer, and eight-layer configurations at various temperatures. The number of G layers can be calculated directly from HRTEM images. Monolayer G is used to layer the buffer layer [123]. Like graphite, bilayer G is AB-stacked (Bernal-stacked) [124]. Further HRTEM observation revealed an ABC-stacking (rhombohedral stacking) of more than three layers [124]. This is a striking contrast to the Bernal bulk stacking graphite. An electric field-induced bandgap is present in the ABC-stacked trilayer G. The ABC-stacked trilayer G demonstrates a ferrimagnetic spin arrangement, an anomalous quantum Hall effect, and an electric field-driven bandgap [125–128]. By using HRTEM observation, the aforementioned atomic-scale growth mechanism was also studied.

2.2.3 Pyrolysis

Using the solvothermal approach, G can be formed chemically using the bottom-up method throughout the pyrolysis technique. For example, during the thermal reaction, the molar ratio of sodium and ethanol is determined 1:1 in the reaction vessel. This process involves heating a 1:1 molar reaction between 2 g sodium and 5 mL ethanol in a sealed reactor vessel at 220°C for 72 hours to produce the G precursor, also known as the solid solvo thermal product. The leftover precursor is then quickly pyrolyzed and washed with 100 mL deionized water. After that, the suspended solid is vacuum-filtered and dried for 24 hours at 100°C. This process yields 0.5 g of G per reaction, or 0.1 g of G every milliliter of ethanol [123]. Another example is the sonication-based pyrolization of sodium ethoxide. A PTFE (Polytetrafluoroethylene) melting pot was placed in an inert atmosphere and filled with 5 ml of ethanol and 2 g of sodium. To produce sodium ethoxide, the tank was firmly sealed and heated to 180°C for 24 hours. This was placed in an ignition dish and ignited in air and left to burn. Keep in mind that handling high pressure, sodium, or open flames requires the utmost caution. After being gathered, the carbonized product was broken up using a pestle, combined with deionized water, and sonated for a number of hours. After that, it was cleaned in deionized water using the method in [123]. The procedure might easily improve the detachment of G sheets. As a result, the produced G sheets have a thickness of 10 μm. To further investigate the graphitic properties, crystalline structure, band structure, and various layers of materials is executed by using Raman spectroscopy, selected area (electron) diffraction, and transmission electron microscopy [129].

2.2.4 Substrate-free gas-phase method

The substrate-free gas-phase (SFGP) approach is a relatively recent technique for producing G materials through gas-phase reactions by eliminating the need for substrates [130]. This process involves introducing liquid ethanol and Ar gas into microwave-generated plasma under atmospheric conditions. As ethanol droplets vaporize and dissociate within the plasma region over a time span of approximately

1 second, G is synthesized. Notably, this method reportedly yielded 2 $mgmin^{-1}$ of G from an input of 164 $mgmin^{-1}$ of ethanol [131]. Dato and Frenklach [132], on the other hand, explored the potential of different carbon precursors using this technique. Isopropyl alcohol and dimethyl ether were identified as possible precursors for generating G nanosheets.

2.2.5 Total organic synthesis

Total organic synthesis involves using polycyclic aromatic hydrocarbons (PAHs) with specific characteristics to create G. PAHs, often described as 2D G segments, composed entirely of sp2 carbons due to their structural resemblance, can be readily modified with aliphatic chains to tailor the solubility of the final product [133]. Crucial to this method is the selection of appropriate precursors that produce high yield quality G through a straight forward reaction pathway. The process results in 2D G nanoribbons (GNRs) up to 12 nm in length, as demonstrated by Yan et al. [134, 135]. However, the relatively narrow size distribution of PAHs might affect the quality of G due to reduced solubility and potential side reactions stemming from their higher molecular weight. Subsequent improvements by Yan et al. addressed these limitations [135].

2.2.6 Template route

The template route employs one-dimensional templates, such as metals, oxides, or polymers, to synthesize G derivatives with high throughput, quality, and well-defined structures [135]. Wei et al. initiated this approach [136]. Through physical vapor deposition (PVD), ZnS ribbons were created as templates for G formation using CH_4 as the carbon source. Subsequent etching with HCl removed residual ZnS nanobelts. Another template-based strategy involves G synthesis through the self-assembly of G on a meso-structured silica template formed from a pyrrole moiety-containing surfactant [137].

2.3 Trending techniques of synthesis of G

2.3.1 Microwave-assisted synthesis

Graphite or GO that has undergone a modified version of Hummer's process is typically used to produce G nanosheets (GNS) [60, 138–140]. Chemical procedures are typically used to rGO in the presence of several hazardous reduction agents, including hydrazine and $NaBH_4$. Thermal treatment, in contrast, uses no toxic reduction chemicals, making it a greener option. The environmentally friendly microwave-assisted method has gained increasing attention as a substitute for the traditional preparation of G in traditional heating systems (furnace or oil bath). In this method, GO [141, 142] or natural graphite [143] can be treated using the microwave-assisted solvothermal/hydrothermal methods in a microwave oven or microwave plasma-enhanced chemical vapor deposition (MPCVD) system.

In this method, nontoxic solvents are used to exfoliate GO within a short reaction period of 1 to 15 min at a relatively low temperature range of 180 to 300°C [138]. According to a study, an easy microwave-assisted solvothermal technique may produce a stable G suspension from a GO suspension in an alkaline medium (pH 10) or in polar solvents such as N,N-dimethylformamide, ethanol, 1-butanol, and water.

Additionally, the water-soluble polymer-grafted G sheets have been produced from GO by preparing them for 4 min at 450 W in a standard household microwave [138]. A residential microwave was used to create a three-dimensional (3D) nanostructure of a "G nano-cup" anchored on a few layers of G substrate [138]. Two stages were reported: the one-pot synthesis of G coated metal nanoparticles anchored on the G sheets and the subsequent etching of metals. This was done under the microwave irradiation in a home microwave oven [138]. More importantly, highly hydrogenated G could be created from GO by a one-step microwave irradiation process in hydrogen plasma, in which the deoxidation and concurrent hydrogenation were both accomplished. Giant G sheets could also be obtained by double microwave-assisted exfoliation of expandable graphite. High local temperatures and pressure are provided by microwave irradiation, and energy is sent directly into the inside of the GO. The polar link of oxygen-containing functional groups on the surface and edge of GO sheets interacts with radiation to produce heat [144]. Additionally, a key element in determining the regularity of deposits is the interaction between polar solvents and the surface oxides on GO sheets. Furthermore, the reduction degree of G sheets is further enhanced, and the functional groups on the surface of GO are successfully lowered.

There are several distinct advantages of utilizing microwave technology to create G. First of all, the microwave-assisted technique is quick and does not require a difficult synthesis process. Second, compared to more traditional methods, this technique is very cost-effective because it uses less chemicals. Third, compared to G made using the traditional heating approach, those produced using microwave-assisted technology may have an average size that is ten times larger. Finally, high-quality G with regulated structure and residual functional groups is produced using microwave-assisted technique [138]. Microwave sources enable localized high temperatures, rapid energy transfer, and efficient precursor disintegration. These attributes lead to homogenous nucleation environments, rapid crystallization, controlled particle size distribution, and precise morphology regulation [145, 146]. This technique facilitates the creation of G-based nanocomposites with adjustable sizes and shapes, enhancing applications like particle/crystal-on-sheet, nanorod/nanofiber-on-sheet, and nanosheet-on-sheet [147].

2.3.2 Electrochemical exfoliation

Electrochemical exfoliation of graphite has become a popular way to make G compounds in recent years. Graphite can be utilized as working electrodes in liquid electrolytes in a range of geometries, such as powders, foils, rods, flakes, or plates [148]. There are two types of exfoliation techniques: cathodic, which applies a negative bias to graphite electrodes, and anodic, which uses a positive bias. In cathodic exfoliation, positively charged electrolyte ions such Li^+ would be drawn to graphite electrodes. In anodic exfoliation, negatively charged ions, such as SO_4^{-2}, may be drawn to electrodes. The van der Waals forces have been hypothesized to be broken by electrochemical reactions, which causes graphite's structural expansion [149]. Additionally, during electrochemical exfoliation, chemical interactions with functionalizing agents can occur simultaneously in order to perform in situ chemical doping (functionalization) of G materials to create a variety of G-based composite materials [150, 151].

2.3.3 Hydrothermal synthesis

Hydrothermal treatment is a thermo-chemical conversion technique that results in efficient hydrolysis, pyrolysis, dehydration, polymerization, and aromatization of organic precursors, as well as high oxygenated functional group content and condensed aromatic structures [152]. The hydrothermal treatment approach, which primarily involves the carbonization and reduction process, can convert biomass and GO to solid char and G, respectively [153]. During the hydrothermal treatment of biomass, small-molecules soluble byproducts such as aromatic compounds, polysaccharides, aldehydic, ketonic, and furan derivatives are also created, from which solid products are produced by further polymerization [154]. Based on pyrolysis, self-assembly, and dehydration, water-soluble and low-molecular weight fulvic acid (FA) can be transformed into G quantum dots. To our knowledge, however, no parallel investigations on water-insoluble, high-molecular-weight humic acid and the associated reaction mechanism have yet been published.

2.3.4 Laser-induced graphenization

Laser-induced graphitization (LIG) is an innovative technique that transforms carbon-rich sources into conductive carbon using lasers [155]. This approach holds promise for creating highly conductive graphenic materials suitable for energy storage devices, sensors, biomedical applications, and hydrogen evolution [156]. LIG has been applied to a variety of bio-based carbon sources, such as paper and wood, to achieve significant graphitization, potentially contributing to their use in various applications.

3. Conclusion

Due to its large surface area, thermal, electrical, and physical properties, the carbon material known as "graphene" has become important in the fields of micro-manufacturing, nanomaterials, biomedical, and composite materials. However, due to its vast range of applications, G must be produced in large quantities, which has recently received a lot of attention from researchers and technologists. Therefore, improving the production process is essential for increasing the yield. Producing high-quality G is also necessary, but the most critical step is to use an easy, affordable, and environmentally friendly method. Future research should therefore focus on the yield measurement, biocompatibility as a result of the use of nontoxic chemicals, high energy, pressure, and poor transfer processes in chemical approaches, all of which have significantly contributed to high production costs, poor yield, and imperfections in the obtained G. The development of new techniques and environmentally friendly materials is crucial for the synthesis of G materials in electronics, nanomaterials, and biomaterials, all other factors being held constant. The size, shape, and optical properties of the resulting material are significantly influenced by the manufacturing processes and the carbon precursor sources used. G is suitable for use in electronics, energy storage, sensors, composites, drug delivery, and biomedical devices, because of its high transparency, thermal and electrical conductivity, and mechanical strength, in addition to its specific area. The improvements in G synthesis have opened up new opportunities for exploiting its extraordinary qualities and developing cutting-edge technology for a variety of industries. In summary, despite extensive study into producing G since its discovery, no method has been found to satisfactorily produce G on an industrial scale. This overview covered prospective uses for G as well as a comparative research and potential approaches.

Acknowledgements

We would like to express our sincere gratitude to Dr. Arshad Jamal Ansari, Postdoctoral Scholar, at University of Southern California and institutions for his valuable suggestions and editing to the completion of this research endeavor.

Competing interests

The authors declare no competing interests.

Author details

Aafreen[1†], Priyanka Verma[1†] and Haris Saeed[2*]

1 Faculty of Pharmacy, Shri Ram Murti Smarak College of Engineering and Technology, Bareilly, Uttar Pradesh, India

2 Keck School of Medicine, University of Southern California, Los Angeles, USA

*Address all correspondence to: haris6229@gmail.com

† These authors are equally contributed.

References

[1] Shinohara H, Tiwari A. Graphene: An Introduction to the Fundamentals and Industrial Applications. Beverly, Massachusetts, USA: John Wiley & Sons; 2015

[2] Allen MJ, Tung VC, Kaner RB. Honeycomb carbon: A review of graphene. Chemical Reviews. 2010;**110**(1):132-145

[3] Brodie BC. On the atomic weight of graphite. Philosophical Transactions of the Royal Society A. 1859;**149**:249-259

[4] Boehm HP, Clauss A, Fischer GO, Hofmann U. Das adsorptionsverhalten sehr dünner kohlenstoff-folien. Zeitschrift für anorganische und allgemeine Chemie. 1962;**316**(3-4):119-127

[5] Wallace PR. The band theory of graphite. Physics Review. 1947;**71**:622-663

[6] Trivedi S, Lobo K, Ramakrishna Matte HSS. Synthesis, properties, and applications of graphene. In: Fundamentals and Sensing Applications of 2D Materials [Internet]. Sawston, United Kingdom: Elsevier; 2019. pp. 25-90

[7] Novoselov KS, Geim AK, Morozov SV, Jiang D, Zhang Y, Dubonos SV, et al. Electric field effect in atomically thin carbon films. Science. 2004;**306**(5696):666-669

[8] Randviir EP, Brownson DAC, Banks CE. A decade of graphene research: Production, applications and outlook. Materials Today. 2014;**17**(9):426-432

[9] Shams SS, Zhang R, Zhu J. Graphene synthesis: A review. Materials Science-Poland. 2015;**33**(3):566-578

[10] Chen D, Feng H, Li J. Graphene oxide: Preparation, functionalization, and electrochemical applications. Chemical Reviews. 2012;**112**(11): 6027-6053

[11] Kumar R, Sahoo S, Joanni E, Singh RK, Yadav RM, Verma RK, et al. A review on synthesis of graphene, h-BN and MoS_2 for energy storage applications: Recent progress and perspectives. Nano Research. 2019;**12**(11):2655-2694

[12] Mas-Ballesté R, Gómez-Navarro C, Gómez-Herrero J, Zamora F. 2D materials: To graphene and beyond. Nanoscale. 2011;**3**(1):20-30

[13] Singh RK, Kumar R, Singh DP. Graphene oxide: Strategies for synthesis, reduction and frontier applications. RSC Advances. 2016;**6**(69):64993-65011

[14] Dong R, Zhang T, Feng X. Interface-assisted synthesis of 2D materials: Trend and challenges. Chemical Reviews. 2018;**118**(13):6189-6235

[15] Singh DP, Herrera CE, Singh B, Singh S, Singh RK, Kumar R. Graphene oxide: An efficient material and recent approach for biotechnological and biomedical applications. Materials Science and Engineering: C. 2018;**86**:173-197

[16] Farooqui UR, Ahmad AL, Hamid NA. Graphene oxide: A promising membrane material for fuel cells. Renewable and Sustainable Energy Reviews. 2018;**82**:714-733

[17] Yadav SK, Kumar R, Sundramoorthy AK, Singh RK, Koo CM. Simultaneous reduction and covalent grafting of polythiophene on graphene oxide sheets for excellent capacitance retention. RSC Advances. 2016;**6**(58):52945-52949

[18] Shuvo MAI, Khan MAR, Karim H, Morton P, Wilson T, Lin Y. Investigation of modified graphene for energy storage applications. ACS Applied Materials & Interfaces. 2013;**5**(16):7881-7885

[19] Tao LQ, Zhang KN, Tian H, Liu Y, Wang DY, Chen YQ, et al. Graphene-paper pressure sensor for detecting human motions. ACS Nano. 2017;**11**(9):8790-8795

[20] Rosli NN, Ibrahim MA, Ahmad Ludin N, Mat Teridi MA, Sopian K. A review of graphene based transparent conducting films for use in solar photovoltaic applications. Renewable and Sustainable Energy Reviews. 2019;**99**:83-99

[21] Ghawanmeh AA, Ali GAM, Algarni H, Sarkar SM, Chong KF. Graphene oxide-based hydrogels as a nanocarrier for anticancer drug delivery. Nano Research. 2019;**12**(5):973-990

[22] Zhang H. Introduction: 2D materials chemistry. Chemical Reviews. 2018;**118**(13):6089-6090

[23] Xia W, Dai L, Yu P, Tong X, Song W, Zhang G, et al. Recent progress in van der Waals heterojunctions. Nanoscale. 2017;**9**(13):4324-4365

[24] Ferrari AC, Bonaccorso F, Fal'ko V, Novoselov KS, Roche S, Bøggild P, et al. Science and technology roadmap for graphene, related two-dimensional crystals, and hybrid systems. Nanoscale. 2015;**7**(11):4598-4810

[25] Singh A, Ahmed A, Sharma A, Arya S. Graphene and its derivatives: Synthesis and application in the electrochemical detection of analytes in sweat. Biosensors. 2022;**12**(10):910

[26] Yu H, Zhang B, Bulin C, Li R, Xing R. High-efficient synthesis of graphene oxide based on improved Hummers method. Scientific Reports. 2016;**6**(1):36143

[27] Singh V, Joung D, Zhai L, Das S, Khondaker SI, Seal S. Graphene based materials: Past, present and future. Progress in Materials Science. 2011;**56**(8):1178-1271

[28] Soldano C, Mahmood A, Dujardin E. Production, properties and potential of graphene. Carbon. 2010;**48**(8):2127-2150

[29] Gutiérrez-Cruz A, Ruiz-Hernández AR, Vega-Clemente JF, Luna-Gazcón DG, Campos-Delgado J. A review of top-down and bottom-up synthesis methods for the production of graphene, graphene oxide and reduced graphene oxide. Journal of Materials Science. 2022;**57**(31):14543-14578

[30] Lu X, Yu M, Huang H, Ruoff RS. Tailoring graphite with the goal of achieving single sheets. Nanotechnology. 1999;**10**(3):269-272

[31] Lang B. A LEED study of the deposition of carbon on platinum crystal surfaces. Surface Science. 1975;**53**(1):317-329

[32] Liang X, Chang ASP, Zhang Y, Harteneck BD, Choo H, Olynick DL, et al. Electrostatic force assisted exfoliation of prepatterned few-layer graphenes into device sites. Nano Letters. 2009;**9**(1):467-472

[33] Ci L, Song L, Jariwala D, Elías AL, Gao W, Terrones M, et al. Graphene shape control by multistage cutting and transfer. Advanced Materials. 2009;**21**(44):4487-4491

[34] Liang X, Fu Z, Chou SY. Graphene transistors fabricated via transfer-printing in device

active-areas on large wafer. Nano Letters. 2007;7(12):3840-3844

[35] Chen JH, Ishigami M, Jang C, Hines DR, Fuhrer MS, Williams ED. Printed graphene circuits. Advanced Materials. 2007;**19**:3623-3627. Available from: https://arxiv.org/abs/0809.1634 [Accessed: December 20, 2023]

[36] Huc V, Bendiab N, Rosman N, Ebbesen T, Delacour C, Bouchiat V. Large and flat graphene flakes produced by epoxy bonding and reverse exfoliation of highly oriented pyrolytic graphite. Nanotechnology. 2008;**19**(45):455601

[37] Song L, Ci L, Gao W, Ajayan P.M. Transfer printing of graphene using gold film. ACS Nano 2009; 3(6): 1353-1356

[38] Casiraghi C, Hartschuh A, Lidorikis E, Qian H, Harutyunyan H, Gokus T, et al. Rayleigh imaging of graphene and graphene layers. Nano Letters. 2007;**7**(9):2711-2717

[39] Stoller MD, Park S, Zhu Y, An J, Ruoff RS. Graphene-based ultracapacitors. Nano Letters. 2008;**8**(10):3498-3502

[40] Brodie BC. Sur le poids atomique du graphite. Annales de Chimie Physique. 1860;**59**:466-472

[41] Hummers WS, Offeman RE. Preparation of graphitic oxide. Journal of the American Chemical Society. 1958;**80**(6):1339-1339

[42] Staudenmaier L. Verfahren zur darstellung der graphitsäure. Berichte der deutschen chemischen Gesellschaft. 1898;**31**(2):1481-1487

[43] Wu YH, Yu T, Shen ZX. Two-dimensional carbon nanostructures: Fundamental properties, synthesis, characterization, and potential applications. Journal of Applied Physics. 2010;**108**(7):071301

[44] Marcano DC, Kosynkin DV, Berlin JM, Sinitskii A, Sun Z, Slesarev A, et al. Improved synthesis of graphene oxide. ACS Nano. 2010;**4**(8):4806-4814

[45] Park S, An J, Jung I, Piner RD, An SJ, Li X, et al. Colloidal suspensions of highly reduced graphene oxide in a wide variety of organic solvents. Nano Letters. 2009;**9**(4):1593-1597

[46] Park S, Suk JW, An J, Oh J, Lee S, Lee W, et al. The effect of concentration of graphene nanoplatelets on mechanical and electrical properties of reduced graphene oxide papers. Carbon. 2012;**50**(12):4573-4578

[47] Paredes JI, Villar-Rodil S, Martínez-Alonso A, Tascón JMD. Graphene oxide dispersions in organic solvents. Langmuir. 2008;**24**(19):10560-10564

[48] Green AA, Hersam MC. Emerging methods for producing monodisperse graphene dispersions. Journal of Physical Chemistry Letters. 2010;**1**(2):544-549

[49] Green AA, Hersam MC. Solution phase production of graphene with controlled thickness via density differentiation. Nano Letters. 2009;**9**(12):4031-4036

[50] Wu J, Pisula W, Müllen K. Graphenes as potential material for electronics. Chemical Reviews. 2007;**107**(3):718-747

[51] Cai J, Ruffieux P, Jaafar R, Bieri M, Braun T, Blankenburg S, et al. Atomically precise bottom-up fabrication of graphene nanoribbons. Nature. 2010;**466**(7305):470-473

[52] Yan X, Cui X, Li B, Li L shi. Large, Solution-processable graphene quantum

dots as light absorbers for photovoltaics. Nano Letters 2010;10(5):1869-1873

[53] Zhi L, Müllen K. A bottom-up approach from molecular nanographenes to unconventional carbon materials. Journal of Materials Chemistry. 2008;**18**(13):1472

[54] Hernandez Y, Nicolosi V, Lotya M, Blighe FM, Sun Z, De S, et al. High-yield production of graphene by liquid-phase exfoliation of graphite. Nature Nanotechnology. 2008;**3**(9):563-568

[55] Hamilton CE, Lomeda JR, Sun Z, Tour JM, Barron AR. High-yield organic dispersions of unfunctionalized graphene. Nano Letters. 2009;**9**(10):3460-3462

[56] O'Neill A, Khan U, Nirmalraj PN, Boland J, Coleman JN. Graphene dispersion and exfoliation in low boiling point solvents. Journal of Physical Chemistry C. 2011;**115**(13):5422-5428

[57] Hernandez Y, Lotya M, Rickard D, Bergin SD, Coleman JN. Measurement of multicomponent solubility parameters for graphene facilitates solvent discovery. Langmuir. 2010;**26**(5):3208-3213

[58] Zheng QF, Guo Y, Liang Y, Shen Q. Graphene nanoribbons from electrostatic-force-controlled electric unzipping of single- and multi-walled carbon nanotubes. ACS Applied Nano Materials. 2020;**3**(5):4708-4716

[59] Chae SJ, Güneş F, Kim KK, Kim ES, Han GH, Kim SM, et al. Synthesis of large-area graphene layers on poly-nickel substrate by chemical vapor deposition: Wrinkle formation. Advanced Materials. 2009;**21**(22):2328-2333

[60] Qian M, Zhou YS, Gao Y, Park JB, Feng T, Huang SM, et al. Formation of graphene sheets through laser exfoliation of highly ordered pyrolytic graphite. Applied Physics Letters. 2011;**98**(17):173108

[61] Zhang Y, Zhou CG, Hua YX, Gao HL, Gao KZ, Cao Y. Synthesis of nafion-reduced graphene oxide/polyaniline as novel positive electrode additives for high performance lead-acid batteries. Electrochimica Acta. 2023;**466**:143045

[62] Stankovich S, Dikin DA, Dommett GHB, Kohlhaas KM, Zimney EJ, Stach EA, et al. Graphene-based composite materials. Nature. 2006;**442**:282-286

[63] Wang X, Zhi L, Mullen K. Transparent, conductive graphene electrodes for dye-sensitized solar cells. Nano Letters. 2008;**8**(1):323-327

[64] Dikin DA, Stankovich S, Zimney EJ, Piner RD, Dommett GHB, Evmenenko G, et al. Preparation and characterization of graphene oxide paper. Nature. 2007;**448**:457-460

[65] Somanathan T, Prasad K, Ostrikov K, Saravanan A, Krishna V. Graphene oxide synthesis from agro waste. Nanomaterials. 2015;**5**(2):826-834

[66] Guo S, Garaj S, Bianco A, Ménard-Moyon C. Controlling covalent chemistry on graphene oxide. Nature Reviews Physics. 2022;**4**(4):247-262

[67] Khanra P, Kuila T, Kim NH, Bae SH, Sheng YD, Lee JH. Simultaneous bio-functionalization and reduction of graphene oxide by baker's yeast. Chemical Engineering Journal. 2012;**183**:526-533

[68] Chua CK, Pumera M. Chemical reduction of graphene oxide: A synthetic chemistry viewpoint. Chemical Society Reviews. 2014;**43**(1):291-312

[69] Kumar PV, Bardhan NM, Chen GY, Li Z, Belcher AM, Grossman JC. New insights into the thermal reduction of graphene oxide: Impact of oxygen clustering. Carbon. 2016;**100**:90-98

[70] Zheng X, Peng Y, Yang Y, Chen J, Tian H, Cui X, et al. Hydrothermal reduction of graphene oxide; effect on surface-enhanced Raman scattering. Journal of Raman Specroscopy. 2017;**48**(1):97-103

[71] Coleman JN. Liquid exfoliation of defect-free graphene. Accounts of Chemical Research. 2013;**46**(1):14-22

[72] Pilli S, Bhunia P, Yan S, LeBlanc RJ, Tyagi RD, Surampalli RY. Ultrasonic pretreatment of sludge: A review. Ultrasonics Sonochemistry. 2011;**18**(1):1-18

[73] Flint EB, Suslick KS. The temperature of cavitation. Science. 1991;**253**(5026):1397-1399

[74] Lotya M, King PJ, Khan U, De S, Coleman JN. High-concentration, surfactant-stabilized graphene dispersions. ACS Nano. 2010;**4**(6):3155-3162

[75] Ciesielski A, Samorì P. Grapheneviasonication assisted liquid-phase exfoliation. Chemical Society Reviews. 2014;**43**(1):381-398

[76] Han JT, Jang JI, Kim H, Hwang JY, Yoo HK, Woo JS, et al. Extremely efficient liquid exfoliation and dispersion of layered materials by unusual acoustic cavitation. Scientific Reports. 2014;**4**(1):5133

[77] Lin Z, Karthik P, Hada M, Nishikawa T, Hayashi Y. Simple technique of exfoliation and dispersion of multilayer graphene from natural graphite by ozone-assisted sonication. Nanomaterials. 2017;7(6):125

[78] Chen X, Dobson JF, Raston CL. Vortex fluidic exfoliation of graphite and boron nitride. Chemical Communications. 2012;**48**(31):3703

[79] Paton KR, Varrla E, Backes C, Smith RJ, Khan U, O'Neill A, et al. Scalable production of large quantities of defect-free few-layer graphene by shear exfoliation in liquids. Nature Materials. 2014;**13**(6):624-630

[80] Chong KY, Chia CH, Chook SW, Zakaria S, Lucas D. Simplified production of graphene oxide assisted by high shear exfoliation of graphite with controlled oxidation. New Journal of Chemistry. 2018;**42**(6):4507-4512

[81] Wang YZ, Chen T, Gao XF, Liu HH, Zhang XX. Liquid phase exfoliation of graphite into few-layer graphene by sonication and microfluidization. Materials Express. 2017;**7**(6):491-499

[82] Pavlova AS, Obraztsova EA, Belkin AV, Monat C, Rojo-Romeo P, Obraztsova ED. Liquid-phase exfoliation of flaky graphite. Journal of Nanophotonics. 2016;**10**(1):012525

[83] Paton KR, Anderson J, Pollard AJ, Sainsbury T. Production of few-layer graphene by microfluidization. Materials Research Express. 2017;**4**(2):025604

[84] Narsimhan G, Goel P. Drop coalescence during emulsion formation in a high-pressure homogenizer for tetradecane-in-water emulsion stabilized by sodium dodecyl sulfate. Journal of Colloid and Interface Science. 2001;**238**(2):420-432

[85] Jafari SM, He Y, Bhandari B. Production of sub-micron emulsions by ultrasound and microfluidization techniques. Journal of Food Engineering. 2007;**82**(4):478-488

[86] Burhan M, Chua KJE, Ng KC. Simulation and development of a multi-leg homogeniser concentrating assembly for concentrated photovoltaic (CPV) system with electrical rating analysis. Energy Conversion and Management. 2016;**116**:58-71

[87] Azoubel S, Magdassi S. The formation of carbon nanotube dispersions by high pressure homogenization and their rapid characterization by analytical centrifuge. Carbon. 2010;**48**(12):3346-3352

[88] Qian J, Yan J, Shen C, Xi F, Dong X, Liu J. Graphene quantum dots-assisted exfoliation of graphitic carbon nitride to prepare metal-free zero-dimensional/two-dimensional composite photocatalysts. Journal of Materials Science. 2018;**53**(17):12103-12114

[89] Santhiran A, Iyngaran P, Abiman P, Kuganathan N. Graphene synthesis and its recent advances in applications—A review. C. 2021;**7**(4):76

[90] Baig Z, Mamat O, Mustapha M, Mumtaz A, Munir KS, Sarfraz M. Investigation of tip sonication effects on structural quality of graphene nanoplatelets (GNPs) for superior solvent dispersion. Ultrasonics Sonochemistry. 2018;**45**:133-149

[91] Jiao L, Zhang L, Wang X, Diankov G, Dai H, et al. Narrow graphene nanoribbons from carbon nanotubes. Nature. 2009;**458**(7240):877-880

[92] Kosynkin DV, Higginbotham AL, Sinitskii A, Lomeda R, Dimiev A, Price BK, et al. Longitudinal unzipping of carbon nanotubes to form graphene nanoribbons. Nature. 2009;**458**(7240):845-846

[93] Abraham GCM, Macias RF, Delgado JC, Gonzalez ECG, Lopez FT, et al. Ex-MWNTs: Graphene sheets and ribbons produced by lithium intercalation and exfoliation of carbon nanotubes. Nano Letters. 2009;**9**(4):1527-1533

[94] Thess A, Lee R, Nikolaev P, Dai H, Petit P, Robert JSG, et al. Crystalline ropes of metallic carbon nanotubes. Science. 1996;**273**(5274):483-487

[95] Liu J, Kim GH, Xue Y, Kim JY, Baek JB, Durstock M, et al. Graphene oxide nanoribbon as hole extraction layer to enhance efficiency and stability of polymer solar cells. Advanced Materials. 2014;**26**(5):786-790

[96] Mahmoud WE, Hazmi FS, Harbi GH. Wall by wall controllable unzipping of MWCNTs via intercalation with oxalic acid to produce multilayers graphene oxide ribbon. Chemical Engineering Journal. 2015;**281**:192-198

[97] Bhuyan MSA, Uddin MN, Islam MM, Bipasha FA, et al. Synthesis of graphene. International Nano Letters. 2016;**6**:65-83

[98] Sood AK, Chakraborty B. In: Rao CNR, Sood AK, editors. Understanding Graphene Via Raman Scattering. Hoboken, NJ, USA: John Wiley & Sons; 2012. pp. 49-90

[99] Mousavi BSM, Sadaf S, Walder L, Gallei M, et al. Poly(vinylferrocene)-reduced graphene oxide as a high power/high capacity cathodic battery material. Advanced Energy Materials. 2016;**1600108**:1-13

[100] Hassanzadeh N, Sadrnezhaad SK, Chen GH. Ball mill assisted synthesis of $Na_3MnCO_3PO_4$ nanoparticles anchored on reduced graphene oxide for sodium ion battery cathodes. Electrochimica Acta. 2016;**220**:683

[101] Lv YY, Yu LS, Jiang CM. Synthesis of graphene nanosheet powder with

layer number control via a soluble salt-assisted route. RSC Advances. 2014;**4**:13350-13354

[102] Kairi MI, Dayou S, Kairi NI. Toward high production of graphene flakes–a review on recent developments in their synthesis methods and scalability. Journal of Materials Chemistry A. 2018;**6**:15010-15026

[103] Jeon IY, Shin YR, Sohn GJ. Edge-carboxylated graphene nanosheets via ball milling. PNAS. 2012;**109**(15):5588-5593

[104] Zhao WF, Fang M, Wu FR. Preparation of graphene by exfoliation of graphite using wet ball milling. Journal of Materials Chemistry. 2010;**20**:5817-5819

[105] Dash P, Dash T, Rout TK, Biswal SK, et al. Preparation of graphene oxide by dry planetary ball milling process from natural graphite. RSC Advances. 2016;**6**:12657-12668

[106] Caicedo CFM, Lopez VE, Agarwal A, Drozd V, et al. Synthesis of graphene oxide from graphite by ball milling. Diamond and Related Materials. 2020;**109**:108064

[107] Lee XJ, Hiew BYZ, Lai KC, Lee YL, et al. Review on graphene and its derivatives: Synthesis methods and potential industrial implementation. Journal of the Taiwan Institute of Chemical Engineers. 2018;**18**:13

[108] Jacobberger RM, Machhi R, Wroblewski J, Taylor B, Gillian-Daniel AL, Arnold MS, et al. Simple graphene synthesis via chemical vapor deposition. Journal of Chemical Education. 2015;**92**:1903-1907

[109] Huang H, Chen S, Wee ATS, Chen W. Epitaxial growth of graphene on silicon carbide (SiC). In: Graphene. Sawston, United Kingdom: Woodhead Publishing; 2014. pp. 3-26

[110] Olatomiwa AL, Adam T, Gopinath SCB, Kolawole SY, Olayinka OH, et al. Graphene synthesis, fabrication, characterization based on bottom-up and top-down approaches: An overview. Journal of Semiconductors. 2022;**43**(6):1-15

[111] Albert D, Michael F. Substrate-free microwave synthesis of graphene: Experimental conditions and hydrocarbon precursors. New Journal of Physics. 2010;**12**:125013

[112] Yang Y, Liu R, Wu J, Jiang X, Cao P, Hu X, et al. Bottom-up fabrication of graphene on silicon/silica substrate via a facile soft-hard template approach. Scientific Reports. 2015;**5**:13480

[113] Nair RR. Fine structure constant defines visual transparency of graphene. Science. 2008;**320**(5881):1308

[114] Brownson DA, Banks CE. The electrochemistry of CVD graphene: Progress and prospects. Physical Chemistry Chemical Physics. 2012;**14**(23):8264-8281

[115] Chen X, Wu B, Liu Y. Direct preparation of high quality graphene on dielectric substrates. Chemical Society Reviews. 2016;**45**(8):2057-2074

[116] Gupta B, Notarianni M, Mishra N, Shafiei M, Iacopi F, Motta N, et al. Evolution of epitaxial graphene layers on 3C SiC/Si (111) as a function of annealing temperature in UHV. Carbon. 2014;**68**:563-572

[117] Jang J, Son M, Chung S, Kim K, Cho C, Lee BH, et al. Low-temperature–grown continuous graphene films from benzene by chemical vapor deposition at ambient pressure. Scientific Reports. 2015;**5**:17955

[118] Wei D, Lu Y, Han C, Niu T, Chen W, Wee AT, et al. Critical crystal growth of graphene on dielectric substrates at low temperature for electronic devices. Angewandte Chemie (International Ed. in English). 2013;**52**(52):14121-14126

[119] Mishra N, Boeckl J, Motta N, Iacopi F, et al. Graphene growth on silicon carbide: A review (Phys. Status solidi a 9 2016). Physica Status Solidi A: Applications and Material Science. 2016;**213**:2269-2289

[120] Riedl C, Coletti C, Starke U. Structural and electronic properties of epitaxial graphene on SiC(0001): A review of growth, characterization, transfer doping and hydrogen intercalation. Journal of Physics D: Applied Physics. 2010;**43**(37):374009

[121] Manukyan KV, Rouvimov S, Wolf EE, Mukasyan AS, et al. Combustion synthesis of graphene materials. Carbon. 2013;**62**:302-311

[122] Norimatsu W, Kusunoki M. Epitaxial graphene on SiC{0001}: Advances and perspectives. Physical Chemistry Chemical Physics. 2014;**16**:3501-3511

[123] Zhang F, Jung J, Fiete GA, Niu Q, MacDonald AH, et al. Physical Review Letters. 2011;**106**:156801

[124] Lui CH, Li Z, Mak KF, Cappelluti E, Heinz TF. Nature Physics. 2011;7:944-947

[125] Bao W, Jing L, Velasco J, Lee Y, Liu G, Tran D, et al. Stacking-dependent band gap and quantum transport in trilayer graphene. Nature Physics. 2011;7:948-952

[126] Zhang L, Zhang Y, Camacho Y, Khodas M, Zaliznyak I, et al. The experimental observation of quantum hall effect of l=3 chiral quasiparticles in trilayer graphene. Nature Physics. 2011;7:953.957

[127] Shioyama H, Sakakihara H, Iwashita N, Tatsumi K, Sawada Y. On the generation of fine metallic particles in graphite matrix. Journal of Materials Science Letters. 1994;**13**:1056

[128] Choi W, Lahiri I, Seelaboyina R, Kang YS, et al. Synthesis of graphene and its applications: A review. Critical Reviews in Solid State and Materials Sciences. 2010;**35**(1):52-71

[129] Jihn YL, Mubarak NM, Abdullah EC, Nizamuddin SK, Mohammad I. Recent trends in the synthesis of graphene and graphene oxide based nanomaterials for removal of heavy metals—A review. Journal of Industrial and Engineering Chemistry. 2018;**66**:29-44

[130] Choucair M, Thordarson P, Stride JA. Gram-scale production of graphene based on solvothermal synthesis and sonication. Nature Nanotech. 2008;**4**(1):30-33

[131] Dato A, Radmilovic V, Lee Z, Phillips J, Frenklach M, et al. Substrate-free gas-phase synthesis of graphene sheets. Nano Letters. 2008;**8**:2012-2016

[132] Dato A, Lee Z, Jeon KJ, Erni R, Radmilovic V, Richardson TJ, et al. Clean and highly ordered graphene synthesized in the gas phase. Chemical Communications. 2009;**40**:6095-6097

[133] Dato A, Frenklach M. Substrate-free microwave synthesis of graphene: Experimental conditions and hydrocarbon precursors. New Journal of Physics. 2010;**12**(12):125013

[134] Choi W, Lee J. Graphene: Synthesis and Applications. Vol. 1. Boca Raton, FL, USA: CRC Press; 2011. pp. 1-394

[135] Zheng Q, Kim J-K. Synthesis, Structure, and Properties of Graphene and Graphene Oxide, in Graphene for Transparent Conductors: Synthesis, Properties and Applications. Vol. 2. New York: Springer New York; 2015. pp. 29-94

[136] Yang X, Dou X, Rouhanipour A, Zhi L, Rader HJ, Müllen K, et al. Two-dimensional graphene nanoribbons. Journal of the American Chemical Society. 2008;**130**(13):4216-4217

[137] Guo S, Dong S. Graphene nanosheet: Synthesis, molecular engineering, thin film, hybrids, and energy and analytical applications. Chemical Society Reviews. 2010;**40**:2644-2672

[138] Wang Y, Sun W, Li H. Microwave-assisted synthesis of graphene nanocomposites: Recent developments on lithium-ion batteries. Reports in Electrochemistry. 2015;**5**:1-19

[139] Murugan AV, Muraliganth T, Manthiram A. Rapid, facile microwave-solvothermal synthesis of graphene nanosheets and their polyaniline nanocomposites for energy strorage. Chemistry of Materials. 2009;**21**:5004-5006

[140] Wang Y, Liu G, An CH, et al. Bimetallic NiCo functional graphene: An efficient catalyst for hydrogen-storage properties of MgH_2. Chemistry, an Asian Journal. 2014;**9**:2576-2583

[141] Chen W, Yan L, Bangal PR. Preparation of graphene by the rapid and mild thermal reduction of graphene oxide induced by microwaves. Carbon. 2010;**48**:1146-1152

[142] Hassan MAH, Abdelsayed V, Khder AERS, et al. Microwave synthesis of graphene sheets supporting metal nanocrystals in aqueous and organic media. Journal of Materials Chemistry. 2009;**19**:3832-3837

[143] Wu ZS, Ren WC, Gao LB, Liu BL, Jiang CB, Cheng HM, et al. Synthesis of high-quality graphene with a pre-determined number of layers. Carbon. 2008;**47**:493-499

[144] Motasemi F, Afzal MT. A review on the microwave-assisted pyrolysis technique. Renewable and Sustainable Energy Reviews. 2013;**28**:317-330

[145] Mutyala S, Fairbridge C, Pare JRJ, Belanger JMR, Ng S, Hawkins R, et al. Microwave applications to oil sands and petroleum: A review. Fuel Processing Technology. 2010;**91**:127-135

[146] Vazquez E, Giacalone F, Prato M. Non-conventional methods and media for the activation and manipulation of carbon nanoforms. Chemical Society Reviews. 2013;**43**:58-69

[147] Gallis KW, Landry CC. Rapid calcination of nanostructured silicate composites by microwave irradiation. Advanced Materials. 2001;**13**:23-26

[148] Nagarajan R, Alan HT. Synthesis, passivation, and stabilization of nanoparticles, nanorods, and nanowires by microwave irradiation. In: Nanoparticles: Synthesis, Stabilization, Passivation, and Functionalization. Vol. 996. Washington, DC, USA: ACS Publication; 2008. pp. 225-247

[149] Panda AB, Glaspell G, El-Shall MS. Microwave synthesis and optical properties of uniform nanorods and nanoplates of rare earth oxides. Journal of Physical Chemistry C. 2007;**111**:1861-1864

[150] Liu F, Wang C, Sui X, Riaz MA, Xu M, Wei L, et al. Synthesis of graphene materials by electrochemical exfoliation:

Recent progress and future potential. Carbon Energy. 2019;**1**(2):173-199

[151] Su CY, Lu AY, Xu Y, Chen FR, Khlobystov AN, Li LJ, et al. High-quality thin graphene films from fast electrochemical exfoliation. ACS Nano. 2011;**5**(3):2332-2339

[152] Abdelkader AM, Cooper AJ, Dryfe RAW, Kinloch IA, et al. How to get between the sheets: A review of recent works on the electrochemical exfoliation of graphene materials from bulk graphite. Nanoscale. 2015;**7**:6944-6956

[153] Huang G, Kang W, Geng Q, Xing B, Liu Q, Jia J, et al. One-step green hydrothermal synthesis of few-layer graphene oxide from humic acid. Nanomaterials. 2018;**8**(4):215

[154] Shi W, Fan H, Ai S, Zhu L, et al. Preparation of fluorescent graphene quantum dots from humic acid for bioimaging application. New Journal of Chemistry. 2015;**1**(10):1-6

[155] Bai Y, Rakhi RB, Chen W, Alshareef HN, et al. Effect of pH-induced chemical modification of hydrothermally reduced graphene oxide on supercapacitor performance. Journal of Power Sources. 2013;**233**:313-319

[156] Edberg J, Brooke R, Hosseinaei O, Fall A, Wijeratne K, Sandberg M, et al. Laser-induced graphitization of a forest-based ink for use in flexible and printed electronics. npj Flexible Electronics. 2020;**4**(1):17

Chapter 2

Recent Progress on Synthesis of 3D Graphene, Properties, and Emerging Applications

Md. Nizam Uddin, Md. Aliahsan Bappy, Md Fozle Rab, Faycal Znidi and Mohamed Morsy

Abstract

Materials based on 3D graphene, such as aerogels, hydrogels, sponges, and foams, are attracting substantial interest due to their superb electrical conductivity, remarkable mechanical properties, and expedited mass and electron transport. These substances preserve the inherent characteristics of 2D graphene sheets and introduce enhanced features like low density, substantial surface area, high porosity, and steadfast mechanical properties. The applications for 3D graphene-based materials are vast, ranging from flexible electronics, sensors, absorbents, and composites to catalysis, energy storage devices, agricultural uses, water purification, biomedical applications, and solar steam generation devices, among others. In this book chapter, we consolidate the latest advancements in the fabrication of 3D graphene-based materials, discussing their properties and the emerging uses in composites and energy storage apparatuses. The synthesis of 3D graphene-based materials on a larger scale poses substantial challenges, the discussion of which might spur innovation and novel approaches in this domain. We aim to provide a comprehensive overview of the contemporary progress in this field, emphasizing the synthesis, properties, and diverse applications of these advanced materials. Our research is anticipated to establish a groundwork for the widespread preparation, understanding of structure–property relationships, and utilization of 3D graphene-based architectures (3DGAs) across various fields, including but not limited to tissue engineering, electronics, supercapacitors, composites, and energy storage devices.

Keywords: 3D graphene, energy storage devices, composites, porous materials, graphene foam

1. Introduction

Carbon is a crucial building block of our universe and plays a significant role in both natural phenomena and technological advancements. In recent years, our understanding of carbon's versatility has grown significantly, especially in the field of sp2-hybridized carbon nanomaterials. Graphene, which is an example of a 2D sp2-hybridized carbon lattice that is only one atom thick, has remarkable physicochemical properties. Its derivatives have found application in various domains like low-dimensional physics, energy storage, electronic devices, catalysis, sensors, and medical equipment. However, a major challenge in harnessing graphene's potential at macroscopic scales is restacking when using 2D graphene sheets, which reduces efficiency and diminishes their unique properties. One solution to this challenge is the conversion of 2D graphene layers into interconnected, 3D frameworks that prevent restacking and retain 2D graphene's exceptional properties. This requires large-scale production and the conversion of individual graphene sheets into multifunctional 3D architectures. A critical step in manufacturing 3D graphene materials, specifically 3D reduced graphene oxide (3D-rGO), involves the reduction of graphene oxide (GO), which is achieved through different chemical methods designed to eliminate oxygen-containing functional groups on the GO plane. While early research focused primarily on synthesizing high-quality graphene sheets from GO, numerous strategies have since emerged, including thermal reduction in inert atmospheres, chemical reduction using various reducing agents, photocatalytic reduction, hydro/solvothermal techniques, laser/flashlight irradiation, electrochemical reduction, hydrogen-plasma/arc-discharge, microwave treatment, and combinations of these methods. Nonetheless, although significant progress has been made in the development of 3D graphene materials in recent years, there is a need for a comprehensive understanding of 3D architectures and their performance in various applications.

This study aims to bridge this gap by presenting advanced fabrication processes and design considerations for 3D graphene-based architectures (3DGAs). It will explore the relationship between 3D graphene properties, formation mechanisms, and key components, ultimately providing an encompassing overview of the 3D graphene family of materials. This will also highlight the significance of these materials in diverse applications and inspire new directions for their development, particularly focusing on simplifying the preparation and functionalization of 3D graphene materials. In the subsequent sections, we will describe the synthesis methods, properties, and potential applications of 3DGAs [1].

2. Synthesis of 3D graphene structure

Over the last decade, numerous techniques for synthesizing 3D graphene have emerged. These methods include electrospraying, supercritical drying, freeze drying, vacuum drying, chemical vapor deposition (CVD), hydrothermal processes, self-assembly, and various other approaches, which this review will explore and discuss the associated distinct advantages and disadvantages of each.

2.1 Template-assisted method

The templates-assisted method involves reducing GO and subsequently extracting the template from the structure. The template-assisted approach, outlined by Ding and Li [2] as well as by Qiu et al. [3], stands out due to its advantages in controlling the formation of 3D graphene structures, offering meticulously crafted morphologies compared to other fabrication methodologies. Within this technique, the chosen template, predominantly constructed from materials such as polystyrene (PS) or silicon dioxide (SiO_2), holds crucial significance, with graphene sheets gathering around it through electrostatic forces. These forces originate from the interaction between negatively charged graphene sheets and the positively charged templates, culminating in the creation of an organized

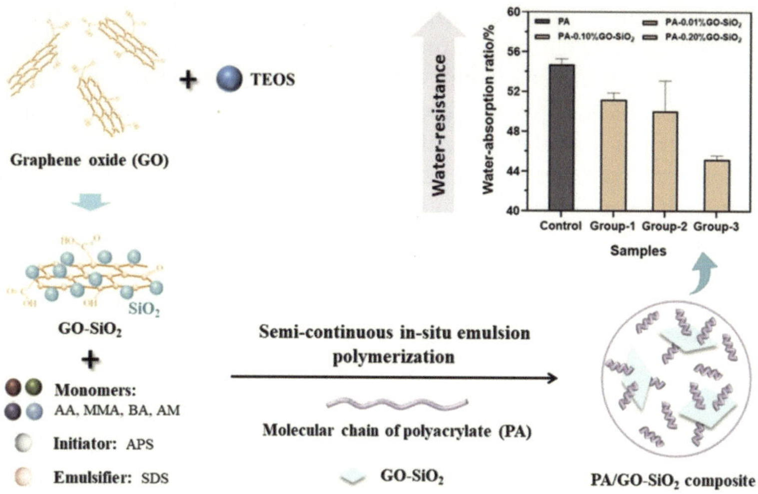

Figure 1.
Synthesis and water resistance characteristics of nano SiO_2 modified graphene oxide composite [5].

composite structure. Compared to self-assembly strategies, this technique ensures superior control over the final product's architecture and morphological attributes.

Several research studies have underscored the efficacy of template-assisted methods in developing 3D graphene-based materials. For instance, positively charged PS spheres were used as templates, coated with GO sheets, which were later reduced to obtain reduced graphene oxide (rGO) using hydrazine. The process was then finalized by calcination to eliminate the PS core, resulting in graphene hollow spheres [4]. This example highlights the extensive potential and adaptability of template-assisted techniques in creating diversified 3D graphene configurations. One remarkable study demonstrated a synthesis pathway for graphene-based hollow spheres, utilizing strong electrostatic interactions between polyethylenimine-functionalized SiO_2 spheres and graphene sheets, forming GO-SiO_2 spherical entities (**Figure 1**). Graphene-based hollow spheres were acquired after reducing and treating with hydrofluoric acid to eliminate the SiO_2 template, showcasing the method's versatility [6]. Additionally, Huang and his team have studied developing nanoporous graphene foams (GFs), emphasizing the role of hydrophobic interactions between GO sheets and modified SiO_2 spherical templates, resulting in nanoporous GF structures with customizable pore sizes [7]. This innovation marks a substantial progression in modifying the structural attributes of 3D graphene materials to accommodate specific needs and applications [8].

The template-assisted fabrication of 3DGAs unveils a multifaceted and regulated strategy to construct graphene-based configurations with optimal morphologies. Employing templates like PS and SiO_2, coupled with specialized electrostatic interactions and reduction phases, allows scientists to innovate in creating 3D graphene materials with a spectrum of applications that range from catalysis, to sensing, and energy storage. The continuous evolution in this domain suggests that template-assisted techniques are important for fostering groundbreaking progressions in materials science and technology.

2.2 Electrospraying

Electrospinning and electrospraying are simple and versatile techniques that can produce graphene-based fibers, spherical structures, and bead-like materials.

The resulting structures can have diameters ranging from a few micrometers to nanometers and have precise control over their shape [9, 10]. The process involves applying a strong electric field between a nozzle containing a graphene-based solution and a grounded metallic collector plate. When the electric field is strong enough, a droplet elongates and forms a continuous jet, ultimately depositing graphene-based fibers or spherical structures on the collector plate. One of the great benefits of electrospinning/electrospraying is its ability to fine-tune the process parameters, allowing for the creation of materials with specific properties. Researchers have made numerous attempts to integrate graphene into fiber structures using both classic and core-shell electrospinning technologies. The core-shell approach has recently gained attention due to its ability to address challenges and enhance control over the resulting structures [11].

Poudeh and colleagues introduced a novel design for creating 3D graphene-based hollow and filled polymeric spheres through a one-step core-shell electrospraying technique (**Figure 2**) [12]. To achieve the desired spherical morphology, the study determined the optimal polymer concentration using Mark–Houwink–Sakurada equations [12]. Proper polymer concentration and solution viscosity are crucial for obtaining the desired spherical shape. In cases where hollowness is desired, the core material should contain a solvent with a higher vapor pressure than the shell solution. By using this innovative approach, researchers have overcome challenges such as crumbling and agglomeration of 2D graphene sheets and ensured better dispersion of graphene layers through the polymer chains. This technique has expanded the potential applications of fabricated structures across various fields, including drug delivery, energy storage, sensors, and nanocomposites. The final morphology of the produced materials is influenced by a combination of solution properties, such as viscosity and electrical conductivity, and process parameters including applied voltage and flow rate.

The interactions that occur between polymeric chains and graphene sheets throughout the sphere formation process are essential in determining the characteristics and capabilities of the subsequently formed materials. The core-shell electrospraying method unveils novel prospects for creating sophisticated graphene-centric structures characterized by customized properties, thus presenting a promising

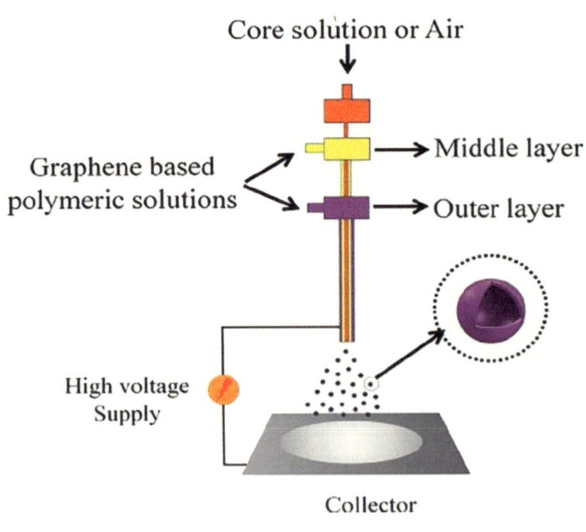

Figure 2.
Preparation of 3D graphene-based spheres by tri-axial electrospraying technique [12].

path for assorted materials science and technology applications. In essence, the electrospinning and electrospraying techniques grant meticulous control in synthesizing graphene-centric materials, enabling the development of structures with the preferred morphologies and attributes. The cutting-edge progress in core-shell electrospraying unveils pathways for creating versatile materials with multifunctional applications spanning various domains. Incorporating graphene into such structures is poised to fuel advancements in sectors like drug delivery, energy storage, sensing technologies, and nanocomposite materials, showcasing immense potential for breakthroughs in innovation.

2.3 Supercritical drying method

Traditional drying techniques, including air drying and vacuum drying, present substantial challenges when employed on graphene hydrogels due to their minimal solid content. The elevated capillary forces induced by solvent evaporation during these methods can disrupt and collapse the graphene network structure, inflicting irreversible damage. To resolve this problem and uphold the network's integrity, advanced drying approaches such as supercritical drying and freeze-drying have been introduced. Supercritical drying emerged as a leading method for synthesizing materials like SiO_2 aerogels. This technique transforms the solvent into a supercritical fluid by meticulously modulating temperature and pressure conditions. At the supercritical state, the lack of a distinct liquid–gas interface effectively eradicates capillary pressure within the network, allowing for the gentle removal of the supercritical fluid through depressurization, yielding aerogels. This method has shown tremendous efficacy in retaining the pristine network structure of SiO_2 aerogels with minimal contraction [13]. However, for graphene-based materials such as 3DGAs, certain constraints exist.

In contrast to SiO_2 aerogels, which possess a network with robust Si-O covalent bonds, 3DGAs have graphene sheets bound by π-π interactions. These sheets are prone to slippage during supercritical drying, causing increased volume and density reduction compared to SiO_2 aerogels. Consequently, supercritical drying is not ideal for fabricating large-scale, low-density 3DGAs. Additionally, the technique's requirement for high pressure and extended durations elevates production costs, making it less advantageous for preparing 3DGAs designated for low-density, large-scale utilization [14]. Therefore, alternative methods, like freeze-drying and air drying, might be more apt for specific graphene-based materials, considering the respective application intents and requisite attributes.

2.4 Freeze drying method

Lyophilization, or freeze drying, is a feasible alternative for mitigating capillary forces during the drying phase of materials like graphene-based hydrogels. This method entails multiple phases to maintain the structural wholeness of the hydrogel, initiated by substituting the solvent in the wet graphene hydrogel with water. The hydrogel is then exposed to freezing at sub-zero temperatures, and the encapsulated water transitions directly to gas via sublimation under vacuum, avoiding the liquid phase completely and thereby preserving the graphene framework. For freeze-drying, specific considerations are essential. When utilized on SiO_2 gels possessing diminutive pore sizes, often around tens of nanometers, the volumetric expansion of water during freezing might compromise the network structure, potentially decreasing the porosity to levels around 80% and halving the material's specific surface area relative to outcomes from supercritical drying. In some instances, the stress induced by freezing might even induce fractures in the sample [15]. Nonetheless, graphene hydrogels, due to their expansive pore dimensions- ranging from several micrometers to tens of micrometers—are more compatible with freeze-drying compared to SiO_2 hydrogels. 3DGAs fabricated via freeze drying exhibit diminished densities, possibly down to 0.16 mg/cm^3, and the method permits structural control by influencing ice crystal formation [16].

However, the process is not devoid of challenges; the water freezing can potentially distort and eject the microstructure of the graphene network, thereby reducing the specific surface area Moreover, freeze drying demands sub-zero temperatures, elevated vacuum levels, and prolonged drying durations, escalating energy usage and preparation expenditures. While lyophilization provides a substantial alternative for contending with capillary forces in the drying of graphene-based hydrogels. It is particularly favorable for graphene hydrogels due to their extensive pore dimensions and potential to produce low-density 3DGAs. However, it involves compromises like potential reductions in specific surface area and heightened energy and cost inputs, which must be weighed against the benefits depending on the intended applications.

2.5 Vacuum drying method

While effective for preserving the structural integrity of 3DGAs, supercritical and freeze drying come with certain drawbacks that hinder their suitability for large-scale production. The main challenges of specific drying methods arise from the requisition of low vacuum, low temperatures, or high-pressure states, which significantly amplify equipment, energy, and time costs. As highlighted earlier, supercritical drying entails modifying a solvent into a supercritical fluid under designated temperature and pressure states, followed by depressurization to yield aerogels. While it proficiently preserves the initial network structure of substances like silica aerogels, it encounters constraints when implemented on 3DGAs. The π–π interactions connecting the graphene sheets tend to displace, causing notable volume reduction in the supercritical drying phase, yielding structures with densities higher than what is optimal for applications necessitating low-density 3DGAs [17]. Furthermore, the time-intensive nature of supercritical drying, coupled with high-pressure conditions, escalates production expenditures. Freeze drying is another alternative, proffering benefits in maintaining the structural soundness of 3DGAs, especially those with expansive pore sizes. Still, it demands elevated vacuum levels, low temperatures, and prolonged durations, elevating energy usage and subsequent costs, especially when juxtaposed with other drying techniques.

In contrast, vacuum drying is a more economically and temporally efficient approach for 3DGAs synthesis, exempting the process from the need for specialized apparatus for temperature, high pressure, or solvent exchange and instead utilizing vacuum conditions for solvent removal. However, the process is challenged by the potential collapse of the delicate network structures of 3D graphene gels, intensifying the density of the final product. While such collapse can be perceived as a constraint, it can also be strategically utilized for the construction of dense 3D graphene network assemblies, marked by their enhanced volumetric/gravimetric energy density, making them ideal for advanced energy storage devices like high-performance batteries. Vacuum drying is a pragmatic and economical technique for developing 3D graphene network assemblies, especially when high volumetric/gravimetric energy density is prioritized. The induced collapse of the 3D network can be optimized to develop materials with properties aligned with the needs of diverse applications, predominantly in advanced energy storage sectors.

2.6 Air drying method

Air drying serves as a simplistic and economical method conducted under atmospheric pressure conditions, presenting a practical solution for drying substances like SiO_2 gels. This method is especially beneficial for the mass production of aerogels compared to intricate techniques such as supercritical drying and freeze-drying. Nonetheless, air drying has challenges; the Young–Laplace equation illustrates that, during the process, the interaction between liquid and air within the pores can create significant capillary pressure due to the minuscule pore size, typically in the tens of nanometers range. This pressure can result in the gradual shrinkage and cracking of the SiO_2 gel skeleton. Various approaches are implemented to counteract these issues, including replacing the solvent in the wet gel with one of lower surface tension to minimize capillary forces and chemically treating the SiO_2 gel skeleton to avoid further condensation of adjacent surface functional groups on the skeleton under capillary stresses. Introducing non-polar groups to the skeleton is vital in maintaining the structural integrity of SiO_2 aerogels throughout the air-drying process [14].

Contrarily, 3DGAs are more adaptable to large-scale production via air drying due to their larger pore sizes, diminishing the capillary forces experienced during the process. Several research endeavors have effectively employed air drying to construct 3DGAs with impressive properties. For instance, durable and flexible 3DGAs have been fabricated by in situ polymerization of polyacrylamide, fortifying the graphene network and allowing it to endure capillary forces during air and vacuum drying while preserving low density. Other investigations have utilized diverse methods, such as GO liquid crystal-stabilized bubbles and ice crystals and air bubbles as templates, to manufacture 3DGAs with the air-drying method, resulting in strong and uniformly porous structures with excellent mechanical properties [18]. Recent research has evidenced the fruitful application of air drying in creating resilient 3DGAs, underscoring the role of aspects like graphene sheet sizes, interconnections between graphene sheets, and the overall graphene network structure in influencing the mechanical characteristics of these materials. Optimization of these elements is paramount for the creation of robust graphene networks with tailored properties suitable for diverse applications.

2.7 Photoinduced reduction

Photoinduced reduction is an innovative process that reduces graphene oxide without the use of chemicals. It delivers high efficiency and rapid processing. This approach harnesses photon energy and has shown remarkable progress as a technique for GO reduction. Researchers began using photon energy for the reduction of GO in solutions as early as 2008 [19]. They employed semiconductor photocatalysts like titanium dioxide (TiO_2) under ultraviolet (UV) light irradiation. This approach paved the way for the development of graphene/semiconductor composites. Later, it was discovered that GO reduction could occur directly under UV light, yielding by-products such as CO, O_2, H_2O, and CO_2 [20]. Additionally, xenon lamp-equipped photographic camera lights and laser lights were found to be capable of thermally deoxygenating GO. In the early days, photoreduction of GO in dilute solutions typically resulted in rGO suspension instead of the desired 3D rGO structures. However, the photoreduction of GO films led to the formation of pore-rich 3D rGO materials characterized by high conductivity and expanded interlayer spacing. Photoinduced reduction of GO occurs through two distinct mechanisms: the photothermal effect and photochemical reduction. Photochemical reduction involves the use of UV light with a wavelength below 390 nm. Wavelengths longer than 390 nm induce the photothermal effect, which involves higher temperatures (approximately 200–230°C) and is highly effective for GO reduction. Laser lights can create intense localized heating by focusing the laser beam on a specific point, leading to rapid reduction [20].

In both mechanisms, strong excitation of the GO surface triggers particle ejection and the formation of plasma plumes. Energy transfer from the plume to the lattice results in the removal of oxygen groups from GO, leading to the formation of CO and CO_2, as well as the evaporation of internal water. This process generates interlayer pressure, causing the rGO layers to expand and form a porous structure. The photoinduced reduction process is also applicable to chemically reduced GO films. Laser lights are widely employed for this purpose. Laser scribing allows for synchronous reduction while patterning GO films. By adjusting laser-processing variables such as intensity, shifting pitch, and scanning pitch, precise control over the degree of reduction in the resulting 3DGAs can be achieved. These 3D graphene patterns, generated through laser-induced reduction, hold significant potential in various biomedical applications, including tissue engineering and cell culture.

The reaction atmosphere plays a crucial role in laser-induced GO reduction, influencing the degree of reduction. The process is notably enhanced in an oxygen-free environment. When the GO precursor is subjected to laser reduction under liquid nitrogen, thermal expansion is suppressed, resulting in 3D rGO films with fewer defects and higher conductivity. Laser intensity can also be directly used to assemble 3D rGO in crystalline GO suspension. By adjusting the focus and intensity of the laser, researchers can create arbitrary homogeneous 3D structures on the inside surface. This approach is particularly favorable for precise 3D localization in electronics and photonics applications. This technique has evolved significantly since its inception, allowing for precise control over the reduction process and opening up a wide range of applications, from composite materials to biomedical scaffolds and electronics.

2.8 Chemical reduction

Chemical reduction is a widely used method to reduce GO to rGO using various chemical routes. Initially, hydrazine was used to reduce dilute GO suspensions, leading to rGO dispersions. Later, researchers focused on synthesizing 3D graphene hydrogels and aerogels using more concentrated GO colloidal solutions and milder reducing agents. The reduction temperature typically ranges from 60 to 100°C, with reagents and absorbed water removed through washing and lyophilization. Several reducing agents, including Na_2S, HI, H_3PO_4/I_2, gelatin, and sodium ascorbate, have been explored to simultaneously reduce and construct 3D GO structures. For instance, sodium ascorbate has been employed for the reduction of GO, resulting in the development of 3D graphene frameworks [21]. The formation of π-π interlinkages among sheets has been identified as a key factor governing the self-assembly of chemically reduced GO. Additionally, the functional groups attached to reducing agents play a crucial role in 3D rGO sheet construction. Covalent bonds form between GO sheets and reducing agents, leading to the substitution of oxygen-containing hydroxyl and epoxy groups and cross-linking of the GO sheets. Carboxylic groups present in reducing agents further promote the development of interlayer hydrogen bonds.

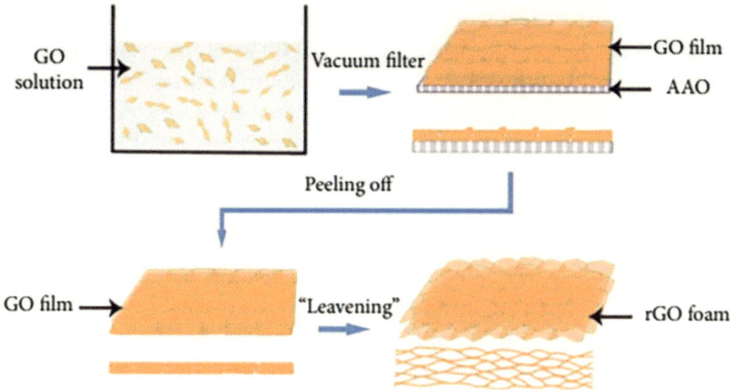

Figure 3.
Schematic of the fermentation process to prepare rGO foam [22].

During the chemical reduction of GO, the evolution of CO and CO_2 gases is common, contributing to the restacking of GO and rGO sheets and modulating the pore structure of 3D materials. Water bubbles can also be introduced through controlled heating, and the number of bubbles plays a critical role. Fewer bubbles may fail to suppress restacking, while an excess can affect the self-assembly of rGO sheets. Hydrazine is considered favorable for reducing GO into 3D structures due to its ability to generate an appropriate number of bubbles. Active metals such as Al, Cu, Al, Fe, and Co can also serve as reducing agents, enabling both reduction and the formation of 3DGAs assemblies on their surfaces. This process relies on redox reactions between metals and GO, occurring at room temperature and accelerating at 60°C. Interestingly, when a conductive substance is placed on the active metal substrate, no reduction in the development of 3D rGO assembly is observed. This unique property has been harnessed to create specialized graphene structures, such as microtubes and micropatterns, with applications in rechargeable Li-ion batteries and flexible rGO 3D thin film devices.

Moreover, vapors produced by reducing agents can also effectively reduce both dried GO films and sponges, leading to side-by-side alterations in 3D structures. For instance, when a dense GO film is used as a precursor, vapor-based reduction significantly alters its 3D configuration. Niu et al. performed the reduction of filtrated GO film by positioning it on the top of a hydrazine monohydrate solution at 90°C for 10 hours (**Figure 3**) [22]. By controlling the concentration of the hydrazine monohydrate solution, the open pore structure can be tuned.

Despite the progress in the chemical reduction of GO, there is still much to learn about the changes that occur during the reduction process. Minimizing the amount of non-carbon impurities in the final product remains a challenge, necessitating further research and optimization of reduction methods. Chemical reduction offers a versatile approach to obtaining 3D graphene structures, with various reducing agents and strategies available to tailor the properties and morphology of the resulting materials.

2.9 Electrochemical reduction

Electrochemical reduction is a prevalent technique for producing 3D graphene constructs, particularly in crafting electrodes for electrochemical devices directly. It permits the formation of resilient 3D graphene frameworks straight on electrode surfaces, optimizing the efficacy of several electrochemical setups. The procedure of electrochemical reduction requires settling GO on a cathode from a GO mixture. Contrasting typical solid graphite strata, the graphene strata in 3D formations are aligned in a fortifying manner, establishing a stable and integrated 3D graphene grid. This method is frequently termed as the immediate expansion of rGO sheets from the electrode interface. The developed 3D graphene lattice presents numerous perks such as superior porosity and escalated electrochemical efficacy.

A crucial consideration in developing 3D rGO substances electrochemically is the selection of the electrode material. Several substances like stainless steel lattice, platinum (Pt) leaf, nickel (Ni) froth, and gold (Au) fiber are feasible as electrode foundations. Utilizing graphene paper as the foundation for the electrode resulted in a closely bonded structure of highly porous rGO layers onto the substrate. Consequently, this yielded a carbon electrode with outstanding versatility for flexible device applications. Employing Ni foam as the electrode base results in the interior vacuities being occupied by rGO, forging a systematic porous framework with diverse pore dimensions. Moreover, hierarchical 3D structures can be synthesized by employing previously acquired 3D rGO substances as electrode foundations [23]. The electrochemical technique is also viable for reducing pre-shaped GO films on electrodes. This method provides superior regulation over the alignment of GO sheets in the casting phase and assures proficient reduction via the electrochemical mechanism. The electrochemical reduction of GO facilitates the production of high-caliber electrodes with customized 3D configurations, enriching their utilization in energy conservation apparatuses, detectors, and other varied electrochemical setups. Electrochemical reduction is a multifaceted and efficient strategy for developing 3D graphene formations, especially significant for the advancement of electrodes in electrochemical appliances. The 3D graphene frameworks that result possess enhanced electrochemical traits, rendering them instrumental in an extensive array of applications.

2.10 Thermal expansion of GO bulk

Expanding GO bulk materials through thermal reduction or annealing is a well-known and efficient method to eliminate oxygen functional groups from GO, yielding 3DGAs with increased bulk volumes. This technique entails exposing GO precursors to elevated temperatures, usually between 800 and 1000°C, inducing a series of structural and chemical modifications. For 3D graphene, high-temperature annealing can trigger thermal exfoliation, causing the expansion of dried and pre-shaped GO bulk precursors. These precursors may appear in several forms, including granular GO, misaligned GO films, tape-cast layers, and bulk GO materials. During the thermal evolution of bulk GO, oxygen-containing functional groups such as carboxyl, hydroxyl, carbonyl, and ether groups are either partially removed or converted into more stable chemical bonds like anhydrides, quinones, and lactones. This leads to the emission of trace gases including CO_2, CO, and H_2O. The degree of purification of the graphene surface is augmented with increased annealing temperatures. At maximum temperatures of 1000°C, unstable oxygen species are efficiently eradicated, enhancing the electrical conductivity of the resultant rGO sheets significantly [24].

Concurrently, the GO mass morphs into honeycomb-esque 3D nanostructures composed of microscopic sheets and numerous pores. This enlargement process contributes to a notable amplification in the material's specific surface area. Depending on the specific precursor and the subsequent reaction, the surface area of 3D rGO structures can generally vary from 400 to 800 m^2/g. Thermal annealing is also utilized for the reduction of partially reduced 3D GO and 3DGAs achieved through dry processes and intense reduction procedures. In such instances, the main goal might not be the modification of 3D structures, but rather the expulsion of oxygen functional groups and conductivity augmentation. Nonetheless, exposing GO precursors to elevated temperatures swiftly, without a gradual heating process, can induce the creation of additional in-wall pores, increasing the surface areas. For instance, subjecting crumpled graphene spheres to thermal shock at 400°C can produce a surface area of roughly 567 m^2/g, whereas maintaining the sample at the same temperature without rapid cooling can yield a lower surface area of 344 m^2/g. Thermal expansion through annealing is a flexible and popular approach to fabricating 3DGAs with enhanced bulk volumes and superior electrical conductivity. This technique involves the elimination of oxygen functional groups and the formation of porous 3D nanostructures, positioning it as a valuable technique for diverse applications in material science and nanotechnology.

2.11 Solvothermal and hydrothermal reduction

The solvothermal and hydrothermal reduction techniques are increasingly recognized for their effectiveness in transforming GO to rGO and crafting 3D graphene structures. These procedures revolve around high-temperature reactions in autoclaves, usually ranging between 100 and 250°C, using water or organic solvents as reducers. In hydrothermal treatments, supercritical water driven by thermal activation serves as the reducer, facilitating the transformation of GO sheets into 3D rGO hydrogels. Early in this process, the surface charges are reduced due to the extraction of carboxylic groups from GO sheets. As the reaction progresses, most carboxylic groups vanish, increasing hydrophobicity and enhancing the attractions and interactions among the rGO sheets. This facilitates the formation of a closely-knit 3D structure. By-products commonly include CO_2, with minor amounts of CO and organic acid remnants. Even if some GO sheets remain intact post-treatment, their oxygen-functional groups help link the 3D rGO sheets, forming a monolithic design. Factors like the initial GO suspension's size, concentration, and surface traits can notably alter the final 3D rGO gel's makeup.

In solvothermal treatments, organic solvents take the place of water, leading to milder reaction temperatures and decreased GO sheet assembly. Given their inherent high pressure and the solvent's low surface energy, solvothermal processes are adept at shaping and reducing GO sheets. This results in 3D rGO products with superior conductivity but diminished surface areas and bulkier walls compared to their hydrothermal counterparts. For example, using ethanol in solvothermal reduction can yield 3D rGO aerogels with distinct features, like near-zero super-elasticity.

Post-reduction drying is vital in maintaining the eventual 3D shape. Traditional drying might induce significant volume contraction and cracks, stemming from substantial capillary pressure during solvent evaporation. To mitigate this, alternative techniques like freeze-drying or supercritical CO_2 drying are adopted. They reduce structural degradation since ice or supercritical CO_2 have minimal interactions with graphene. The capillary pressure experienced during solvent evaporation is influenced by aspects like solvent surface tension, contact angle, and pore size. Several methods are explored to achieve ambient drying while preserving the 3D structure and elevating material rigidity. While the degrees of reduction achieved in solvothermal and hydrothermal reactions are somewhat moderate due to their temperature constraints, further treatments like annealing or chemical reduction can be applied to amplify reduction and reinstate π-conjugation. These augmentations also bolster electrical conductivity. Moreover, these processes can be employed to manufacture 3D graphene-infused composites by embedding nanoparticles like Fe_3O_4 or CO_3O_4 into the 3DGAs. Solvothermal treatments especially lead to a more evenly spread nanoparticle distribution within the 3D rGO, creating consistent composites. The solvothermal and hydrothermal reduction techniques are potent tools for devising 3D graphene structures, aerogels, and composites. The selection of solvents and drying methods remains essential in shaping the final material's attributes and structure.

2.12 Freeze-casting process (pre- or post-reduction)

The freeze-casting process is a well-established solution-phase technique widely employed for the fabrication of 3D structures from GO or partially reduced GO suspensions or gels. This method leverages the freezing point of water and capitalizes on ice crystallization to arrange GO or partially reduced GO sheets into a continuous 3D framework. For this to occur, it is crucial to exceed the percolation threshold by concentrating GO or partially reduced rGO sheets. After freeze-casting, a subsequent reduction step is usually necessary to transform the porous GO monolith into a 3D rGO framework, thereby modifying the surface properties of the sheets. However, it should be noted that the freeze-casting process and subsequent reduction may have a marginal effect on the microscopic morphology of the final product. In the freeze-casting of GO suspensions, the chemical characteristics of GO sheets play a pivotal role. When GO suspensions are frozen directly, they produce monoliths that are brittle and randomly oriented. In contrast, when freeze-drying is applied to GO-filtered gels or specific partially reduced GO dispersions, it results in super-elastic structures with a honeycomb-like cellular orientation. In part, this contrast in monolith structure can be explained by the heightened attraction between partially reduced GO sheets. Furthermore, the capacity of partially reduced GO sheets to adsorb onto ice surfaces, owing to their hydrophobic properties and the presence of abundant oxygen-containing groups, facilitates the development of these super-elastic configurations [25].

The term "super-elasticity," as applied to freeze-cast partially reduced GO monoliths, characterizes their exceptional ability to endure substantial deformation and rapidly recover, rendering them exceptionally resilient materials. These materials exhibit remarkable load-bearing capabilities, sustaining loads up to 450,000 times their weight and rebounding from 480% compression quite quickly. Moreover, the freezing process's temperature holds a significant role in shaping the 3D structure during freeze-casting. Various freezing temperatures affect ice crystal growth dynamics, resulting in different wall thicknesses, overall dimensions, and pore architectures. Through precise control of the freezing temperature, researchers gain the ability to tailor the freeze-cast material's properties to precise specifications. In recent years, bidirectional freezing technologies have also emerged as valuable tools for controlling the freezing process precisely. This innovative approach facilitates the creation of distinctive structures, such as fan-shaped arrangements of GO, which find applications across diverse fields, including water purification. Moreover, freeze-casting can use alternative solvents, such as those characterized by high vapor pressures and melting points slightly higher than room temperature. This variation, known as room-temperature freeze gelation, presents significant energy-saving advantages compared to freeze-drying with water. However, it's worth noting that the cost associated with these organic solvents may present challenges for large-scale applications. The freeze-casting method stands as a versatile and efficient technique for producing

3D structures from GO or partially reduced GO suspensions or gels. The method allows the fabrication of super-elastic, highly resilient materials that can be tuned by manipulating freezing parameters and applying bidirectional freezing methodologies. This approach holds great promise across a broad spectrum of applications, encompassing domains like water purification and the development of lightweight structural materials.

2.13 3D printing before reduction

The extrusion-based 3D printing method, also known as robot-assisted deposition, robocasting, direct ink writing, or continuous extrusion, stands out for its ability to fabricate complex 3D rGO structures that are challenging to achieve using alternative manufacturing techniques. This technique involves the layer-by-layer deposition of GO or partially reduced GO inks, enabling the creation of intricate 3D structures. The effectiveness of extrusion-based 3D printing depends on the rheological properties of the GO ink. Specifically, the ink should exhibit shear-thinning behavior and viscoelastic characteristics to ensure proper adhesion between layers while maintaining the desired print shape. The rheological properties of GO inks are influenced by various factors, including flake size and GO concentration. Smaller lateral sizes of GO sheets, typically in the range of 150–400 nm, display non-Newtonian fluid behavior at a concentration of 20 mg/mL. Higher concentrations result in increased viscosity and improved printability. To enhance printability further, modified GO suspensions are often used, incorporating additives such as pH-sensitive polymers, hydrophobic fumed silica powder, and basic compounds. These additives help regulate viscosity and shear-yield stress, making it possible to 3D print the inks effectively. In addition to extrusion-based 3D printing, optical 3D printing has emerged as a precise technique for manipulating micrometer-scale structures. This approach combines photo-excited polymerization with layer-by-layer 3D printing to generate complex macro-3D rGO architectures. The photocurable resin used in this process is created by blending diluted GO dispersion with photocurable acrylates and a photoinitiator. This resin can rapidly solidify through light-initiated polymerization. During optical 3D printing, thin layers of the resin are deposited, cross-linked, and solidified using patterned light irradiation. This process is repeated for subsequent layers, resulting in the gradual construction of 3D structures. A light source with a wavelength of 405 nm is typically employed, along with a spatial light modulator, to enable precise control over 3D patterns and resolutions. This technology allows for the fabrication of intricate 3D structures with controlled micron-scale dimensions, maintaining a high resolution of approximately 10 μm (**Figure 4**) [26].

2.14 Wet-spinning before reduction

Wet-spinning is a widely employed technique for the fabrication of 3D rGO structures, encompassing a range of forms such as fabrics, films, cylinders, spheres, and fibers. This method offers continuous alignment of rGO materials, resulting in the development of materials with distinctive properties. The wet-spinning process involves extruding a GO suspension through a nozzle into a coagulation bath. The concentration of GO within the suspension plays a crucial role in the successful formation of fibers. High concentrations of GO in the suspension foster strong interactions among GO sheets, promoting alignment and coagulation. Conversely, low concentrations can lead to the formation of brittle fibers and collapsed structures. It's important

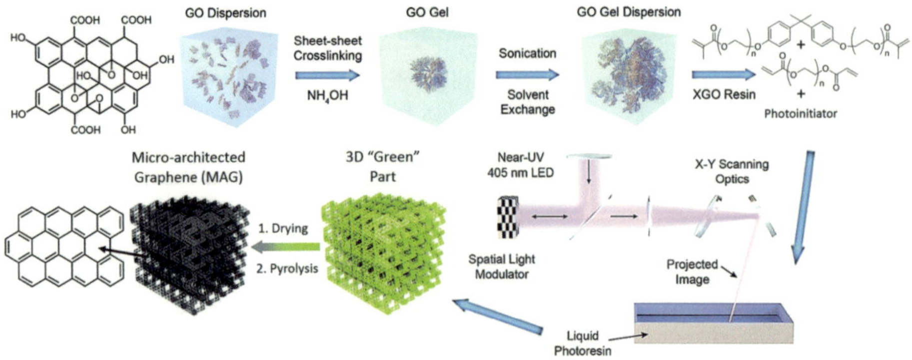

Figure 4.
Illustration of the optical 3D printing process [26].

to note that the formation of graphene fibers through wet-spinning follows a multi-step mechanism. Initially, a multilayer GO film is generated as negatively charged GO sheets repel each other. As the charge neutralizes within the coagulation bath, the film undergoes bending and folding, resulting in highly aligned fine particles. The versatility of wet-spinning allows for the production of 3D graphene materials with diverse shapes and structures by controlling the rotation process. For instance, the wet-spinning of liquid GO crystals into a rotating coagulation bath can yield super-elastic graphene aerogel millispheres. These millispheres possess continuous shell and core structures, exceptional mechanical strength, and unique jumping properties. Wet-spinning, when followed by a reduction step, emerges as a versatile and efficient approach for manufacturing a wide array of 3D rGO structures with tailored properties. The selection of GO concentration, spinning conditions, and parameters for the reduction process can be finely tuned to achieve specific characteristics in the final material. This method finds applications across various domains that demand high-performance graphene-based materials, showcasing its versatility and adaptability.

3. Properties of 3D graphene

Since their initial creation in 2011, 3DGAs have consistently showcased exceptional properties. The Table below provides an overview of the physical characteristics that define various 3DGAs, encompassing attributes such as specific surface area (SSA), electrical conductivity, pore structure, density, and mechanical properties. It highlights the remarkable qualities inherent to three-dimensional graphene networks, with a specific focus on GFs, graphene sponges (GSs), and graphene aerogels (GAs). These materials stand out for their impressive features, including substantial surface areas and pore volumes, reduced densities, noteworthy electrical conductivities, and robust mechanical performance. In a general context, it is evident that 3DGAs characterized by chemically bonded structures exhibit notably superior properties in comparison to those relying on physically assembled structures. As a result of these chemically bonded variants, they consistently demonstrate advantages such as reduced contact resistance, enhanced electrical conductivity, and superior mechanical properties, such as heightened strength, toughness, and flexibility. As illustrated in the **Table 1**, chemically bonded GAs have a bulk electrical conductivity of approximately 1 S/cm. This level of conductivity outperforms what has been reported for

Synthesis method	Specific surface area (m^2/g)	Pore structure	Density (mg/cm^3)	Electrical conductivity (S/cm)	Mechanical property	Ref.
GF/Ni Foam template-assisted CVD growth	~ 850 (~3 Layer graphene)	Porosity ~ 99.7% Pore size ~400 μm	~5 (~3 Layer graphene)	~7 (~3 Layer graphene)	Flexible	[27]
GS/freezing drying of GH	53.4	Pore size 2–100 nm	9	0.54	Good Flexibility, Young's Moduli 7.56 kPa	[27]
GS/Self-assembly of GO, hydrothermal freeze-drying	423	Long axis ~570–620 μm short axis ~150–300 μm	12 ± 5			[28]
GA/ supercritical drying or freeze drying of GH	512	Pore volumes 2.48 cm^3/g, pore size 2–100 nm	12–96	~1	Young's moduli 1.2–6.2 MPa	[14]
GA/organic cross-linker	584	Pore volume 2.96 cm^3/g, pore size 1–100 nm	10	~1		[29]
GA/organic cross-linker		Porosity 99.7–99.8%	3–5		Highly compressible	[16]
GA/Chemical reduction			15	0.87		[21]
GH/ hydrothermal		Pore size ~ 0.1–7 μm		5×10^{-3}		[30]
3D RGO/ hydrothermal freeze-drying			30	2.5×10^{-3}	Compressive strength 0.042 MPa compress modulus 0.26 MPa	[31]

Table 1.
The physical characteristics of various 3DGAs have been synthesized through diverse methods.

macroscopic 3DGAs exclusively constructed using physical cross-linkers or partial chemical bonding. For instance, GO hydrogel, with a conductivity of 5×10^{-3} S/cm, and 3D RGO, with a conductivity of 2.5×10^{-3} S/cm, significantly lag in this regard. In contrast, GFs exhibit significantly higher electrical conductivity due to their continuous and interconnected networks. Notably, GAs consistently demonstrate high electrical conductivities of around 1 S/cm, regardless of whether or not an organic cross-linker is employed.

This phenomenon may be attributed to the cross-linking interactions that occur among functional groups present on the surfaces and edges of the graphene oxide during the sol–gel process. Alterations in fabrication conditions lead to variations in structural features, including the orientation of graphene sheets, physical or chemical interconnections between these sheets, pore size, porosity, and the number of graphene sheet layers. Consequently, these structural changes affect the properties of the material. For example, GFs possess an isotropic structure due to the absence of specific graphene sheet orientation, while GSs exhibit anisotropic structures. In the case of GSs, large graphene films align nearly parallel to one another, resulting in an anisotropic arrangement. The adaptable manufacturing process employed for GFs provides precise control over both their macrostructure and microstructure [31]. The pore structure within Ni foam can be carefully manipulated to tailor the pore size and porosity to meet specific requirements. Simultaneously, the adjustment of CH_4 concentration plays a vital role in influencing several key characteristics of GFs, including the average number of graphene layers, surface area, and overall density. The concentration of CH_4 enables more graphene layers to be formed, resulting in a substantial change in SSA, density, and electrical conductivity. Intriguingly, it is noteworthy that the electrical conductivity of GFs exhibits an initial increase followed by a subsequent decrease as the number of graphene layers escalates.

In a similar vein, the microstructure and characteristics of graphene sponges (GSs) can be finely tailored by making adjustments to the synthesis conditions. Particularly, when GSs are fabricated through the freeze-drying process using graphene hydrogels (GH), the freezing temperature emerges as a crucial factor (**Figure 5**). Changing freezing temperatures cause significant pore size and wall thickness changes, changes in pore morphology from anisotropic lamellar to uniform cellular structures, and changes in Young's modulus. It's worth emphasizing that the mean pore size exerts a direct influence on water absorption properties, where larger pores impart water resistance to the sponge, while smaller pores facilitate water absorption. This tunable aspect of GS properties, contingent on freezing temperature, offers remarkable versatility for an array of applications [32].

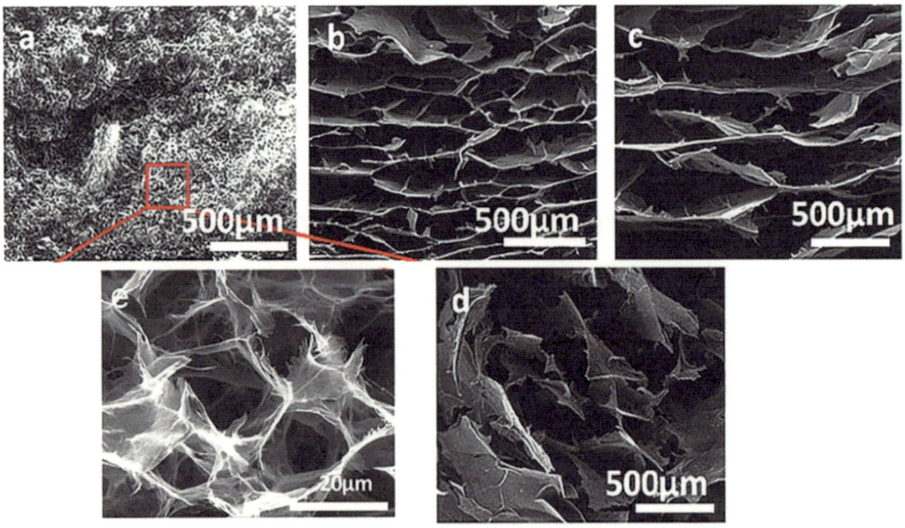

Figure 5.
(a–d) SEM images of the porous structures of four 3DGAs fabricated at different freezing temperatures of −170, −40, −20, and − 10°C, respectively. (e) Enlargement of the square area in image (a) [32].

4. Applications of 3D graphene

4.1 Drug delivery and cancer treatment

In conventional drug delivery methods, a range of challenges can arise, including low solubility, insufficient selectivity, and toxicity concerns. To address these limitations, researchers have increasingly used carbon nanotubes (CNTs), a subgroup of graphene, as drug delivery platforms. CNTs offer a substantial surface-to-volume ratio, enabling the incorporation of significant drug volumes, encompassing both hydrophilic and lipophilic drugs. Drugs can be loaded either on the exterior or inside of CNTs, making this a viable option for multidrug therapies. Of notable importance, drugs attached to the surface of CNTs possess the capability to recognize specific cell receptors and traverse cellular barriers without inducing toxicity. This innovative approach has demonstrated significant potential in the delivery of medications such as methotrexate, epirubicin, and doxorubicin, particularly in the context of cancer treatment [33].

4.2 Tissue engineering

Tissue engineering constitutes the development of functional tissues for purposes such as transplantation and regenerative medicine. Materials based on graphene have proven valuable in the design of scaffolds in this domain. The scaffolds promote crucial cellular processes, such as adhesion, proliferation, and differentiation. As a noteworthy example, the utilization of graphene oxide-modified 3D acellular cartilage extracellular matrix (ACM) scaffolds has yielded substantial advancements in the realm of cartilage regeneration. These scaffolds were found to promote good cell behavior and biocompatibility, making them promising for repairing cartilage

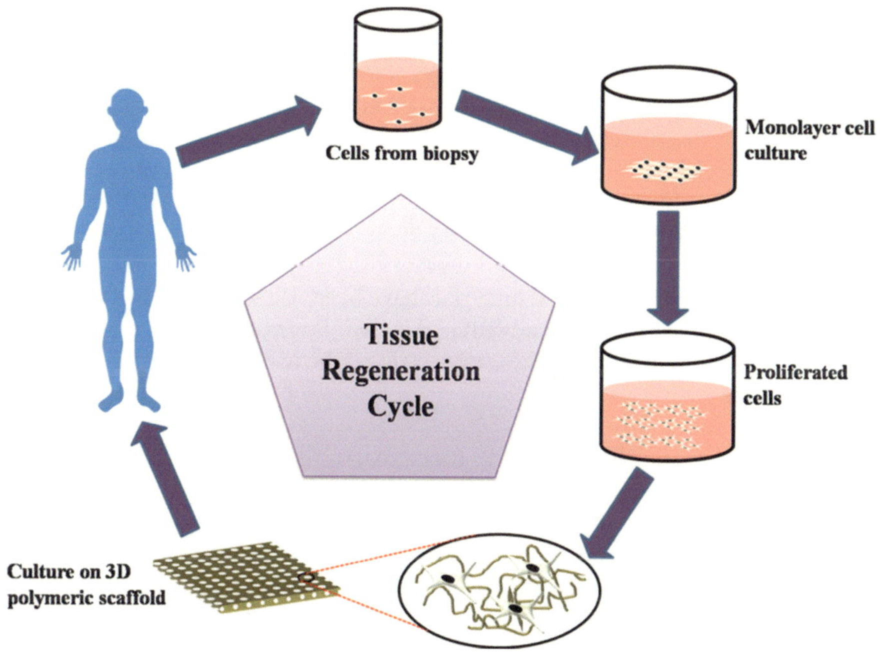

Figure 6.
Graphene-based scaffolds have been explored for bone tissue engineering [34].

injuries [34]. Additionally, as shown in **Figure 6**, graphene-based scaffolds have been extensively investigated in various tissue engineering applications, including bone tissue engineering, cardiac tissue engineering, neural tissue engineering, and skin tissue engineering. The versatility and potential of graphene-based scaffolds in tissue engineering are underscored by this multifaceted application.

4.3 Sensors for magnetic resonance imaging (MRI)

Graphene's unique properties have made it suitable for use in medical MRI sensors. Traditional sensors made from metals can introduce distortion and inhomogeneity to the magnetic field, leading to misdiagnosis. Graphene's high conductivity, strain resilience, and non-toxic nature make it an excellent candidate for replacing traditional metallic conductors in piezoelectric sensors used in MRI. Graphene-based sensors have shown promising results in providing clearer images with minimal distortion during MRI scans [35].

4.4 Stem cell-based transplant

Transplant therapy based on stem cells holds significant promise for the restoration of damaged tissues and organs, particularly within the realm of neural regeneration. Neural stem cells (NSCs) can self-renew and differentiate into a variety of neural cell types, making them invaluable for treating nerve damage, promoting neurogenesis, and stimulating axonal growth. Within the sphere of regenerative medicine, carbon-based materials, including CNTs, carbon nanofibers, and 3DGAs, have attracted considerable attention owing to their distinctive electrical, mechanical, and biological properties [36].

4.5 Dental and oral application

There has been considerable interest in graphene-based materials in the field of dentistry due to their exceptional properties, particularly their potential antibacterial properties [37]. These materials have demonstrated their effectiveness against a broad spectrum of bacterial pathogens, including both gram-positive and gram-negative bacteria [38]. The utilization of 3D GAs in dentistry presents various applications and advantages, as outlined below:

a. Tooth filling: Graphene-based materials have been utilized in tooth-filling procedures. One of their notable advantages is their low cytotoxicity, making them safe for dental applications. These materials can be used to fill cavities and repair damaged teeth [39].

b. Antibacterial properties: Graphene-based materials exhibit robust antibacterial properties, contributing to the prevention of bacterial growth in the oral cavity. This, in turn, reduces the risk of dental infections and enhances overall oral health. The effectiveness of this antibacterial action has been extensively investigated and documented in studies [39].

c. Biocompatibility: Graphene materials are biocompatible with oral tissues. This means they are well-tolerated by the body and do not induce adverse reactions or inflammation when in contact with oral tissues [40].

d. Enhanced imaging: Graphene oxide materials have also been investigated for their potential in enhancing dental imaging techniques such as X-rays. They can improve the contrast and quality of dental radiographs, aiding in more accurate diagnoses [41].

e. Tooth whitening: Graphene-based materials have been explored for their teeth-whitening properties [42]. They can be incorporated into dental products to help whiten teeth effectively.

f. Bioactive coatings: Graphene coatings can be applied to dental implants and orthodontic devices to improve their biocompatibility and reduce the risk of infections or complications [43].

Thus, a variety of dental procedures may benefit from graphene-based materials, including antibacterial properties, biocompatibility, and imaging enhancement. These properties make graphene-based materials promising candidates for improving oral health and dental treatments [42].

4.6 Supercapacitors

Due to their distinctive properties, such as high power density and extended cycle life, supercapacitors have received substantial attention as energy storage devices. These devices are categorized into two primary groups: pseudo-capacitors and electrochemical double-layer capacitors (EDLCs). Pseudo-capacitors, such as transition metal oxides and conducting polymers, store charges via chemical reactions on their surfaces. In contrast, EDLCs, frequently incorporating carbon-based materials such as graphene, store energy through ion adsorption at the electrode-electrolyte interface. Because of its large surface area and varied dimensions, graphene has gained widespread use in EDLC electrodes [44]. Recently, 3D graphene structures have emerged as compelling candidates for supercapacitors, thanks to their porous nature, expansive surface area, and interconnected networks, which enhance electrolyte ion accessibility and bolster electrical conductivity. The following 3D graphene configurations have been investigated for applications in supercapacitors:

a. Graphene spheres: Hollow micro/nanostructured graphene spheres deliver advantages such as a remarkable surface area and abbreviated diffusion length for charge and mass transport. A graphene-wrapped polyaniline hollow sphere demonstrated an exceptional specific capacitance of 614 F/g and maintained over 90% capacitance after 500 charging/discharging cycles.

b. Mesoporous graphene nanoballs: These nanoballs, synthesized via CVD, exhibited a high specific capacitance of 206 F/g at a scan rate of 5 mV/s and retained 96% of their capacitance after 10,000 charging and discharging cycles [44].

c. 3D graphene networks: Structures like GFs, sponges, and hydrogels have garnered attention owing to their porous architectures, which facilitate the movement of electrolyte ions within the graphene frameworks. The high energy density of spongelike graphene nanostructures enables them to retain 98% of their capacitance after 10,000 cycles, for instance [45].

d. Flexible supercapacitors: Flexible supercapacitors have been designed for wearable and portable devices. Graphene hydrogel films have been employed as electrodes in flexible solid-state supercapacitors, exhibiting high gravimetric capacitance and excellent cycling capability [45]. The films are mechanically robust and are well suited to flexible applications.

e. The graphene/MnO_2 composite network has been used to manufacture ultralight, freestanding, flexible supercapacitors. Using CVD on nickel foam and electrochemically depositing MnO_2, these networks achieved a high specific capacitance of 130 F/g and demonstrated low resistance variations upon vending [46].

So, 3DGAs have shown promise in supercapacitor applications, offering enhanced performance, high energy density, and excellent cycling capability. These structures, whether in the form of spheres, networks, or films, hold great potential for energy storage in various electronic devices and portable applications. The utilization of graphene in flexible supercapacitors further expands their range of applications, making them suitable for emerging technologies like electronic textiles and wearable devices [44].

4.7 Lithium-ion batteries

In recent years, 3DGAs have become the focus of extensive research due to their potential application as active electrodes in batteries, particularly in lithium-ion batteries (LIBs). These 3DGAs offer numerous advantages, including enhanced lifetime, higher energy density, and improved electrochemical performance, addressing some of the challenges typically associated with batteries, such as low reversible capacity and limited cyclic life. In the design of batteries, such as LIBs, it is crucial to consider the role of different components, including electrodes and electrolytes, in enhancing overall battery performance. In addition to its exceptional properties, 3D graphene is a highly promising candidate for achieving high-performance LIBs due to its large surface area, porous structure, rapid mass and charge transfer, and interconnected network. Among the key strategies for harnessing 3D graphene's potential for LIBs is to incorporate metal or metal oxides such as Sn, NiO, Fe_3O_4, and $LiFePO_4$, along with CNTs into graphene sheets. 3D graphene composites were formed as a result of these efforts, which exhibit remarkable electrochemical properties and significantly improve the performance of LIB electrodes. The incorporation of conductive nanomaterials such as graphene into Fe_3O_4 has attracted considerable attention, primarily due to the appealing characteristics of Fe_3O4 as a prospective electrode material for LIBs. Aside from its high theoretical capacity, Fe_3O_4 has the advantage of being cost-effective and non-toxic.

Stable electrode performance has been hindered by previously encountered challenges, such as Fe_3O_4's low conductivity and high volume expansion. To tackle these issues, researchers have pioneered the development of 3D GFs-supported Fe_3O_4 LIBs. This innovative approach has yielded LIB electrodes with an impressive capacity of 785 mA h/g at a 1 C charge–discharge rate, demonstrating stable performance over 500 cycles [47]. 3D graphene's incorporation into LIB electrodes confers several advantages in all the aforementioned studies. Firstly, it establishes a shorter path length for lithium ion transport, expediting charge and discharge processes. Secondly, it enhances electron transport and electrode conductivity. Lastly, it curbs the agglomeration of active materials, ensuring uniform and consistent electrode performance over extended cycling. The 3D graphene structures can significantly enhance LIB performance. Through the integration of diverse metal or metal oxide composites and CNTs, researchers have harnessed the distinctive attributes of 3D graphene, resulting in improved electrode materials characterized by enhanced capacity, stability, and rate capability. These advancements pave the way for the development of high-performance batteries, with applications spanning from portable electronics to electric vehicles [48].

4.8 Sensors

In recent years, graphene-based materials adorned with metals and metal oxides have emerged as highly promising contenders for a diverse array of sensing devices, encompassing electrochemical sensing and biosensing. As a result of graphene's exceptional optoelectronic attributes and the remarkable catalytic activity conferred by metals and metal oxides, graphene is attracting a growing amount of attention. Several applications involving these materials have yielded impressive results in terms of sensitivity, selectivity, and detection limits. Presented below are noteworthy instances of graphene-based sensing materials and their respective applications [49–52].

a. Ammonia (NH_3) and nitrogen dioxide (NO_2) detection utilizing 3D graphene networks: Yavari et al. developed a 3D graphene network tailored for the detection of NH_3 and NO_2 under standard room temperature and atmospheric pressure conditions [53]. Due to its high sensitivity, this sensor was able to detect these gases at minute concentrations within the ppm (parts per million) range. The sensor's exceptional performance can be attributed to the outstanding optoelectronic properties inherent to graphene, complemented by its intricate 3D structural design. This work holds substantial significance for applications such as air quality monitoring and environmental sensing.

b. Detection of hydrogen peroxide (H_2O_2) utilizing platinum-ruthenium bimetallic nanocatalyst-integrated 3D GFs: Kung et al. designed a sensor specifically for the detection of H_2O_2, featuring the integration of a platinum-ruthenium bimetallic nanocatalyst into 3DGFs [54]. Through this innovative approach, the surface area of the electrochemical reaction was increased, and efficient transport was facilitated. The resultant material exhibited remarkable performance in terms of electrochemical oxidation of H_2O_2, showcasing both high sensitivity and an impressively low detection limit. The application of these sensors is wide, ranging from healthcare to environmental monitoring to food safety.

c. Utilizing 3D graphene interconnected with polyaniline (PANI) nanofibers for the determination of guanine and adenine: Yang et al. fabricated 3D graphene structure intricately connected with GO by self-doped PANI nanofibers through a simple adsorption and intercalation route *via* sonication of the mixed dispersions of both components [55]. The distinctive free-standing, three-dimensional interconnected nanostructure arises from the robust π–π^* stacking interaction between PANI backbones and GO basal planes, coupled with electrostatic repulsion between negatively charged PANI and GO sheets. This nanocomposite exhibits a significant specific surface area, maintaining a uniform and stable dispersion with PANI. Consequently, it acquires high conductivity and excellent electrocatalytic activity. The nanocomposite's negative charge and specific structure facilitate the adsorption of positively charged guanine and adenine through strong π–π^* interactions or electrostatic adsorption. Leveraging these properties, the hybrid serves as an outstanding sensing platform for the highly sensitive determination of guanine and adenine. Beyond its applicability to guanine and adenine, this versatile platform lends itself to the determination of a multitude of other analytes, particularly within the realm of bioanalytical and clinical applications. These exemplars underscore the adaptability and potential of graphene-based materials to diverse sensing applications. Due to their unique properties, such as extensive surface areas, excellent electrical conductivity, and amenable functionalization, they are ideal candidates for sensor development. Furthermore, the integration of metals and metal oxides serves to bolster their catalytic prowess, broadening the horizon of possibilities in environmental monitoring, healthcare, and beyond. A graphene-based sensor can detect a wide range of analytes and has several advantages, such as fast response times, greater sensitivity, selectivity, and a wide range of analytes. As continued research in this domain advances, the anticipation of even more intricate and efficient sensing devices, spanning industries from electronics to biomedicine, becomes increasingly foreseeable.

4.9 Fuel cells

In the pursuit of sustainable energy sources and the transition away from finite fossil fuels, scientists and researchers have been exploring various avenues to develop renewable energy technologies. Among these, 3DGAs have garnered significant attention for their potential applications in fuel cells and microbial fuel cells (MFCs), where they serve as catalysts or catalyst carriers. In particular, these graphene structures improve the performance of oxygen reduction reactions (ORRs) in these energy conversion devices. In addition, they have shown promise for improving the power density and efficiency of MFCs, which can be applied to both energy production and environmental bioremediation [56–60].

a. Fuel cells: enhancing catalyst performance with 3D graphene.

Fuel cells serve as electrochemical devices, directly converting the chemical energy stored in fuels like hydrogen or methanol into electricity. Fundamentally, they operate by catalyzing reactions: anodes oxidize fuel, and cathodes reduce oxygen. The efficacy and performance of fuel cells heavily depend on the catalyst materials employed. Researchers are actively studying the potential of 3DGAs as catalyst supports or even as catalysts. One illustrative case involves the integration of 3D graphene as an anode electrode within microbial fuel cells (MFCs). MFCs represent devices designed to harness the chemical energy embedded in biodegradable organic compounds through a bio-oxidation process, simultaneously yielding electricity and contributing to environmental bioremediation. Nevertheless, conventional MFCs have issues such as low power density and low bacterial adhesion to electrodes. To tackle these issues, Yong et al. introduced an innovative anode electrode featuring a macroporous and monolithic structure, constructed from a hybrid material merging PANI with 3D graphene. As a result of graphene's generous surface area, it is more robustly integrated with bacterial films, resulting in enhanced electron transfer along multiple conductive pathways.

b. Microbial fuel cells (MFCs) offer an innovative avenue for converting organic substances, including wastewater, into electricity via microbial-driven processes. There are, however, several challenges associated with MFCs, including low power density and limited bacterial adhesion to electrode surfaces. Recent investigations have delved into the utilization of 3DGAs to tackle these hurdles and enhance MFC performance. To illustrate, researchers developed freestanding anodes for MFCs by adorning 3DGA with platinum (Pt) nanoparticles. This inventive configuration yielded an exceptional power density of 1460 mW/m^2. The superior performance of these MFCs can be attributed to several factors, including a heightened bacterial loading capacity, streamlined electron transfer between bacteria and the 3D graphene/Pt anode, and expedited ion diffusion facilitated by the porous 3D structure.

c. Applications of 3DGAs in MFCs related to the environment and energy.

The integration of 3DGAs in microbial fuel cells holds significant promise for both energy generation and environmental bioremediation. These advancements not only enhance the efficiency of MFCs but also contribute to sustainable energy production while mitigating environmental issues through the treatment of organic waste. As a result, the use of 3D graphene in fuel cells and microbial fuel cells represents an important advancement in the field of renewable energy and environmental technology. These graphene structures contribute to improving catalyst performance, power density, and overall efficiency in energy conversion devices. As research continues in this area, we can anticipate further innovations and applications of 3DGAs in addressing global energy and environmental challenges.

5. Conclusion

3DGAs are noble materials for their distinctive 3D porous architectures, expansive specific surface areas, remarkable adsorption capabilities, exceptional electrical conductivity, mechanical robustness, and swift mass and electron transport properties. 3DGAs represent a transformative technology with far-reaching implications across various facets of human life. This review provides a comprehensive overview of recent advancements in 3DGA synthesis and their burgeoning applications in fields such as sensors, fuel cells, lithium-ion batteries, supercapacitors, dental materials, tissue engineering, drug delivery systems, and many other domains. The synthesis of 3DGAs has evolved significantly, embracing methods like electrospraying, 3D printing, and chemical and electrochemical reduction, among others. Efforts aimed at achieving cost-effective large-scale production and utilization of various 3D graphene materials are essential for their widespread adoption in industrial settings. It also holds the promise of mitigating toxicity concerns and ensuring safer applications by prioritizing the development of biocompatible 3DGAs.

Author details

Md. Nizam Uddin[1*], Md. Aliahsan Bappy[2], Md Fozle Rab[3], Faycal Znidi[1] and Mohamed Morsy[1]

1 Department of Engineering and Physics, Texas A & M University-Texarkana, Texarkana, TX, USA

2 Department of Industrial Engineering, Lamar University, TX, USA

3 Bangladesh Railway, Rajshahi, Bangladesh

*Address all correspondence to: muddin@tamut.edu

References

[1] Sun Z, Fang S, Hu YH. 3D graphene materials: From understanding to design and synthesis control. Chemical Reviews. 2020;**120**(18):10336-10453. DOI: 10.1021/acs.chemrev.0c00083

[2] Ding M, Li C. Recent advances in simple preparation of 3D graphene aerogels based on 2D graphene materials. Frontiers in Chemistry. 2022;**10**:1-9. DOI: 10.3389/fchem.2022.815463

[3] Qiu L, He Z, Li D. Multifunctional cellular materials based on 2D nanomaterials: Prospects and challenges. Advanced Materials. 2018;**30**(4):1704850. DOI: 10.1002/adma.201704850

[4] Shao Q, Tang J, Lin Y, Zhang F, Yuan J, Zhang H, et al. Synthesis and characterization of graphene hollow spheres for application in supercapacitors. Journal of Materials Chemistry A. 2013;**1**:15423-15428

[5] Liu H, Pang X, Ding W, Guo S, Ding Z. Preparation of nano-SiO2 modified graphene oxide and its application in polyacrylate emulsion. Materials Today Communications. 2021;**27**:102245. DOI: 10.1016/j.mtcomm.2021.102245

[6] Wu L, Feng H, Liu M, Zhang K, Li J. Graphene-based hollow spheres as efficient electro-catalysts for oxygen reduction. Nanoscale. 2013;**5**:10839-10843

[7] Huang X, Qian K, Yang J, Zhang J, Li L, Yu C, et al. Functional Nanoporous graphene foams with controlled pore sizes. Advanced Materials. 2012;**24**(32):4419-4423. DOI: 10.1002/adma.201201680

[8] Uddin MN, Huang ZD, Mai YW, Kim JK. Tensile and tearing fracture properties of graphene oxide papers intercalated with carbon nanotubes. Carbon. 2014;**77**:481-491. DOI: 10.1016/j.carbon.2014.05.053

[9] Uddin MN, Desai FJ, Rahman MM, Asmatulu R. Highly efficient fog harvester of electrospun permanent Superhydrophobic-hydrophilic polymeric nanocomposite Fiber Mats. Nanoscale Advances. 2020;**2**:4627-4638. DOI: 10.1039/D0NA00529K

[10] Uddin MN, Desai FJ, Asmatulu E. Biomimetic electrospun nanocomposite Fibers from recycled polystyrene foams exhibiting Superhydrophobicity. Energy, Ecology, and Environment. 2020;**5**(1):1-11. DOI: 10.1007/s40974-019-00140-7

[11] Bhuyan MSA, Uddin MN, Islam MM, Bipasha FA, Hossain SS. Synthesis of graphene. International Nano Letters. 2016;**6**:65-83. DOI: 10.1007/s40089-015-0176-1

[12] Poudeh LH, Okan BS, Zanjani SMJ, Yildiz M, Menceloglu YZ. Design and fabrication of hollow and filled graphene-based polymeric spheres via core-shell electrospraying. RSC Advances. 2015;**5**:91147-91157

[13] Zhang Z, Scherer GW. Evaluation of drying methods by nitrogen adsorption. Cement and Concrete Research. 2019;**120**:13-26. DOI: 10.1016/j.cemconres.2019.02.016

[14] Zhang F, Pant D, Logan BE. Long-term performance of activated carbon air cathodes with different diffusion layer porosities in microbial fuel cells. Biosensors and Bioelectronics. 2011;**30**(1):49-55. DOI: 10.1016/j.bios.2011.08.025

[15] Fang Q, Shen Y, Chen B. Synthesis, decoration and properties of three-dimensional graphene-based macrostructures: A review. Chemical Engineering Journal. 2015;*264*:753-771. DOI: 10.1016/j.cej.2014.12.001

[16] Hu H, Zhao Z, Wan W, Gogotsi Y, Qiu J. Ultralight and highly compressible graphene aerogels. Advanced Materials. 2013;**25**(15):2219-2223. DOI: 10.1002/adma.201204530

[17] Li H, Jia D, Ding M, Zhou L, Wang K, Liu J, et al. Robust 3D graphene/cellulose nanocrystals hybrid lamella network for stable and highly efficient solar desalination. Solar RRL. 2021;**5**(8):2100317. DOI: 10.1002/solr.202100317

[18] Zhang X, Sui Z, Xu B, Yue S, Luo Y, Zhan W, et al. Mechanically strong and highly conductive graphene aerogel and its use as electrodes for electrochemical power sources. Journal of Materials Chemistry. 2011;**21**(18):6494. DOI: 10.1039/c1jm10239g

[19] Williams G, Seger B, Kamat P. TiO_2-graphene Nano-composites. UV-assisted photocatalytic reduction of graphene oxide. ACS Nano. 2008;**2**(7):1487-1491

[20] Smirnov VA, Arbuzov AA, Shulga YM, Baskakov SA, Martynenko VM, Muradyan VE, et al. Photoreduction of graphite oxide. High Energy Chemistry. 2011;*45*(1):57-61. DOI: 10.1134/S0018143911010176

[21] Chen W, Yan L. In situ self-assembly of mild chemical reduction graphene for three-dimensional architectures. Nanoscale. 2011;**3**:3132-3137

[22] Niu Z, Chen J, Hng H, Ma J, Chen X. A leavening strategy to prepare reduced graphene oxide foams. Advanced Materials. 2012;**24**:4144-4150

[23] Chen K, Chen L, Chen Y, Bai H, Li L. Three-dimensional porous graphene-based composite materials: Electrochemical synthesis and application. Journal of Materials Chemistry. 2012;**22**:20968-20976

[24] Wu Z-S, Ren W, Gao L, Zhao J, Chen Z, Liu B, et al. Synthesis of graphene sheets with high electrical conductivity and good thermal stability by hydrogen arc discharge exfoliation. ACS Nano. 2009;**3**(2):411-417. DOI: 10.1021/nn900020u

[25] Cote LJ, Cruz-Silva R, Huang J. Flash reduction and patterning of graphite oxide and its polymer composite. Journal of the American Chemical Society. 2009;**131**(31):11027-11032. DOI: 10.1021/ja902348k

[26] Hensleigh RM, Cui H, Oakdale JS, Ye JC, Campbell PG, Duoss EB, et al. Additive manufacturing of complex microarchitected graphene aerogels. Materials Horizons. 2018;**5**(6):1035-1041. DOI: 10.1039/C8MH00668G

[27] Chen Z, Ren W, Gao L, Liu B, Pei S, Cheng H-M. Three-dimensional flexible and conductive interconnected graphene networks grown by chemical vapour deposition. Nature Materials. 2011;**10**(6):424-428. DOI: 10.1038/nmat3001

[28] Bi H, Xie X, Yin K, Zhou Y, Wan S, He L, et al. Spongy graphene as a highly efficient and recyclable sorbent for oils and organic solvents. Advanced Functional Materials. 2012;**22**(21):4421-4425. DOI: 10.1002/adfm.201200888

[29] Worsley MA, Pauzauskie PJ, Olson TY, Biener J, Satcher JH, Baumann TF. Synthesis of graphene aerogel with high electrical conductivity. Journal of the American Chemical Society. 2010;**132**(40):14067-14069. DOI: 10.1021/ja1072299

[30] Xu Y, Sheng K, Li C, Shi G. Self-assembled graphene hydrogel via a one-step hydrothermal process. ACS Nano. 2010;**4**(7):4324-4330. DOI: 10.1021/nn101187z

[31] Tang Z, Shen S, Zhuang J, Wang X. Noble-metal-promoted three-dimensional macroassembly of single-layered graphene oxide. Angewandte Chemie International Edition. 2010;**49**(27):4603-4607. DOI: 10.1002/anie.201000270

[32] Xie X, Zhou Y, Bi H, Yin K, Wan S, Sun L. Large-range control of the microstructures and properties of three-dimensional porous graphene. Scientific Reports. 2013;*3*(1):2117. DOI: 10.1038/srep02117

[33] Dhanasekaran PS, Uddin MN, Wooley P, Asmatulu R. Fabrication and biological characterization of highly porous PEEK bio-nanocomposites incorporated with carbon and hydroxyapatite nanoparticles for scaffold applications. Molecules. 2020;**25**(16):3572. DOI: 10.3390/molecules25163572

[34] Ambekar RS, Kandasubramanian B. Progress in the advancement of porous biopolymer scaffold: Tissue engineering application. Industrial & Engineering Chemistry Research. 2019;**58**(16):6163-6194. DOI: 10.1021/acs.iecr.8b05334

[35] Raslan A, Saenz del Burgo L, Ciriza J, Pedraz JL. Graphene oxide and reduced graphene oxide-based scaffolds in regenerative medicine. International Journal of Pharmaceutics. 2020;**580**:119226. DOI: 10.1016/j.ijpharm.2020.119226

[36] Zhu W, Ye T, Lee S-J, Cui H, Miao S, Zhou X, et al. Enhanced neural stem cell functions in conductive annealed carbon nanofibrous scaffolds with electrical stimulation. Nanomedicine:Nanotechnology, Biology, and Medicine. 2018;**14**:2485-2494

[37] Karahan HE, Wiraja C, Xu C, Wei J, Wang Y, Wang L, et al. Graphene materials in antimicrobial Nanomedicine: Current status and future perspectives. Advanced Healthcare Materials. 2018;**7**(13):e1701406. DOI: 10.1002/adhm.201701406

[38] Shahnaz T, Hayder G. Exploring graphene's antibacterial potential for advanced and sustainable solutions in water treatment. Journal of Water Process Engineering. 2023;**56**:104530. DOI: 10.1016/j.jwpe.2023.104530

[39] Aati S, Chauhan A, Shrestha B, Rajan SM, Aati H, Fawzy A. Development of 3D printed dental resin nanocomposite with graphene nanoplatelets enhanced mechanical properties and induced drug-free antimicrobial activity. Dental Materials. 2022;**38**:1921-1933. DOI: 10.1016/j.dental.2022.10.001

[40] Dreanca A, Sarosi C, Parvu AE, Blidaru M, Enacrachi G, Purdoiu R, et al. Systemic and local biocompatibility assessment of graphene composite dental materials in experimental mandibular bone defect. Materials. 2020;**13**(11):2511. DOI: 10.3390/ma13112511

[41] Antoine C, Pijeira MSO, Ricci-Junior E, Alencar LMR, Santos-Oliveira R. Graphene quantum dots as bimodal imaging agent for X-ray and computed tomography. European Journal of Pharmaceutics and Biopharmaceutics. 2022;**179**:74-78. DOI: 10.1016/j.ejpb.2022.08.020

[42] Li X, Liang X, Wang Y, Wang D, Teng M, Xu H, et al. Graphene-based nanomaterials for dental applications: Principles, current advances, and future Outlook. Bioengineering and

Biotechnology. 2022;**10**:804201. DOI: 10.3389/fbioe.2022.804201

[43] Mobarak MH, Hossain N, Hossain A, Mim JJ, Khan F, Rayhan MT, et al. Advances of graphene nanoparticles in dental implant applications-a review. Applied Surface Science Advances. 2023;**18**:100470. DOI: 10.1016/j.apsadv.2023.100470

[44] Cao X, Yin Z, Zhang H. Three-dimensional graphene materials: Preparation, structures and application in supercapacitors. Energy & Environmental Science. 2014;**7**(6):1850-1865. DOI: 10.1039/c4ee00050a

[45] Lei S, Liu X, Zhang X, Huang Z, He C, Yang Y. One-pot hydrothermal approach to graphene/poly(3,4-ethylenedioxythiophene) composites for high-capacitance supercapacitors. Materials Today Communications. 2019;**20**:100549. DOI: 10.1016/j.mtcomm.2019.100549

[46] He Y, Chen W, Li X, Zhang Z, Fu J, Zhao C, et al. Freestanding three-dimensional graphene/MnO2 composite networks as ultralight and flexible supercapacitor electrodes. ACS Nano. 2013;**7**(1):174-182. DOI: 10.1021/nn304833s

[47] Luo J, Liu J, Zeng Z, Ng CF, Ma L, Zhang H, et al. Three-dimensional graphene foam supported Fe_3O_4 lithium battery anodes with long cycle life and high rate capability. Nano Letters. 2013;**13**(12):6136-6143. DOI: 10.1021/nl403461n

[48] Bagoole O, Rahman M, Younes H, Shah S, Al Ghaferi A. Three-dimensional graphene interconnected structure, fabrication methods and applications: Review. Journal of Nanomedicine & Nanotechnology. 2017;**8**(2):438-445. DOI: 10.4172/2157-7439.1000438

[49] Agudosi ES, Abdullah EC, Numan A, Mubarak NM, Aid SR, Benages-Vilau R, et al. Fabrication of 3D binder-free graphene NiO electrode for highly stable supercapattery. Scientific Reports. 2020;**10**(1):1-13. DOI: 10.1038/s41598-020-68067-2

[50] Uddin MN, Nageshkar VN, Asmatulu R. Improving water-splitting efficiency of electrolysis process via highly conductive nanomaterials at lower voltages. Energy, Ecology and Environment. 2020;**5**:108-111. DOI: 10.1007/s40974-020-00147-5

[51] Asmatulu R, Veisi Z, Uddin MN, Mahapatro A. Highly sensitive and reliable electrospun polyaniline nanofiber based biosensor as a robust platform for COX-2 enzyme detections. Fibers and Polymers. 2019;**20**(5):966-974. DOI: 10.1007/s12221-019-1096-x

[52] Salahuddin M, Uddin MN, Hwang G, Asmatulu R. Super-hydrophobic PAN nanofibers for gas diffusion layers of proton exchange membrane fuel cells for Cathodic water management. International Journal of Hydrogen Energy. 2018;**43**(25):11530-11538. DOI: 10.1016/j.ijhydene.2017.07.229

[53] Yavari F, Castillo E, Gullapalli H, Ajayan PM, Koratkar N. High sensitivity detection of NO2 and NH3 in air using chemical vapor deposition grown graphene. Applied Physics Letters. 2012;**100**:203120. DOI: 10.1063/1.4720074

[54] Kung CC, Lin PY, Buse FJ, Xue Y, Yu X, Dai L, et al. Preparation and characterization of three-dimensional graphene foam supported platinum–ruthenium bimetallic nanocatalysts for hydrogen peroxide based electrochemical biosensors. Biosensors and Bioelectronics. 2014;**52**:1-7. DOI: 10.1016/j.bios.2013.08.025

[55] Yang T, Guan Q, Li Q, Meng L, Wang L, Liu C, et al. Large-area, three-dimensional interconnected graphene oxide intercalated with self-doped polyaniline nanofibers as a free-standing electrocatalytic platform for adenine and guanine. Journal of Materials Chemistry B. 2013;**1**(23):2926-2933. DOI: 10.1039/c3tb20171f

[56] Kumar SSA, Uddin MN, Rahman MM, Asmatulu R. Introducing graphene thin films into carbon Fiber composite structures for lightning strike protection. Polymer Composites. 2018;**40**(S1):E517-E525. DOI: 10.1002/pc.24850

[57] Bhuyan MSA, Uddin MN, Bipasha FA, Islam MM, Hossain SS. A review of functionalized graphene. Properties and its Applications, International Journal of Innovation and Scientific Research. 2015;**17**(2):303-315

[58] Pang Y, Jian J, Tu T, Yang Z, Ling J, Li Y, et al. Wearable humidity sensor based on porous graphene network for respiration monitoring. Biosensors and Bioelectronics. 2018;**116**:123-129. DOI: 10.1016/j.bios.2018.05.038

[59] Uddin MN, Le L, Zhang B, Nair R, Asmatulu R. Effects of graphene thin films and nanocomposite coatings on fire Retardancy and thermal stability of aircraft composites: A comparative study. Journal of Engineering Materials and Technology. 2019;**141**(3):031004-1-7. DOI: 10.1115/1.4042663

[60] Xu Z, Peng M, Zhang Z, Zeng H, Shi R, Ma X, et al. Graphene-assisted electrochemical sensor for detection of pancreatic cancer markers. Frontiers in Chemistry. 2021;**9**:733371. DOI: 10.3389/fchem.2021.733371

Section 2

Surface Reactivities

Chapter 3

Surface Functionalization Reactions of Graphene-Based Nanostructure and Their Practical Application

Neeraj Kumari and Meena Bhandari

Abstract

Graphene (G) has captured the attention of scientists and researchers due to its remarkable electronic, structural, optical, and mechanical properties. While pristine G has been used for various desirable applications requiring high electrical conductivity, there is also a demand for altered or functionalized versions of G, such as G oxide, reduced G, and other functionalized variants, in numerous other applications. The structural alteration of G through chemical functionalization unveils a multitude of possibilities for adjusting its configuration, and various chemical and physical functionalization techniques have been explored to enhance G's stability and adaptability. Functionalization allows the customization of graphene's properties, such as electronic, chemical, and mechanical characteristics, to suit specific applications. This chapter highlights the functionalization of graphene-based nanostructure, encompassing both covalent and non-covalent approaches, for a wide range of applications as well as for addressing current challenges and for outlining potential future research directions concerning surface functional modification for G and graphene oxide (GO).

Keywords: graphene, functionalization, surface modification, hydrophilic, doping

1. Introduction

The well-known materials of the carbon family prior to the 1980s were graphite and diamond. The discovery of molecular carbon allotropes, including fullerenes, carbon nanotubes (CNT), and most recently, 2-D graphene (G), has completely altered the landscape of the inorganic chemistry of carbon [1]. Graphene possesses a 2D honeycomb-like lattice structure, made up of single-layer sp^2-bonded carbon atoms (**Figure 1**) [2].

Single-layer G exhibits a unique characteristic of having a zero-band gap, leading to its exceptional optical transparency of 97.7%. Furthermore, G has an impressive specific surface area of 2600 m^2/g [3], which is greater than that of CNTs yet is less reactive because it lacks the binding stress that CNTs' curvature produces.

Figure 1.
Structure of graphene.

Therefore, G is used by researchers and scientists in a large-scale application in different fields including drug delivery systems, biosensing, polymer compositing, and fabrication of liquid crystal devices. Other exceptional qualities displayed by G are its electrical, mechanical, optical, and transport nature and the G bipolar field effect [4–6]. These unique properties of G have opened up extensive opportunities for surface chemical applications and sparked intense interest among researchers and technologies [7].

Despite its advantageous properties, G sheets tend to restack and aggregate because of van der Waals interactions between the layers. This propensity poses significant challenges in applications linked to nanomaterials and biomaterials-based technologies, where maintaining the desired structural arrangement is crucial. The attractive van der Walls forces found between G sheets prevent the G dispersion in various solvents. Moreover, due to these interactions, the monolayer G tends to re-aggregate after exfoliation and dispersion. These additional limitations of the single-component materials, including challenging processing, impose significant constraints on its practical applications.

However, as compared to G, graphene oxide (GO), a derivative material obtained by the partial oxidation of G is gaining more attention as it can be synthesized on a large scale with more ease and cost-effectiveness. Thus, GO has experienced a remarkable increase in attention in recent times. The surfaces of GO sheets are rich in oxygen-containing functional groups that may include hydroxyl, epoxide, diol, ketone, and carboxyl sites (**Figure 2**). These groups play a crucial role in altering the van der Waals interactions, leading to a diverse range of solubilities in both water and organic solvents [8].

The functional modifications are essential to expanding the potential use of both G and its derivatives including (GO). Thus, the fundamental basis for achieving functionalization lies in effecting changes to the intrinsic structure of G and its oxide. The two faces, edges, and defect sides of G can all be functionalized through surface or substitutional doping [9]. The existence of structural imperfections and the number of layers have an impact on G's properties [10]. For example, G's electrical conductivity tends to decrease with the introduction of defects during covalent functionalization though the sp^2 structure is preserved under non-covalent functionalization [11].

With improved synthetic techniques, G is made more usable in electronics and other fields that rely on quick electron transfer mechanisms including photocatalytic applications for renewable energy production. Another significant problem that needs to be addressed is the need for conversion of the 2-D G material into 3-D, as the creation of higher-order nanostructures from G continues to show

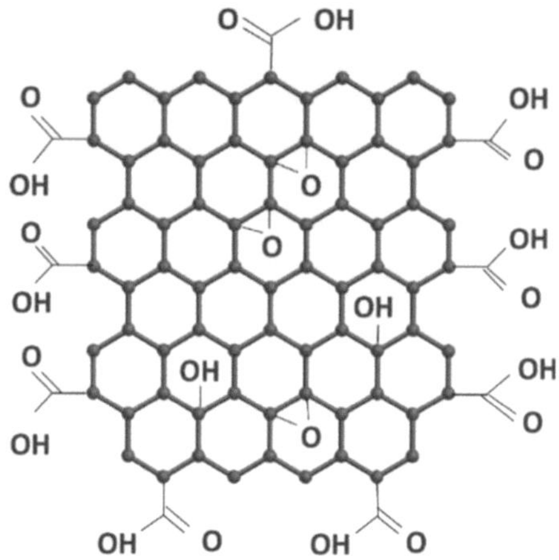

Figure 2.
G oxide containing various functional groups like hydroxyl, carboxyl, and epoxy.

promise in applications like supercapacitors, fuel cells, water purification, drug delivery, photovoltaics, catalysis, gas adsorption, sensing [12], touch screens, spintronic devices, high-frequency circuits, toxic material removal, and flexible electronics [13–15].

The ability of the G surface to be altered and functionalized has opened a plethora of possibilities to develop specialized practical compounds [10]. The bandgap of single-layer G is altered for microelectronic devices [11]. Moreover, highly porous 3D structures can be engineered from the naturally non-porous 2D G and used in gas sorption, storage, separation, sensing, and electrochemical devices like batteries, fuel cells, and supercapacitors. Nevertheless, the vast array of potential applications can be expanded through numerous functionalization techniques for G and its derivatives. These diverse functionalization techniques provide numerous opportunities to enhance the present applications of G in other areas, such as bioimaging or increasing band gaps for electronic use. In this chapter, methods for functional modification of G and GO have been discussed focusing on the essential chemical bonds and functional groups that affect their structural integrity. The chapter is divided into two sections. Firstly, the synthesis and structure of G-based materials are discussed. Then, G functionalization and applications of the derived materials are explored. The methods of G modification discussed include reactions with organic and inorganic molecules, as well as chemical modifications through covalent and non-covalent interactions with G [16].

2. Synthesis of graphene-based nanostructure

2.1 Pristine graphene

The basic structure of all graphitic materials including charcoal, CNT, and graphite is based on G which is an indeterminately large atomic molecule. There are several approaches that range from mechanical exfoliation of high-quality graphite to direct growth on carbides by bottom-up and top-down techniques which are used to synthesize G [17].

G sheets of different thicknesses can be synthesized through mechanical exfoliation using a simple peeling process. The natural graphite or single-crystal graphite is engraved in oxygen plasma to create deep mesas followed by wedging and peeling off layers. The collected flakes of G are washed off and transferred to a substrate. Another approach to synthesizing defect-free monolayer G is physical exfoliation through ultrasonication using a high-boiling-point solvent [18].

One more promising method for synthesizing mono- or few-layer G is the chemical vapor deposition method (CVD), which also allows for the thickness and crystallinity of the G layer to be controlled using this method [19].

An alternative approach for G synthesis is plasma-induced CVD where the synthesis process takes place at low temperature. This method is more commonly used to synthesize G as compared to CVD as during synthesis, low temperature and less deposition time are required [20].

The chemical exfoliation of graphite is one of the well-established variant methods to produce G. In this method, the increment of interlayer spacing of graphite is achieved by using active functional moieties, which results in the weakening of the van der Wall forces. In this technique, the intercalated G compounds undergo exfoliation either through rapid heating, reduction process, or ultrasonication [21].

2.2 Graphene oxide

GO is highly hydrophilic and can easily swell after dispersion in water due to extra carbonyl and carboxyl groups that are present at the edges of the GO sheets. Due to this hydrophilicity, ultrasonication treatment is done to introduce a single or few layers of G which are exceedingly steady in deionized water and other solvents. This is necessary to understand that graphite oxide and GO are different [22].

Graphite oxide is a multi-layered system while GO is a few or single-layer system. On the basis of oxygen functionalities, numerous models related to GO structure have been studied. GO can be synthesized through the oxidation of graphite using an oxidizing agent in the presence of concentrated inorganic acids. Thus, GO is widely synthesized through various methods like Hummer's method where potassium permanganate, as an oxidizing agent, concentrated sulfuric (VI), and nitric (V) acids are used to oxidize the G-to-GO [23].

Another method used to synthesize GO is the Staudenmaier method which involves the use of fuming nitric acid and $KClO_3$ as oxidants [24], and the Tour method using concentrated phosphoric acid with potassium permanganate oxidant [25]. All these methods are known as chemical oxidation methods where various oxygen-containing functional groups are created on the edges and surface of G (**Figure 3**). These functional groups are responsible for the hydrophilic nature of GO as they break van der Wall forces. The characteristics of GO can be modified through functionalization to tailor the materials for specific applications.

2.3 Reduced graphene

When GO undergoes reduction, G is produced due to the removal of oxygen-containing functional groups. Various chemical, thermal, and electrochemical methods are used for the reduction of GO. The chemical reduction has been done using various reducing agents like hydrazine [26], sodium borohydride [27], hydroquinone [28], alkaline solution [29], ascorbic acid [30], and glucose [31].

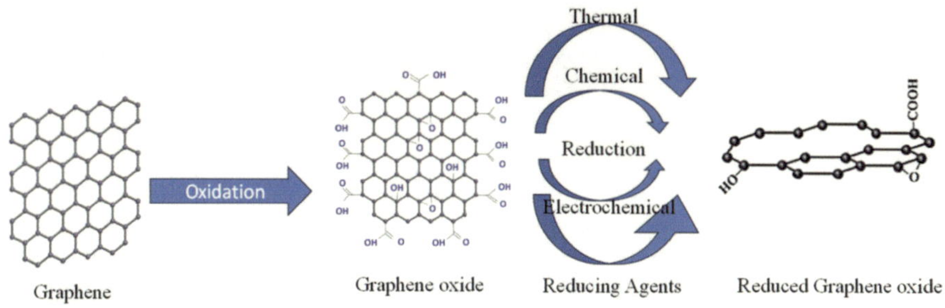

Figure 3.
Reduction of GO through different methods.

During the reduction method, a brown precipitate of G, which turns black during dispersion, results in aggregation and precipitation of reduced sheets of G oxide. During the thermal reduction process, G oxide is heated to remove various oxide groups resulting in exfoliations with the evolution of CO_2. The electrochemical reduction method can also be initiated at −0.8 V and completed at −1.5 V with the formation of black precipitates. There are some other methods like photochemical and photothermal methods which are used for the formation of reduced G oxide [32].

3. Surface functionalization

Surface functionalization is one of the most important processes that enable the use of G-based materials for various applications. The functionalization affects the hydrophobicity and surface charge of G by changing the ionization of G. G's surface can be modified via doping with chemical agents that include elements, compounds, polymers, and nanoparticles which enhance its dispersion, stability, and tribological properties. Through surface modification, stable G-based hybrid materials can be formed. When GO and its composites are used in a water-borne coating system, the main problem is to achieve homogenous dispersion of G. Therefore, to overcome this problem, the surface functionalization of GO is done by covalent/non-covalent reactions involving hydroxyl, carboxyl, and epoxy group using chemical moieties such as organic and inorganic, polymer, and nanocomposites (**Figure 4**). Among

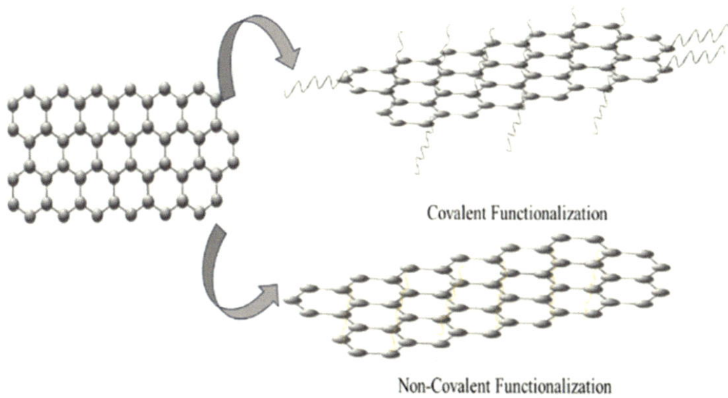

Figure 4.
Covalent and non-covalent functionalization of G.

the groups, the epoxy sites react with the edges and defect sites of G to enhance the dispersion ability of G by forming a stable crosslinker between the epoxy binders and functional G. To enhance practical nano-electronic devices and sensors, band gap widening of G can be achieved through doping, intercalation, and striping processes [15, 33].

The appropriate functionalization of G and GO allows to protect their natural properties by preventing their aggregation during reduction steps [32] and it forms further functional groups to impart additional properties to the materials. As a result, the functionalization processes of G materials have been divided into four categories: that include: (1) covalent, (2) non-covalent functionalization, (3) substitutional doping, and (4) hybridization. With organic groups such as epoxide, carboxylic groups, amino, and hydroxyl groups, G can be functionalized directly through covalent bonding onto the surface sites. To increase dispersibility, for instance, carboxylic acid groups at the GO edges are then swapped out for amine groups such as ethylenediamine and ethanolamine.

3.1 Covalent functionalization

For the purpose of enhancing G's performance, covalent bonds are used to join G with newly added groups. When an appropriate functional group establishes a covalent bond with the sp^2 carbon structure of the G surfaces, at the edges or basal planes, covalent functionalization takes place. The aromatic nature of G can be improved by adding functional groups via the covalent method, which alters the electrical properties of materials while boosting its solubility and stability and widening the bandgap [34].

Covalent organic functionalization offers several advantages, allowing the combination of G's properties with other functional materials like chromophores or polymers, and it enhances the dispersibility of the material in organic solvents and water [35]. Surface functionalization at the edges and flaw functionalization are all possible. The resulting structural defects, which include structural flaws, atoms, and solvent molecules randomly adsorbed onto the materials, are brought on by the chemical processes used to produce G [5]. These defects also include damage to the carbon lattice and structural flaws. Covalent modification can, however, be categorized into different types, such as free radical addition, atomic radical addition, nucleophilic addition, cycloaddition, electrophilic substitution reactions, carboxyl, carbon-carbon skeleton functionalization, and hydroxyl functionalization.

3.1.1 Free radical addition

Functionalization of G based on the free radical addition involves the formation of covalent bonds through the interaction of free radicals with the G surfaces through thermal treatment, photochemical, or chemical treatments (**Figure 5**) [36]. One of the common radical reactions is carried out with aryl diazonium salts. Functionalization of G through free radical addition was performed by Tour and coworkers where aryl radical forms were formed after the removal of nitrogen from aryl diazonium ion. It was found that the addition of free radical moiety took place on the sp^2 surface of G by donating an electron [37]. The organic free radicals generated during the reaction then reacted with G through the covalent formation and were found to be accountable for self-polymerization of the materials. The free radical addition

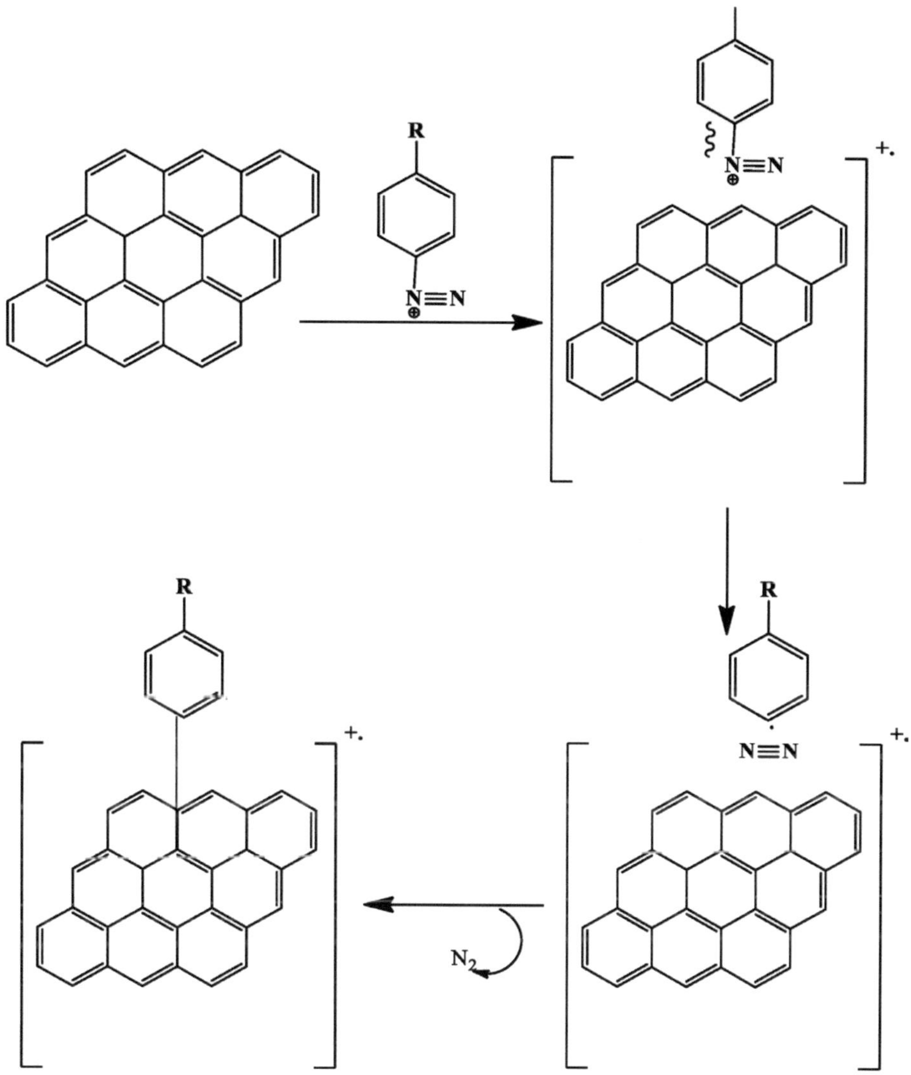

Figure 5.
Free radical addition with aryl diazonium salt.

reaction is controlled by the G layers, and it is observed that G with a single layer is 10 times more reactive than bi or multi-layered G [38].

This type of reaction is employed to study the antimicrobial characteristics of synthesized G and its composites using chlorophenyl groups. On functionalization with chlorophenyl groups, G has been found to be more effective as an antimicrobial agent. Numerous studies demonstrated the generation of free radicals used to induce the denaturation of DNA/RNA, leading to the inactivation of microbial cells.

An alternative method for introducing free radicals involves the reaction of benzoyl peroxide with graphene sheets, which is initiated through photochemical means. This process entails focusing an Ar-ion laser beam onto the graphene sheets immersed in a solution of benzoyl peroxide and toluene [39].

3.1.2 Nucleophilic addition reactions

In these reactions, G always behaves as an electron acceptor. For instance, when poly-9,9′-dihexyfluorene carbazole reacts with G, a base is used to initially form an anionic moiety, leading to the generation of nitrogen anions on carbazole (**Figure 6**). Subsequently, this anion reacts with the G surface, resulting in the formation of a covalent bond [40].

The primary sites of reactivity in the nucleophilic addition reaction are the epoxy groups found in GO. The amine (-NH_2) functionality present in the organic modifiers, possessing a lone pair of electrons, initiates an attack on the epoxy groups of GO. Notably, nucleophilic addition takes place with remarkable ease, even at room temperature and in an aqueous environment. Consequently, this method has garnered significant attention as a promising approach for the large-scale production of functionalized graphene when compared to other techniques [41].

3.1.3 Cycloaddition reactions

There are different types of cycloaddition reactions like [2 + 1], [2 + 2], [3 + 2], and [4 + 2]. The most prominent reaction is [3 + 2] cycloaddition reaction including 1,3-dipolar cycloaddition which was performed by Trapalis and his co-workers (**Figure 7**). G can be used on a large scale in the Diels-Alder reaction as it behaves both as a diene and a dienophile [42].

Due to its gentle reaction condition, the Bingel reaction is one of the exceptionally valuable reactions for functionalizing carbon nanomaterials, such as graphene. This

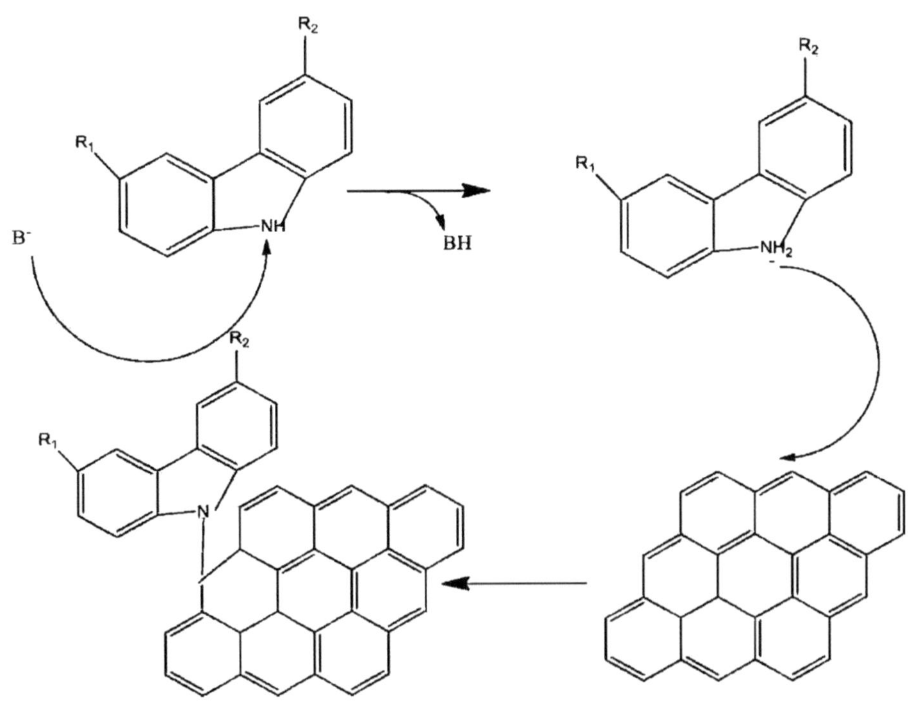

Figure 6.
Nucleophilic addition reaction with carbazole.

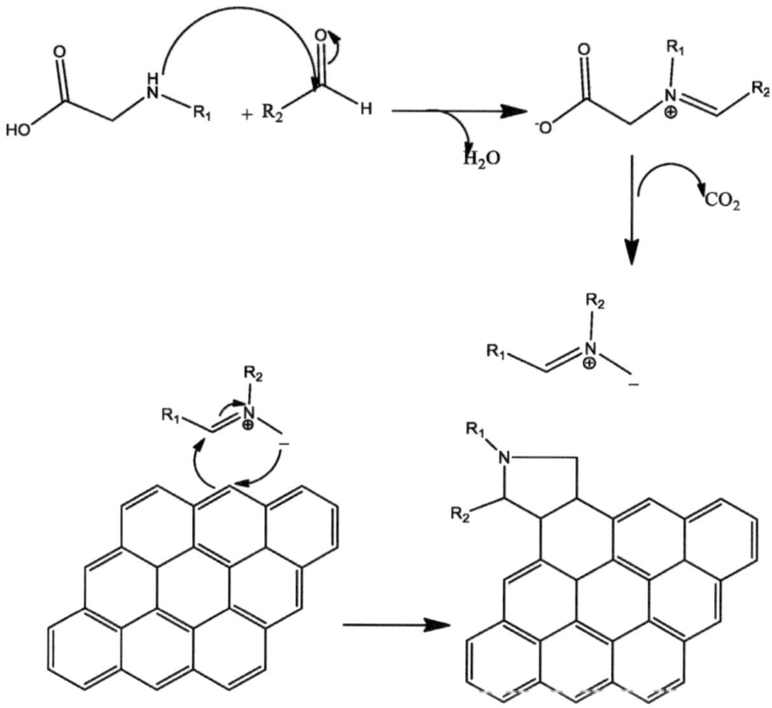

Figure 7.
1,3-Dipolar cycloaddition reaction.

reaction entails the use of a halide derivative of a malonate ester group in the presence of a base. The base extracts a proton from this derivative, leading to the formation of an enolate that subsequently attacks the C=C bonds within the graphene structure. The resulting carbanion undergoes a nucleophilic substitution, displacing the halide and resulting in the formation of a cyclopropane ring [43].

Arynes are well-known reactive intermediates in nucleophilic aromatic substitution reactions, and they are used on a large scale in various reactions. In recent times, there have been several attempts to employ aryne cycloaddition reactions for carbon nanomaterials, demonstrating its successful application in functionalizing fullerenes and their derivatives. Notably, various research groups have reported the chemical modification of graphene through aryne cycloaddition. Based on weight loss analysis, it is estimated that the degree of functionalization is approximately more than one functional group per 17 carbon atoms. The resulting graphene exhibits exceptional solubility and thermal stability, remaining stable even at temperatures as high as 500°C. Additionally, similar reactions using 1,3-dipolar cycloaddition of azomethine ylides and cyclopropanated malonate have also yielded successful chemical functionalization of G and its derivatives [44, 45].

3.1.4 Reaction with atomic radicals

When the reaction with organic free radicals and atomic species like hydrogen, and fluorine is compared, the possibility of side chain reaction is less in case of reaction with atomic radicals due to which uniform and homogenous functionalization of G occurs (**Figure 8**) [46]. During hydrogenation, deformation of lattice occurs on attaching of

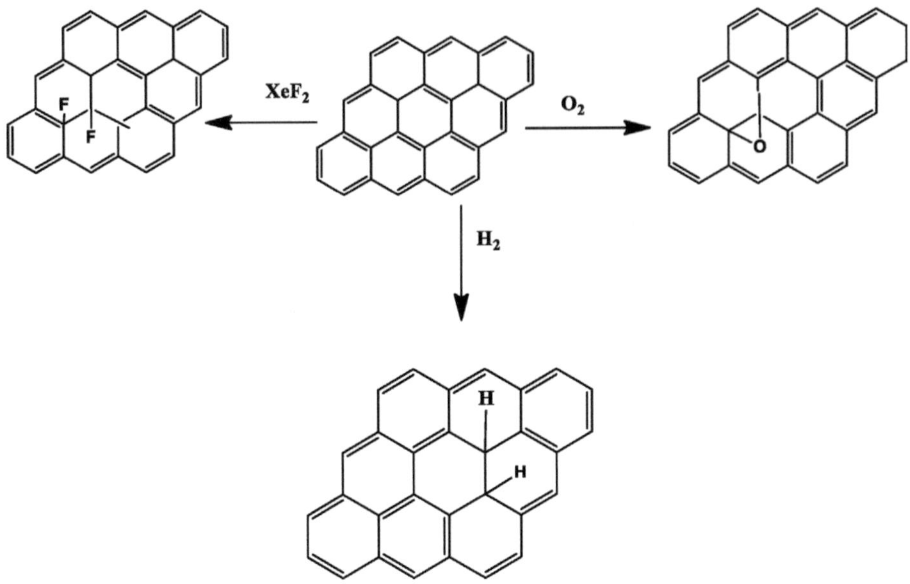

Figure 8.
Reaction with gas phase atomic radicals.

atomic hydrogen resulting in easy formation of a second C-H bond. G in its hydrogenated form is known as graphane [47]. The fluorination reaction of G is found to be comparable with hydrogenation as fluorine attached to the carbon atoms of G through a single bond with enhanced binding strength results in high functionalization. Generally, there are three methods to synthesize fluoroG i.e. through: (1) exposition to XeF, (2) etching fluorinated complexes, and (3) graphite fluoride exfoliation [48].

Oxygenated G produced by applying Hummer's method, exhibits significant heterogeneity. In this reaction, it is assumed that oxygen atoms will attach to graphene, resulting in the formation of epoxide groups. Researchers have successfully generated these epoxide groups by exposing graphene to oxygen plasmas and atomic oxygen beams [49].

3.1.5 Electrophilic substitution reactions

The functionalization of graphene (G) is made possible through electrophilic substitution reactions, taking into account its electron-rich structure. Various reactions like Friedel-Craft acylation and hydrogen-lithium exchange can be included in electrophilic substitution reactions. In the Friedel-Craft reaction, acyl cation is formed as a reactive intermediate after introducing ketone moieties. The reactive intermediate is an acyl cation generated in the presence of a Lewis acid. This reaction was employed to produce brominated flame-retardant high-density polyethylene composites containing graphene nanoplatelets. In the hydrogen-lithium exchange reaction, the first deprotonation or carbometallation of G takes place using butyl lithium (**Figure 9**). The derivative of lithium-G reacts with an electrophile resulting in the formation of a covalent bond [50].

Covalent bond functionalization is simpler on GO as compared to G due to oxygen-containing groups on its surface. Common chemical processes involving these groups include addition reactions, carboxylic acylation, epoxy ring opening, isocyanation, and diazotization [7].

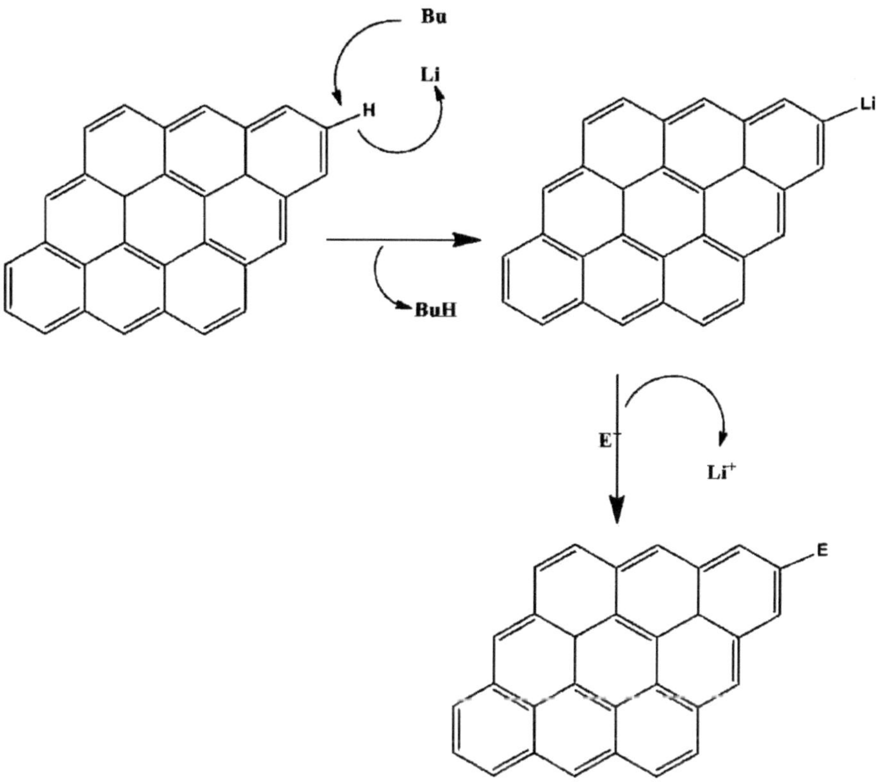

Figure 9.
Mechanism of electrophilic reaction with BuLi.

By employing polystyrene particles that have been "armored" using nanoscale GO sheets made via aqueous mini-emulsion polymerization of styrene, Wang et al. [51] took advantage of GO's amphiphilic characteristics without the use of conventional surfactants. From graphite nanofibers, a unique technique was used to manufacture the nanoscale GO sheets of 100 nm diameter. Increased processability and new functions are brought about via covalent bond modification [51].

It is possible to covalently functionalize GO using a variety of other methods. GO nanosheets were chemically modified to include a sulfanilic acid group, which enhanced water dispersibility due to ionic repulsion [52].

Covalent bond modification facilitated increased processability and introduced new functionalities. It has been reported that chemical modification of GO nanosheets with a sulfanilic acid group improved water dispersibility through ionic repulsion. Moreover, covalent functionalization with polyaniline (PANI) not only increased the surface area of reduced GO aerogel (rGOA) but also enhanced electrical conductivity and prevented G nanosheet aggregation. The functionalization started with a free radical reaction where GO donated an electron to aryl diazonium ion resulting in the formation of aryl radical. GO attached to aryl group through covalent bonding after cleavage from the N of PANI-grafted rGOA as high-performance supercapacitor electrodes [53].

The reaction was further preceded by amidation. Lastly, amino group act as active sites for polymerization of aniline on rGOA in an acidic medium using ammonium persulfate as oxidant resulting in the formation of PANI-grafted

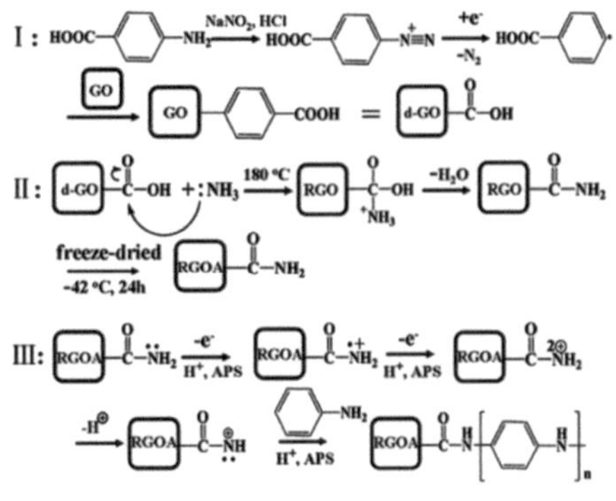

Figure 10.
Covalent functionalization of rGOA with PANI.

rGOA as shown in **Figure 10**. Consequently, the PANI-grafted rGOA exhibited superior capacitance performance (396 F/g at 10 A/g) compared to rGOA (183 F/g at 10 A/g).

Shin et al. [54] prepared GO-LMB 21 (locked nucleic acid molecular beacon) using amine functionalized DNA, LMB, and carboxylate GO. The synthesized GO-LMB enhanced the detection limit of miRNA sensing to the picomolar level by covalently coupling fluorescence-labeled dsDNA probes onto the GO. This approach was adopted because interactions between GO and DNA molecules were obstructed by the presence of small molecules like lipids, proteins, and nucleic acids, which caused nonspecific probe desorption [54].

A porphyrin-G nanohybrid was created by Choi et al. [55]. By adding polymers such as polyethylene glycol, dextran, and chitosan, the amidation process has frequently been employed to produce biocompatible GO [55]. The synthesis of chitosan (CS) modified graphene nanosheets under microwave irradiation in N,N-dimethylformamide medium is the best example of amidation reaction which involved the reaction between the carboxyl groups of graphene oxide nanosheets (GONS) and the amido groups of chitosan followed by the reduction of graphene oxide nanosheets into graphene nanosheets using hydrazine hydrate. The results showed that chitosan was covalently grafted onto the surface of graphene nanosheets via amido bonds. Solubility measurements indicated that the resultant nanocomposites dispersed well in aqueous acetic acid.

Four esterification methods, namely direct, carbodiimide activated, oxalyl chloride acylation, and via an acid-functionalized GO intermediate, were investigated for the preparation of surface-functionalized GO nanosheets using tannic acid (TA). In the first approach, direct esterification of GO was done with TA in an acidic medium. In the second approach, acid-functionalized GO (GO-COOH) was prepared through the activation of GO with $C_2H_3ClO_2$ in a basic medium. In the third experiment, the aqueous solution of GO-COOH was ultrasonicated followed by the addition of TA under ambient conditions to form the final product GO-COOH-g-TA-2. In the last stage, the formation of acyl chloride derivative (GO-CO-Cl) takes place through the conversion of carboxylic acid into acyl chloride. The covalent grafting of TA onto the GO surface renders it more hydrophobic, leading to enhanced organic solvent dispersion. Additionally, TA acts as a crosslinker between the GO nanosheets, thereby increasing its thermal resistance. Furthermore, the combined effect of GO and TA results in the suppression of bacterial growth. Among the examined methods, esterification using carbodiimide showed the highest degree of grafting, maximum thermal stability, and strongest antibacterial activity [56].

3.1.6 Carboxyl functionalization

The functionalization of GO has been extensively explored because of the abundance of carboxyl groups near the edge of the material which are highly reactive groups [57]. The reactions are often started by the carboxyl functionalization step, after which dehydration of an amino group and a hydroxyl group takes place to create an ester/amide bond. The chemicals usually employed for carboxyl activation include thionyl chloride ($SOCl_2$), 2-(7-aza-1H-benzotriazole-1-yl)-1,1,3,3-tetramethyluronium hexafluorophosphate, N,N-dicyclohexyl carbodiimide (DCC), and 1-ethyl-3-(3-dimethylaminopropyl)-carbodiimide (EDC) [58].

Functionalization of semiconducting G nanoribbons (GNRs) with Stone-Wales (SW) imperfections was demonstrated using carboxyl (COOH) group. A significant change in the properties of GNR was observed due to the SW defect as the electrons of unsaturated carbon atoms in COOH interact directly with donor state of GNR nanoribbons. The significant rehybridization interaction results in noticeable downward shifts of the states originating from the two bonding carbon atoms to a lower energy range. This phenomenon helps to clarify why the defect band shifts downward after the adsorption of the COOH group. Essentially, the alteration in electrical conductivity with the density of single-wall defective carbon nanotubes (SWDCPs) can be attributed to the redistribution of electronic states. When SWDCPs axial concentration increases, the behaviors of the system shift from semiconducting to p-type metallic. Consequently, G nanoribbons (GNRs) offer promising prospects for applications in nano-electronics and chemical sensors [59]. Following the addition of CRGO-CN (chemically reduced G oxide having CN group) to a solution of sodium hydroxide and methanol, a hydrolysis reaction produced CRGO that was more abundant in carboxyl groups (CRGO-COOH) [60]. A poly(3-hexylthiophene) molecule was grafted onto the carboxylic acids to activate them in order to create heterojunction photovoltaic devices [61].

3.1.7 Carbon-carbon skeleton functionalization

The aromatic basal plane of GO can be directly functionalized using highly reactive intermediates including nitriles, carbenes, and aryl diazonium salts. This process could change the basal plane's sp^2 hybridization to sp^3 after functionalization. In comparison to non-functionalized GO, GO functional groups considerably increased dispersion stability in water, dimethyl sulfoxide, and N,N-dimethylformamide (DMF). In the aromatic ring of G or GO, the C=C bond is mostly used for the functional alteration of the carbon skeleton. Both the Diels-Alder reaction and the GO diazotization reaction have been documented [62]. The functionalization of G sheets was done through the one-pot process using anthraquinone molecules. In this process, the G electrode undergoes oxidative electrochemical exfoliation in the presence of 0.1 M H_2SO_4 solution containing anthraquinone diazonium ions. Functionalization takes place through the spontaneous reaction of newly formed graphene sheets with diazonium ions (**Figure 11**) [63].

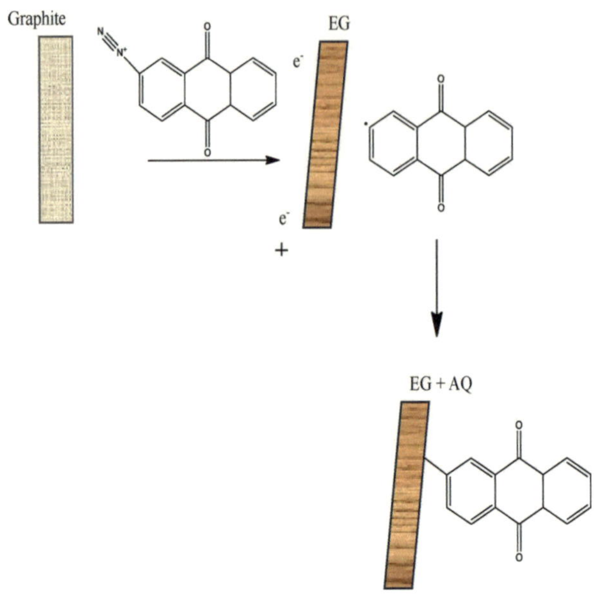

Figure 11.
The one-pot electrochemical exfoliation of graphite leads to the spontaneous functionalization of graphene sheets (EG) with anthraquinone.

Xiong et al. [62] produced 4-propargyloxyphenyl G (GCCH) through a two-step process. They mixed solution-phase G with 4-propargyloxydiazobenzenetetrafluoroborate at 45°C for 8 hours [62]. Then, the G carbon skeleton was treated and further functionalized using a click chemical reaction with azido polyethylene glycol carboxylic acid [64]. This approach proved to be versatile and practical, allowing for the creation of biosensors and composite materials with G by modifying the functional groups connecting the layers.

The underlying mechanism involves the formation of a diazonium salt or a diazo compound through the diazotization of an aromatic amine-containing material with a reactive functional group, leading to the generation of a free radical during deaeration [65]. The benzene derivative linked through sigma bond with the reactive functional group undergoes an addition reaction with a carbon double bond (C=C) to form a new single C-C bond. Eventually, G with the reactive functional groups undergoes functional modification with G oxide.

3.1.8 Hydroxy functionalization

Another significant method of modifying G is through hydroxy functionalization, in which the occurrence of several hydroxyl groups present on the GO surface enables the functionalization of hydroxyl-based via amidation reaction or esterification. The ester produced by GO is then further modified using other functional groups [66]. The hydroxyl groups were substituted after esterification in order to create aziolated GO. Stirring GO with 2-bromoisobutyryl bromide for 2 days proceeded by dispersion of the esterified GO in dimethylformamide and sodium azide for 24 hours produced azido functionalized GO resulting in improvement in the solubility of the functionalized GO in polar solvents like chloroform and tetrahydrofuran [67].

The researchers utilized tris buffer (TB) and diethanolamine (DEA) which are water-soluble, non-toxic, and biocompatible to easily produce alcohol-functionalized GO material. This alcohol functionalization on GO resulted in the expansion of inter-layer space and specific surface areas, allowing for the molecular grafting of amines to G. In a 60% KOH solution with a density of 1 A g^{-1}, in a 6% KOH solution with a density of 1 A g^{-1}, the synthesized materials exhibited high specific capacitance values of 372 and 252 F g^{-1} [68].

Erythritol, a promising medium-temperature candidate, was combined with GO nanosheets as an additive to increase the dispersion stability of the composites. GO nanosheets treated with hydroxyl groups were used for this purpose. Moreover, the application of functionalized GO nanosheets as an additive has proven effective in enhancing erythritol performance, which is a promising medium-temperature candidate. To optimize overall performance, a trade-off evaluation on the loading of the additive is necessary. By combining GO nanosheets with erythritol, the dispersion stability improved significantly while using GO nanosheets after treatment with hydroxyl-containing compound, the highest loading of 1.0 wt% GO nanosheets led to a twofold increase in thermal conductivity and a reduction in supercooling from 64°C to 48°C. It has been demonstrated that functionalized GO nanosheets are effective in enhancing the performance of erythritol, however, to obtain the greatest overall performance, a trade-off analysis on the loading would be necessary [69].

In other words, the hydroxyl groups were selectively functionalized using a variety of chemical strategies. The Williamson reaction with an amino-terminated linker derivatized the hydroxyl groups, resulting in the production of ether bonds. In contrast to most derivatization methods for hydroxyl groups, this reaction occurred under benign conditions at ambient temperature, preventing the reduction of GO. Then, effective esterification of the hydroxyls using aminocaproic acid at room temperature has been carried out. Overall, the hydroxyls' reactivity in the Williamson reaction and esterification was quite similar [70].

When dispersing gold nanoparticles (GNPs) in pyridine, Georgakilas et al. [39] employed 1,3-dipolar cycloaddition of azomethine ylide to add the hydroxyl group to G surfaces. For 30 days, the functionalized GNPs could persistently be disseminated in ethanol [39]. Using a ball mill to exfoliate graphite and potassium hydroxide, the authors were able to produce hydroxyl-functionalized G. The G with the hydroxyl functional is very electroactive, hydrophilic, and water dispersible. The functionalized G is made up of single- to few-layer GNPs. Esterification has been reported by the addition of poly(vinyl) alcohol to GNPs [71]. The carboxylic group was added to the pre-functionalized GNPs by oxidizing them in a cold nitric/sulfuric acid solution while subjecting them to extended sonication. Amiri et al. [70] functionalized GNPs with ethylene glycol via a microwave-aided electrophilic process. In water/ethylene glycol conditions, the functionalized GNPs exhibit good dispersibility [70].

3.2 Non-covalent functionalization

The stable dispersion of G materials in organic solvents can be achieved through covalent functionalization using aromatic compounds [72–73] isocyanates, and aliphatic amines [74]. However, this covalent approach comes at the cost of reducing many of the advantageous characteristics of G-based materials (GBMs), such as their barrier, electrical, and mechanical properties. In contrast, the non-covalent approach focuses on achieving stable dispersion by relying on physical adsorption and/or enfolding of molecules or polymers through weak interactions like hydrogen bonding, van der Waals forces, and H-π, π-π, cation-π, and anion-π interactions. This non-covalent method allows for the preservation of the electrical properties of GO [32].

Non-covalent functionalization procedures have a significant benefit over covalent functionalization approaches since they lessen the impact on the G structure and its inherent material qualities [74]. GBMs can be non-covalently functionalized using a variety of techniques, including polymeric grafting, interaction of small molecules with aromatic rings, or use of surfactants [75].

The G structure suffers little harm from the non-covalent change, and maximum properties can be preserved [13, 76]. Previously, conjugated molecules such as 1-pyrenebutyrate, sulfonated polyaniline, dendronized perylene bisimides, polyacetylenes, and carboxylated oligoanilines have been used to disperse G [39, 77]. Through dispersant, van der Waals interactions or electrostatic repulsion prevails which may aid in G stability in solution.

3.2.1 π-π interactions

G interaction with other compounds or nanomaterials is influenced by two different forms of π-π interactions that take place between the electron-rich and electron-poor areas. This is frequently observed in C_6R_6 (R is substituents like hydrogen or other groups altering the ring's polarity) where C and R arrange through face and edges interaction. Beyond benzene, these interactions also occur with biologically significant compounds such as DNA and porphyrins. These interactions, which are also present in small molecules, can be used to functionalize GO and G systems for processing and property modification.

A novel approach to non-covalently functionalized GO involved using hyperbranched polyesters with terminal carboxyl (HBP) to produce HBP-GO through strong π-π coupling between hyperbranched polyesters and GO nanosheets. This method was aimed at enhancing the interfacial characteristics between GO and epoxy resin (EP). The presence of hyperbranched polyesters embedded within the GO layer created a steric hindrance effect, effectively preventing the aggregation of GO nanosheets, and greatly improving their dispersibility. Additionally, this non-covalent functionalization increased surface energy, interfacial energy, and adhesion work, reduced the contact angle of HBP-GO with EP, and significantly enhanced the wetting property of HBP-GO [78].

The G compounds with non-covalent functionalization spread easily in polar aprotic liquids. According to Marcia et al., non-covalent pyrene derivatives were applied to GBMs to prevent agglomeration and improve compatibility with the polymer matrix [79]. Layek et al. [80] found that functional groups on the pyrene basal plane edge also improved the interface between G and the polymeric matrix [79]. Similar to this, non-covalently functionalized G composites with enhanced barrier characteristics and G dispersion utilizing pyrene derivates and polyketones [80]. The fluorescence qualities of the dispersant with conjugated structure are typically present; however, once the G is combined, the fluorescence capabilities will be significantly quenched [81].

The reduction of GO through the wet chemical method produced a G-based material having aggregation-induced emission (AIE) properties. During the GO reduction process, TPEP, a conjugated molecule containing tetraphenylethylene (TPE) and pyrene, was used as a stabilizer. The resulting rGO-TPEP exhibited AIE property and high dispersion in solution compared to TPEP alone. The fluorescence intensity of rGO-TPEP was 2.23 times stronger. rGO-TPEP proved to be a sensitive chemical sensor for explosive detection in both aggregated and solid states, owing to its unique optical features and AIE effect. In the aggregated state, rGO-TPEP demonstrated the ability to detect even small concentrations of 2,4-dinitrotoluene (DNT) at 0.91 ppm, with a high quenching constant of 2.47 × 104 M^{-1} [82].

3.2.2 Functionalization with polymers, drugs, and biomolecules

The π-π interactions between G derivatives and polymers containing aromatic rings provide an excellent example of how tight binding can result in highly homogeneous polymer composites with improved mechanical, electrical, and thermal properties [81, 82]. For instance, Kevlar, an aromatic ring-containing polymer, can strongly interact with G through stacking, and when incorporated into a G nanoribbon (GNR) composite, it enhances the mechanical properties significantly. For example, the addition of 1 wt% GNR to Kevlar/PVC composite increased the Young's modulus and yield strength by approximately 72.3% and 106%, respectively.

Non-covalent functionalization of G has been explored in other applications. One method involves non-covalently functionalizing G with amine-terminated polystyrene, which enhances its dispersibility in organic environments [71]. Another approach uses self-assembled monolayers of 1,4-benzenedimethanethiol (BDMT) to anchor gold nanoparticles on the G surface, enabling highly sensitive electrochemical detection of H_2O_2 [83].

A novel method for modifying G in field-effect transistors (FETs) application was developed, involving positive and negative doping effects induced by sequentially treating G with gold nanoparticles (AuNPs) and thiol-SAM molecules [82]. This technique demonstrated a Dirac voltage switcher on a G FET using heavy metal ions.

Similar to this, a G structure that self-assembled and adorned with glucose oxidase was employed to support the sensing of glucose within the 20 nm detection limit [84–86]. To detect nitrate, copper nanoparticles and adorned self-accumulated G was synthesized by Wang, Kim, and Cui [87]. Non-covalently functionalized G with 60 nm thickness and 300 nm diameter was obtained with dextran and chitosan via layer assembly on the GO surface for an anticancer function [88–89].

Jung et al. [90] introduced a novel G probe functionalized with single-stranded DNA (ssDNA) for the accurate detection of H_2S and NH_3 in exhaled breathing. The functionalization of G with ssDNA creates an ion-conducting channel, enabling efficient proton hopping at humidity levels above 80%. This process results in excellent carrier density modulation, which has the potential to detect biomarkers for illnesses such as kidney diseases and halitosis. The chemiresistive probe was fabricated using a hybrid configuration, where non-covalent π-stacking interactions between ssDNA and G play a crucial role in facilitating the detection process [90].

Concha et al. [91] utilized self-limiting monolayers of ammonium-substituted pyrenes to impart a general positive charge to the G surface and sulfonate-substituted pyrenes to impart a general negative charge. Both types of pyrenes resulted in a stable hydrophilic surface, which allowed for the specific immobilization of macromolecules that carried either negative or positive charges. This straightforward and versatile non-covalent approach is applicable to G on various substrates, including Cu, SiO_2, suspended G, and graphite. By transforming G from hydrophobic to hydrophilic, this method facilitates the use of electrostatic interactions to control the adsorption of macromolecules [91].

Gan et al. [92] developed a nanocomposite to functionalize G with D-glucose using poly (vinyl alcohol) (PVA) and poly (methyl methacrylate) (PMMA) as matrices. The interaction between the polymer blend and D-glucose moieties through hydrogen bonds connected to the fillers led to the uniform diffusion of functionalized G within the matrices. This resulted in a significant improvement in the thermomechanical properties of the nanocomposite [92].

Chhetri et al. [89] used a catalyst based on 3-amino-1,2,4-triazole (TZ) in KOH to functionalize GO nanoparticles and then incorporate them into epoxy resin resulting in better thermal and mechanical resistance. More specifically, compared to composites containing pure GO, both the tensile strength and elastic modulus increased by roughly 30%, while fracture toughness increased more than a magnitude [89].

Functionalization of high-density polyethylene (HDPE) with maleic anhydride (GO-g-MA), GO with ethylenediamine (GO-EDA), and oxidized CNTs (MWCNTs-COOH) were used in a complicated system that Bian et al. [93] developed. To connect GO-EDA and MWCNTs-COOH, L-aspartic acid was employed, creating a hybrid network. This hybrid network was then melted into HDPE-g-MA [93].

The packaging business is a significant additional area where the utilization of G and its derivatives may be of tremendous relevance. Given the growing environmental concerns around waste disposal, two significant areas of application are food and electronic packaging, both requiring excellent barrier properties against gases, particularly water vapor. Functionalization of G has also been employed in packaging materials, where it can enhance mechanical qualities, chemical endurance, and barrier properties [86]. Examples include G functionalized with D-glucose to improve thermomechanical properties in polyvinyl alcohol and poly(methyl methacrylate) matrices [93], as well as GO functionalized with 3-amino-1,2,4-triazole to enhance thermal and mechanical resistance in epoxy resin [94].

G's utilization in electronic packaging holds particular promise due to its ability to enhance mechanical properties, chemical resistance, and barrier performance [79].

Nanocomposites were formed using vinyl silicone resin prepolymer. The addition of 1% of functionalized G nanoplatelets (GNP) resulted in a substantial improvement in mechanical properties, increasing the tensile strength by approximately 500% and the elastic modulus by up to 1000%. Moreover, at higher percentages of functionalized GNP (10–15%), there was a significant enhancement in thermal conductivity, reaching 16 to 38 times the initial conductivity of the resin [94].

Hierarchical structures were created by growing *in situ* SiO_2 nanoparticles on reduced G oxide (rGO) nanoplatelets that had been non-covalently functionalized. These were discovered to serve as a reinforcing agent for the matrix made of hydrogenated nitrile butadiene rubber, significantly resulting in a significant improvement in both the static and dynamic mechanical properties [95–96].

Ou et al. [97] developed an eco-friendly polyurethane comprising of poly (L-lactic acid) as flexible fragments and G. The process involved initiating the ring-opening polymerization of L-lactic acid with phenol-derivatized G, resulting in a polymer grafted with G. Subsequently, the polymer endured condensation polymerization with diphenylmethane-diisocyanate to produce the polyurethane. This polyurethane exhibited superior hydrolysis and antifouling behavior compared to the clean polyurethane, thus making it a suitable coating for flat surfaces [97].

Layer-by-layer assembly propelled by electrostatic contact was used to alternately deposit polyethyleneimine and functionalized GO (FGO) on poly (vinyl alcohol) (PVA) film surface, giving the coated PVA film remarkable flame retardancy. The PVA matrix was enclosed in a shield of defense created by the multilayer FGO-based coating, which successfully stopped the mass and heat transfer that occurs during combustion. Compared to plain PVA, coated PVA shows an initial decomposition temperature of 260°C and a nearly 60% lower total heat output [98].

3.3 Nanoparticle functionalization

Another common type of functionalization is the addition of various nanoobjects, such as nanoparticles (NPs), nanowires, nanorods, and nanospheres (NSs), to G heterostructures and nanocomposites. In contrast to pure G, nanoobjects are equipped with a variety of capabilities based on the inherent features of materials. To expand G's potential in a variety of electronic and optoelectronic processes, the nanoobject-based G nanocomposites in particular combine specialized optical and electrical capabilities. The semiconductor nanoobjects that can outperform the low-absorption behavior of pure G have certain exceptional optical properties. For instance, when coupled with G to create nanocomposites, the semiconductor cadmium sulfide quantum dots/nanoparticles (CdS QDs/NPs) have significantly enhanced the photo-absorption as well as photoelectrical reactions [99].

When paired with G/ZnO heterostructures, ZnO nanowires/nanorod semiconductors with UV activity have a wider bandgap, which can enhance G's UV responsiveness. Likewise, when mixed with G to create nanocomposites, the TiO_2 NPs demonstrated better photocatalytic and photoelectric activity [100–101]. The G/nanoobjects also demonstrate exceptional activity in electrical and electrochemical fields; according to one study, in a hydrogen evolution reaction, MoS_2 NSs/G nanocomposites displayed superior electrocatalytic properties to pure MoS_2 [102]. Additionally, Ni $(OH)_2$, the NSs/G nanocomposites have shown improved performance in investigations pertaining to electrochemical capacitors [103].

Pure Co_3O_4 NPs and Co_3O_4 NPs/G nanocomposites demonstrated significantly higher oxygen reduction than a C/Pt catalyst. During the oxidation of methanol, the metal NPs, which are only Pt-functionalized with G, demonstrate excellent electro-catalytic activity [104]. Shahid et al. demonstrated the functionalization of G with nanoscale particles which show selectivity for hydrazine [105].

G-gold nanostructures (G-AuNS) have emerged as highly effective sensing substrates, enabling the development of electrochemical biosensors that are affordable, dependable, rapid, and sensitive. These electrochemical devices have found significant applications in the biomedical domain, for detecting glucose through enzymatic or catalytic approaches, as well as H_2O_2, biomolecules (DNA, protein), small molecules (dopamine), microorganisms, foodborne pathogens, environmental pollutants, and a wide range of other analytes. In general, electrochemical biosensors present remarkable advantages, including customization, miniaturization, and rapid analysis capabilities. Nevertheless, they also come with certain common analytical limitations, such as susceptibility to interferences from complex biological sample matrices and the inability to simultaneously detect multiple analytes [106].

3.4 Plasma hydrogenation

G is an example of a 2D material with exceptional qualities that can be used for huge on/off ratio devices and hydrogen storage. G's outstanding electrical, optical, mechanical, and thermal capabilities have propelled it to the forefront of cutting-edge research and industry. However, a significant barrier to its use has been G's lack of a sizable band gap. Various methods have therefore been developed to open and control a band gap in functionalized G.

The most straightforward chemical modification of G involves hydrogenation. In this process, each lattice carbon can bond with only one hydrogen atom, assuming no carbon-carbon bonds are broken. Surprisingly, due to the similar electronegativities of carbon and hydrogen, the C-H bond is expected to be nonpolar, not significantly impacting the G's doping. Based on our knowledge of organic chemistry, the C-H bond in organic molecules is relatively unreactive, leading to the anticipation that hydrogenated G would be chemically stable and inert. This hydrogenated form of G was envisaged as a chemically stable, nonpolar, two-dimensional material with potential applications in electronics, and possibly hydrogen storage (**Figure 12**) [106].

On the other hand, hydrogen plasma treatment for surface modification of single-layered G has garnered significant scholarly interest due to its potential for conventional wafer-scale production. Researchers have successfully created a monolayer chemical-vapor-deposited G, hydrogenated by an indirect hydrogen plasma, with no structural defects. They observed that adjusting the hydrogen coverage allows for precise control of the band gap, achieving values of up to 3.9 eV [107].

Plasma containing H_3^+ ions having 3.45, 5.35, and 7.45 eV energies was used to expose G sheets. Only the specimen subjected to the lowest energy plasma could be thermally annealed back into G; the other specimens had irreversible features as a result of vacancy defects caused by ions having high energies. By employing plasma with the proper ion energy and Joule heating, the alterable property in G FETs has been demonstrated, proving that the damage caused by plasma was minimal [108].

The photoluminescence of hydrogenated G exhibits an intriguing optical characteristic. This phenomenon arises from the creation of electronically disconnected conjugated polycyclic regions, exhibiting a diverse range of absorption outlines and remarkably fluorescent emission features as atoms randomly populate the G lattice [109]. These luminous regions possess functionality comparable to carbon nanodots. Consequently, highly hydrogenated G has garnered interest for applications involving white light fluorescence, optoelectronic characteristics, and imaging capabilities, akin to quantum dots [110]. Elsewhere, Elias et al. discovered that the hydrogenation had a strong p-doping behaviour. However, it was discovered that once the sample

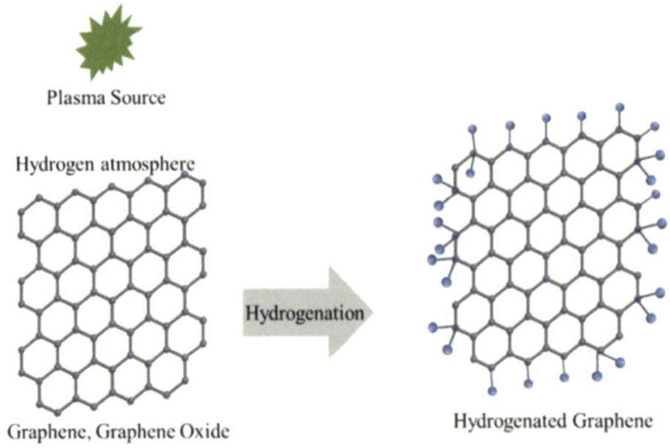

Figure 12.
Plasma hydrogenation of G.

was dried, hydrogen showed the n-doped behaviour in the G and p-doping effect was initiated by adsorbed water molecules [111].

These studies highlight the importance of surface modification of GO in tuning its properties and enabling novel applications in analytical chemistry. Fortunately, GO possesses numerous active surface functional groups, which allow for surface modification through various interactions, including covalent and non-covalent interactions. Additionally, successful surface modification of GO has been achieved through the doping of heteroatoms or nanocomposites.

3.5 Substitutional doping of graphene

The most realistic and appropriate approach to amend the band structure of G is doping where G changes from semi-metal to n or p-type semiconductor. This is accomplished by substituting or replacing carbon atoms from G lattice with foreign elements, a process known as substitutional doping of G. Substitutional doping is going to be interesting as it introduces the charge in the structure of G.

In this type of doping, G's carbon atom is replaced by another atom like nitrogen, phosphorous, boron, sulfur, etc., near the opening (**Figure 13**). In a pristine G, the unpaired electrons are tightly bonded and passivated within its delocalized structure, rendering it chemically unreactive and limiting its energy absorption capacity [112]. However, G can be easily p-type doped through surface absorption. When pristine G is exposed to molecules with electron-withdrawing groups (such as H_2O, O_2, N_2, NO_2, PMMA, etc.), noticeable p-type doping occurs, but it can quickly return to its original state once the doping molecules are removed.

In contrast, achieving stable n-type doping in G presents more challenges. Although some electron-donating molecules like ammonia, potassium, phosphorus, hydrogen, and poly(ethyleneimine) (PEI) can induce n-type doping through surface electron transfer, these doping effects often prove to be unstable. As an alternative approach, introducing nitrogen-containing precursors during the growth process can partially replace lattice carbon atoms with nitrogen atoms, leading to effective n-doping.

Combining both p-type and n-type doping methods enables the creation of p-n junctions in mono- or bi-layer G. These hetero-doped G p-n junctions have paved the way for novel functional devices like photothermoelectric devices [113].

Heteroatom insertion can produce extraordinarily abundant active sites in G. The addition of a heteroatom with a different electronegativity from the carbon atom could break the electroneutrality of G resulting in the generation of unstable charged zones and such zones might act as active sites. These active sites may exist as

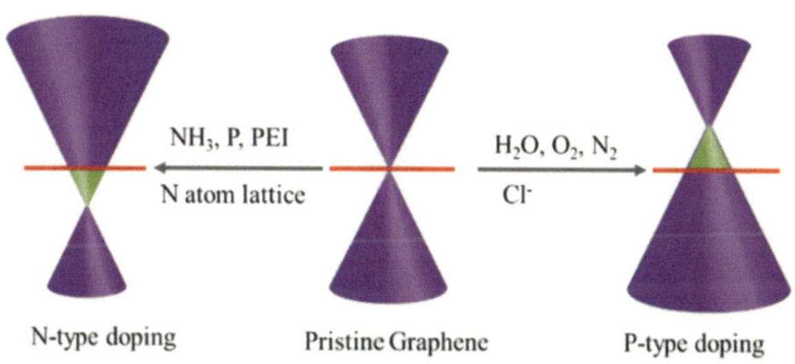

Figure 13.
Doping mechanism in G layer.

structural defects (arising from the lattice strain). The lattice strain is mainly due to the difference in size of dopant and carbon atom. The base of these active sites introduces the band gap and semiconducting properties of G by improving their chemical properties. The heteroatom can be substituted into G through various methods like solid-phase synthesis, liquid-phase synthesis, and direct synthesis [114].

CVD and segregation-growth approach are the direct synthesis methods. The most prominent method to introduce the heteroatom in G is the CVD method where heteroatoms are directly incorporated into lattice of G [114–116]. This method is used to synthesize n-type semiconductors using G and nitrogen element as dopant accompanied by alteration of charge mobilization and transference of electron results in the transition of G from metal to semiconductor. As a result, nitrogen-doped graphene finds applications in both the electronic and optoelectronic realms [117].

Another method used to synthesize doped G is a segregation-growth method which is also a direct synthesis method. In this method, some selective doping is possible as a heteroatom incorporated into some selective surface of G. Recently, Wang et al. confirmed the fabrication of nitrogen-doped G via this method exhibited a bandgap of 0.16 eV. This method is used to control the concentration and position of doping materials hence the synthesized G used for FET application [118].

4. Conclusion, challenges, and future prospectives

The chapter delves into the ability of G materials for various applications such as energy storage, conversion devices, and many more. To leverage the unique properties of G, it can be combined with other nanomaterials like metal, metal oxide, magnetic nanoparticles, quantum dots, etc., using surface functionalization and nanoparticle functionalization, along with plasma dehydrogenation and substitutional doping. The appropriate modification of the surface through doping and/or functionalization creates promising opportunities for the use of these materials in device applications.

Surface functionalization can be achieved through covalent and non-covalent functionalization. Non-covalent functionalization, while easy and rapid, relies on hydrophobic, van der Waals, and electrostatic interactions, making it susceptible to leaching out functions from the G sheets during application. On the other hand, covalent functionalization capitalizes on oxygen functional groups present on G surfaces, such as carboxylic acid groups at the edges and epoxy/hydroxyl groups on the basal plane, to modify the surface functionality of G. Covalent functionalization addresses the drawbacks of non-covalent functionalization and allows for the precise decoration of desired functions on suitable platforms.

Regarding surface transfer doping, its long-term stability is lacking as adsorbed species can desorb from the G surface and react with reactive molecules. Substitutional doped G, where metal or heteroatoms are attached to the carbon linkage of G, offers improved stability. However, it faces critical challenges in terms of large-scale production, doping controllability, and mechanisms. The production of G indeed faces several challenges, including the need for specific conditions and high temperatures in some methods such as the CVD method where high temperatures are required (typically above 1000°C) to create the necessary conditions for graphene growth. These high temperatures can be energy-intensive and may limit the scalability and cost-effectiveness of production. Developing a controlled synthesis of doped

G could provide a desirable solution. As the field of novel materials and innovative applications continuously advances, understanding the mechanisms for reactions in many electrochemical systems remains complex. Further investigations into doped G materials hold promise for contributing to energy systems. Controlled reduction offers a viable pathway for mass-producing semiconducting G.

GO has proven its versatility in various applications, including optoelectronics, drug delivery materials, biodevices, and polymer composites. Consequently, G emerges as a promising candidate for immobilizing various substances, such as metals, biomolecules, fluorescent molecules, drugs, and inorganic nanoparticles.

Acknowledgements

The authors are thankful to K. R. Mangalam University for providing technical support.

Conflict of interest

The authors declare no conflict of interest.

Abbreviations

AIE	aggregation-induced emission
BDMT	1,4-benzenedimethanethiol
CdS QDs/NPs	cadmium sulfide quantum dots/nanoparticles
CVD	chemical vapor deposition
DA	dodecyl amine
DCC	N,N-dicyclohexyl carbodiimide
DEA	diethanolamine
DMF	N,N-dimethylformamide
DNT	2,4-dinitrotoluene
EDA	ethylenediamine
EDC	1-ethyl-3-(3-dimethylaminopropyl)-carbodiimide
FET	field-effect transistor
FGO	functionalized GO
G	graphene
GBMs	G-based materials
GO	graphene oxide
G-AuNS	G-gold nanostructures
GNPs	gold nanoparticles
GNS	graphene nanosheets
GNRs	G nanoribbons
HBP	hyperbranched polyesters with terminal carboxyl
HDPE	high-density polyethylene
LMB	locked nucleic acid molecular beacon
MA	maleic anhydride
MWCNT	multi-walled carbon nanotube
NPs	nanoparticles

NSs	nanospheres
ODA	octadecylamine
PANI	polyaniline
Py-PGMA	poly (glycidyl methacrylate) with localized pyrene groups
PMMA	poly (methyl methacrylate)
PNVP	poly-N-vinyl-2-pyrrolidone
PVA	poly (vinyl alcohol)
rGO	reduced graphene oxide
rGOA	reduced GO aerogel
ssDNA	single-stranded DNA
SW	Stone-Wales
SWDCPs	SW defect COOH pairs
TA	tannic acid
TB	tris buffer
TPE	tetraphenylethylene
thiol-SAM	thiol self-assembled monolayer

Author details

Neeraj Kumari and Meena Bhandari*
Department of Chemistry, School of Basic and Applied Sciences, K.R. Mangalam University, Gurugram, India

*Address all correspondence to: meena.bhandari@krmangalam.edu.in

References

[1] Park MJ, Lee JK, Lee BS, Lee YW, Choi IS, Lee S. Covalent modification of multiwalled carbon nanotubes with imidazolium-based ionic liquids: Effect of anions on solubility. Chemistry of Materials. 2006;**18**:1546-1551. DOI: 10.1021/cm0511421

[2] Liu Y, Zheng J, Zhang X, Li K, Du Y, Yu G, et al. Recent advances on graphene microstructure engineering for propellant-related applications. Journal of Applied Polymer Science. 2021;**138**: 1-22. DOI: 10.1002/app.50474

[3] Sun Y, Wu Q, Shi G. Graphene based new energy materials. Energy and Environmental Science. 2011;**4**: 1113-1132. DOI: 10.1039/C0EE00683A

[4] Novoselov KS, Geim AK, Morozov SV, Jiang D, Katsnelson MI, Grigorieva IV, et al. Two-dimensional gas of massless Dirac fermions in graphene. Nature. 2005;**438**:197-200. DOI: 10.1038/nature04233

[5] Schedin F, Geim AK, Morozov SV, Hill EW, Blake P, Katsnelson MI, et al. Detection of individual gas molecules adsorbed on graphene. Nature Materials. 2007;**6**(9):652-655. DOI: 10.1038/nmat1967

[6] Lee C, Wei XD, Kysar JW, Hone J. Measurement of the elastic properties and intrinsic strength of monolayer graphene. Science. 2008;**321**(5887): 385-388. DOI: 10.1126/science.1157996

[7] Boukhvalov DW, Katsnelson MI. Chemical functionalization of graphene. Journal of Physics: Condensed Matter. 2009;**21**(34):344205. DOI: 10.1088/0953-8984/21/34/344205

[8] Zhang S, Yang K, Feng L, Liu Z. In vitro and in vivo behaviors of dextran functionalized graphene. Carbon. 2011;**49**(12):4040-4049. DOI: 10.1016/j.carbon.2011.05.056

[9] Ma R, Zhou Y, Bi H, Yang M, Wang J, Liu Q, et al. Multidimensional graphene structures and beyond: Unique properties, syntheses and applications. Progress in Materials Science. 2020;**113**:100665. DOI: 10.1016/j.pmatsci.2020.100665

[10] Pappas GS, Ferrari S, Wan C. Recent advances in graphene-based materials for lithium batteries. Current Organic Chemistry. 2015;**19**:838-1849. DOI: 10.2174/1385272819666150526010249

[11] Worku AK, Ayele DW. Advances and challenges in the fabrication of porous metal phosphate and phosphonate for emerging applications. In: Gupta RK, editor. Metal Phosphates Phosphonates. Germany: Springer International Publishing; 2023. pp. 393-407. DOI: 10.1007/978-3-031-27062-8_22

[12] Chandra V, Park J, Chun Y, Woo Lee J, Hwang IC, Kim KS. Water-dispersible magnetite-reduced graphene oxide composites for arsenic removal. ACS Nano. 2010;**4**:3979. DOI: 10.1021/nn1008897

[13] Lee WH, Park J, Kim Y, Kim KS, Hong BH, Cho K. Control of graphene field-effect transistors by interfacial hydrophobic self-assembled monolayers. Advanced Materials. 2011;**23**:3460. DOI: 10.1002/adma.201101340

[14] Wang Y, Li Z, Wang J, Li J, Lin Y. Graphene and graphene oxide: Biofunctionalization and applications in biotechnology. Trends in Biotechnology. 2011;**29**(5):205-212. DOI: 10.1016/j.tibtech.2011.01.008

[15] Pumera M. Graphene-based nanomaterials and their electrochemistry. Chemical Society Reviews. 2010;**39**: 4146-4157. DOI: 10.1039/C002690P

[16] Novoselov KS, Geim AK, Morozov SV, Jiang D, Zhang Y, Dubonos SV, et al. Electric field effect in atomically thin carbon films. Science. 2004;**306**:666-669. DOI: 10.1126/science.1102896

[17] Mattevi C, Kim H, Chhowalla MA. Review of chemical vapour deposition of graphene on copper. Journal of Materials Chemistry. 2011;**21**:3324-3334. DOI: 10.1039/C0JM02126A

[18] Varykhalov A, Rader O. Graphene grown on Co(0001) films and islands: Electronic structure and its precise magnetization dependence. Physical Review B. 2009;**80**:035437. DOI: 10.1103/PhysRevB.80.035437

[19] Park S, Ruoff RS. Chemical methods for the production of graphenes. Nature Nanotechnology. 2009;**4**:217-224. DOI: 10.1038/nnano.2009.58

[20] Park S, An J, Jung I, Piner RD, An SJ, Li XS, et al. Colloidal suspensions of highly reduced graphene oxide in a wide variety of organic solvents. Nano Letters. 2009;**9**:1593-1597. DOI: 10.1021/nl803798y

[21] Hummers WS, Offeman RE. Preparation of graphitic oxide. Journal of the American Chemical Society. 1958;**80**:1339-1339. DOI: 10.1021/ja01539a017

[22] Staudenmaier L. Verfahren zur darstellung der graphitsäure. Berichte der Deutschen Chemischen Gesellschaft. 1898;**31**:1481-1487. DOI: 10.1002/cber.18980310237

[23] Marcano DC, Kosynkin DV, Berlin JM, Sinitskii A, Sun ZZ, Slesarev A, et al. Improved synthesis of graphene oxide. ACS Nano. 2010;**4**: 4806-4814. DOI: 10.1021/nn1006368

[24] Stankovich S, Dikin DA, Piner RD, Kohlhaas KA, Kleinhammes A, Jia YY, et al. Synthesis of graphene-based nanosheets via chemical reduction of exfoliated graphite oxide. Carbon. 2007;**45**:1558-1565. DOI: 10.1016/j.carbon.2007.02.034

[25] Shin HJ, Kim KK, Benayad A, Yoon SM, Park HK, Jung IS, et al. Efficient reduction of graphite oxide by sodium borohydride and its effect on electrical conductance. Advanced Functional Materials. 2009;**19**:1987-1992. DOI: 10.1002/adfm.200900167

[26] Sheshmani S, Fashapoyeh MA. Suitable chemical methods for preparation of graphene oxide, graphene and surface functionalized graphene nanosheets. Acta Chimica Slovenica. 2013;**60**:813-825

[27] Fan XB, Peng WC, Li Y, Li XY, Wang SL, Zhang GL, et al. Deoxygenation of exfoliated graphite oxide under alkaline conditions: A green route to graphene preparation. Advanced Materials. 2008;**20**:4490-4493. DOI: 10.1002/adma.200801306

[28] Zhou XJ, Zhang JL, Wu HX, Yang HJ, Zhang JY, Guo SW. Reducing graphene oxide via hydroxylamine: A simple and efficient route to graphene. Journal of Physical Chemistry C. 2011;**115**:11957-11961. DOI: 10.1021/jp202575j

[29] Zhang JL, Yang HJ, Shen GX, Cheng P, Zhang JY, Guo SW. Reduction of graphene oxide via L-ascorbic acid. Chemical Communications. 2010;**46**:1112-1114. DOI: 10.1039/B917705A

[30] Liu HT, Ryu S, Chen ZY, Steigerwald ML, Nuckolls C, Brus LE. Photochemical reactivity of graphene.

Journal of the American Chemical Society. 2009;**131**:17099-17101. DOI: 10.1021/ja9043906

[31] Joshi DJ, Koduru JR, Malek NI, Hussain CM, Kailasa SK. Surface modifications and analytical applications of graphene oxide: A review. TrAC Trends in Analytical Chemistry. 2021;**144**:116448. DOI: 10.1016/j.trac.2021.116448

[32] Park J, Jo SB, Yu YJ, Kim Y, Yang JW, Lee WH, et al. Single-gate bandgap opening of bilayer graphene by dual molecular doping. Advanced Materials. 2012;**24**:407-411. DOI: 10.1002/adma.201103411

[33] Georgakilas V, Otyepka M, Bourlinos AB, Chandra V, Kim N, Kemp KC, et al. Functionalization of graphene: Covalent and non-covalent approaches derivatives and applications. Chemical Reviews. 2012;**112**:6156-6214. DOI: 10.1021/cr3000412

[34] Lomeda JR, Doyle CD, Kosynkin DV, Hwang W-F, Tour JM. Diazonium functionalization of surfactant-wrapped chemically converted graphene sheets. Journal of the American Chemical Society. 2008;**130**(48):16201-16206. DOI: 10.1021/ja806499w

[35] Ma X, Li F, Wang Y, Hu A. Functionalization of pristine graphene with conjugated polymers through diradical addition and propagation. Chemistry – An Asian Journal. 2012;**7**(11):2547-2550. DOI: 10.1002/asia.201200520

[36] Jaworski S, Wierzbicki M, Sawosz E, Jung A, Gielerak G, Biernat JK, et al. Graphene oxide-based nanocomposites decorated with silver nanoparticles as an antibacterial agent. Nanoscale Research Letters. 2018;**13**:1-17. DOI: 10.1186/S11671-018-2533-2

[37] Ioniţa M, Vlăsceanu GM, Watzlawek AA, Voicu SI, Burns JS, Iovu H. Graphene and functionalized graphene: Extraordinary prospects for nanobiocomposite materials. Composites Part B: Engineering. 2017;**121**:34-57. DOI: 10.1016/j.compositesb.2017.03

[38] Wang G, Wang B, Park J, Yang J, Shen X, Yao J. Synthesis of enhanced hydrophilic and hydrophobic graphene oxide nanosheets by a solvothermal method. Carbon. 2009;**47**:68-72. DOI: 10.1016/j.carbon.2008.09.002

[39] Georgakilas V, Bourlinos AB, Zboril R, Steriotis TA, Dallas P, Stubos AK, et al. Organic functionalisation of graphene. Chemical Communications. 2010;**46**:1766-1768. DOI: 10.1039/B922081J

[40] Chua CK, Pumera M. Covalent chemistry on graphene. Chemical Society Reviews. 2013;**42**(8):3222-3233. DOI: 10.1039/c2cs35474h

[41] Zhang X, Jin J, Li S. Aryne cycloaddition: Highly efficient chemical modification of graphene. Chemical Communications. 2010;**46**(39):7340-7342. DOI: 10.1039/C0CC02389B

[42] Quintana M, Spyrou K, Grzelczak M. Functionalization of graphene via 1,3-dipolar cycloaddition. ACS Nano. 2010;**4**(6):3527-3533. DOI: 10.1021/nn100883p

[43] Economopoulos S, Rotas G, Miyata Y. Exfoliation and chemical modification using microwave irradiation affording highly functionalized graphene. ACS Nano. 2010;**4**(12):7499-7507. DOI: 10.1021/nn101735e

[44] Johns JE, Hersam MC. Atomic covalent functionalization of graphene. Accounts of Chemical Research.

2013;**46**(1):77-86. DOI: 10.1021/ar300143e

[45] Wojtaszek M, Tombros N, Caretta A, van Loosdrecht PHM, van Wees BJ. A road to hydrogenating graphene by a reactive ion etching plasma. Journal of Applied Physics. 2011;**110**(6):063715. DOI: 10.1063/1.3638696

[46] Medeiros PVC, Mascarenhas AJS, de Brito MF, de Castilho CM. A DFT study of halogen atoms adsorbed on graphene layers. Nanotechnology. 2010;**21**(48):485701. DOI: 10.1088/0957-4484/21/48/485701

[47] Barinov A, Malcioglu OB, Fabris S, Sun T, Gregoratti L, Dalmigilo M, et al. Initial stages of oxidation on graphitic surfaces: Photoemission study and density functional theory calculations. Journal of Physical Chemistry C. 2009;**113**(21):9009-9013. DOI: 10.1021/jp902051d

[48] Yuan C, Chen W, Yan L. Amino-grafted graphene as a stable and metal-free solid basic catalyst. Journal of Materials Chemistry. 2012;**22**(15):7456-7460. DOI: 10.1039/c2jm30442b

[49] Yu W, Sisi L, Haiyan Y, Jie L. Progress in the functional modification of graphene/graphene oxide: A review. RSC Advances. 2020;**10**:15328-15345. DOI: 10.1039/D0RA01068E

[50] Wei Y, Liu Q, Lian Q, Wang Y, Zhu Y, Zhang P, et al. Insight into the separation mechanism of graphene oxide membrane by designing dual layered structure. Desalination. 2022;**532**:115687. DOI: 10.1016/j.desal.2022.115687

[51] Wang Y, Wang F, Dong S, He H, Lu Y, Shi J, et al. Ultra-small SiO_2 nanoparticles highly dispersed on non-covalent functionalized reduced graphene oxide nanoplatelets for high-performance elastomer applications. Composites Science and Technology. 2020;**198**:108297. DOI: 10.1016/j.compscitech.2020.108297

[52] Man SHC, Thickett SC, Whittaker MR, Zetterlund PB. Synthesis of polystyrene nanoparticles "armoured" with nano dimensional graphene oxide sheets by miniemulsion polymerization. Journal of Polymer Science Part A: Polymer Chemistry. 2013;**51**:47-58. DOI: 10.1002/pola.26341

[53] Xu Y, Liu Z, Zhang X, et al. A graphene hybrid material covalently functionalized with porphyrin: Synthesis and optical limiting property. Advanced Materials. 2009;**21**(12):1275-1279. DOI: 10.1002/adma.200801617

[54] Shin B, Kim WK, Yoon S, Lee J. Duplex DNA-functionalized graphene oxide: A versatile platform for miRNA sensing. Sensors and Actuators B: Chemical. 2020;**305**:127471. DOI: 10.1016/j.snb.2019.127471

[55] Choi HK, Oh Y, Jung H, Hong H, Ku BC, You NH, et al. Influences of carboxyl functionalization of intercalators on exfoliation of graphite oxide: A molecular dynamics simulation. Physical Chemistry Chemical Physics. 2018;**20**(45):28616. DOI: 10.1039/C8CP05436C

[56] Cao Y, Lai Z, Feng J, et al. Graphene oxide sheets covalently functionalized with block copolymers via click chemistry as reinforcing fillers. Journal of Materials Chemistry. 2011;**21**(25):9271. DOI: 10.1039/C1JM10420A

[57] Ouyang F, Huang B, Li Z, et al. Chemical functionalization of graphene nanoribbons by carboxyl groups on stone-wales defects. The Journal of Physical Chemistry C. 2008;**112**(31):12003-12007. DOI: 10.1021/jp710547x

[58] Bonanni A, Chua CK, Pumera M. Rational design of carboxyl groups perpendicularly attached to a graphene sheet: A platform for enhanced biosensing applications. Chemistry: A European Journal. 2014;**20**(1):217-222. DOI: 10.1002/chem.201303582

[59] Yu D, Yang Y, Durstock M, Baek J-B, Dai L. Soluble P_3HT-grafted graphene for efficient bilayer–heterojunction photovoltaic devices. ACS Nano. 2010;**4**(10):5633-5640. DOI: 10.1021/nn101671t

[60] Xia Z, Leonardi F, Gobbi M, et al. Electrochemical functionalization of graphene at the nanoscale with self-assembling diazonium salts. ACS Nano. 2016;**10**(7):7125-7134. DOI: 10.1021/acsnano.6b03278

[61] Ossonon BD, Belanger D. Functionalization of graphene sheets by the diazonium chemistry during electrochemical exfoliation of graphite. Carbon. 2017;**111**:83-93. DOI: 10.1016/j.carbon.2016.09.063

[62] Xiong Y, Xie Y, Zhang F, Ou E, Jiang Z, Ke L, et al. Reduced graphene oxide/hydroxylated styrene–butadiene–styrene tri-block copolymer electroconductive nanocomposites: Preparation and properties. Materials Science and Engineering B. 2012;**177**(14):1157-1163. DOI: 10.1016/j.mseb.2012.05.012

[63] Yang X, Ma L, Wang S, Li Y, Tu Y, Zhu X. "Clicking" graphite oxide sheets with well-defined polystyrenes: A new strategy to control the layer thickness. Polymer. 2011;**52**(14):3046-3052. DOI: 10.1016/j.polymer.2011.04.062

[64] Farquhar AK, Dykstra HM, Waterland MR, et al. Spontaneous modification of free-floating few-layer graphene by aryldiazonium ions: Electrochemistry, AFM and infrared spectroscopy from grafted films. Journal of Physical Chemistry C. 2016;**120**(14):7543. DOI: 10.1021/acs.jpcc.5b11279

[65] Ramesh P, Sambathkumar S, Jagatheesan R, Vasanth KS. A facile synthesis of hydroxyl functionalized graphene oxide as an electrode material for energy storage devices. Materials Letters. 2019;**256**:126639. DOI: 10.1016/j.matlet.2019.126639

[66] Shao XF, Lin JC, Teng HR, Yang S, Fan LW, Chiu JNW, et al. Hydroxyl group functionalized graphene oxide nanosheets as additive for improved erythritol latent heat storage performance: A comprehensive evaluation on the benefits and challenges. Solar Energy Materials and Solar Cells. 2020;**215**:110658. DOI: 10.1016/j.solmat.2020.110658

[67] Isabella A, Vacchi JR, Bianco A, Ménard-Moyon C. Controlled derivatization of hydroxyl groups of graphene oxide in mild conditions. 2D Materials. 2018;**5**:035037. DOI: 10.1088/2053-1583/aac8a9

[68] Yan L, Lin M, Zeng C, Chen Z, Zhang S, Zhao X, et al. Electroactive and biocompatible hydroxyl- functionalized graphene by ball milling. Journal of Materials Chemistry. 2012;**22**:8367-8371. DOI: 10.1039/C2JM30961K

[69] Veca LM, Lu F, Meziani MJ, Cao L, Zhang P, Qi G, et al. Polymer functionalization and solubilization of carbon nanosheets. Chemical Communications. 2009:2565-2567. DOI: 10.1039/B900590K

[70] Amiri A, Sadri R, Shanbedi M, Ahmadi G, Kazi SN, Chew BT, et al. Synthesis of ethylene glycol-treated graphene nanoplatelets with one-pot,

microwave-assisted functionalization for use as a high-performance engine coolant. Energy Conversion and Management. 2015;**101**:767-777. DOI: 10.1016/j.enconman.2015.06.019

[71] Su Q, Pang S, Alijani V, Li C, Feng X, Müllen K. Composites of graphene with large aromatic molecules. Advanced Materials. 2009;**21**:3191. DOI: 10.1002/adma.200803808

[72] Ren PG, Wang H, Huang H, Li ZM. Characterization and performance of dodecyl amine functionalized graphene oxide and dodecyl amine functionalized graphene/high-density polyethylene nanocomposites: A comparative study. Journal of Applied Polymer Science. 2014;**131**(2):39803. DOI: 10.1002/app.39803

[73] Georgakilas V, Tiwari JN, Kemp KC, Perman JA, Bourlinos AB, Kim KS, et al. Noncovalent functionalization of graphene and graphene oxide for energy materials, biosensing, catalytic, and biomedical applications. Chemical Reviews. 2016;**116**:5464. DOI: 10.1021/acs.chemrev.5b00620

[74] Mann JA, Dichtel WR. Noncovalent functionalization of graphene by molecular and polymeric adsorbates. The Journal of Physical Chemistry Letters. 2013;**4**:2649-2657. DOI: 10.1021/jz4010448

[75] Choi E-Y, Han TH, Hong J, Kim JE, Lee SH, Kim HW, et al. Noncovalent functionalization of graphene with end-functional polymers. Journal of Materials Chemistry. 2010;**20**(10):1907-1912. DOI: 10.1039/b919074k

[76] Xu Y, Bai H, Lu G, Li C, Shi G. Flexible graphene films via the filtration of water-soluble noncovalent functionalized graphene sheets. Journal of the American Chemical Society. 2008;**130**(18):5856-5857. DOI: 10.1021/ja800745y

[77] Kozhemyakina NV, Englert JM, Yang G, Spiecker E, Schmidt CD, Hauke F, et al. Non-covalent chemistry of graphene: Electronic communication with dendronized perylene bisimides. Advanced Materials. 2010;**22**(48):5483-5487. DOI: 10.1002/adma.201003206

[78] Gu L, Liu S, Zhao H, Yu H. Facile preparation of water-dispersible graphene sheets stabilized by carboxylated oligoanilines and their anticorrosion coatings. ACS Applied Materials & Interfaces. 2015;7(32):17641-17648. DOI: 10.1021/acsami.5b05531

[79] Marcia M, Hirsch A, Hauke F. Perylene-based non-covalent functionalization of 2D materials. FlatChem. 2017;**1**:89-103. DOI: 10.1016/j.flatc.2017.01.001

[80] Layek RK, Samanta S, Nandi AK. The physical properties of sulfonated graphene/poly (vinyl alcohol) composites. Carbon. 2012;**50**:815-827. DOI: 10.1016/j.carbon.2011.09.039

[81] Cho I, Jeon SY, Kim S, Lim JY. Improving dispersion and barrier properties of polyketone/graphene nanoplatelet composites via noncovalent functionalization using aminopyrene. ACS Applied Materials and Interfaces. 2017;**9**:27984. DOI: 10.1021/acsami.7b10474

[82] Zhang Y, Li H, Wu Q-Y, Gu L. Non-covalent functionalization of graphene sheets by pyrene-endcapped tetraphenylethene: Enhanced aggregation-induced emission effect and application in explosive detection. Frontiers in Chemistry. 2022;**10**:970033. DOI: 10.3389/fchem.2022.970033

[83] Lian M, Fan J, Shi Z, Li H, Yin J. Kevlar-functionalized graphene

nanoribbon for polymer reinforcement. Polymer. 2014;**55**:2578-2587. DOI: 10.1016/j.polymer.2014.03.059

[84] Zhang J, Xu Y, Cui L, Fu A, Yang W, Barrow C, et al. Mechanical properties of graphene films enhanced by homo-telechelic functionalized polymer fillers via p–p stacking interactions. Composites Part A: Applied Science and Manufacturing. 2015;**71**:1-8. DOI: 10.1016/j.compositesa.2014.12.013

[85] Shin D, Kim HR, Hong BH. Gold nanoparticle-mediated non-covalent functionalization of graphene for field-effect transistors. Nanoscale Advances. 2021;**3**:1404. DOI: 10.1039/d0na00603c

[86] Liu S, Tian J, Wang L, Luo Y, Lu W, Sun X. Self-assembled graphene platelet–glucose oxidase nanostructures for glucose biosensing. Biosensors and Bioelectronics. 2011;**26**:4491-4496. DOI: 10.1016/j.bios.2011.05.008

[87] Wang L, Kim J, Cui T. Self-assembled graphene and copper nanoparticles composite sensor for nitrate determination. Microsystem Technologies. 2018;**24**(9):3623-3630. DOI: 10.1007/s00542-018-3792-7

[88] Xie M, Lei H, Zhang Y, Xu Y, Shen S, Ge Y, et al. Non-covalent modification of graphene oxide nanocomposites with chitosan/dextran and its application in drug delivery. RSC Advances. 2016;**6**:9328. DOI: 10.1039/C5RA23823D

[89] Chhetri S, Adak NC, Samanta P, Mallisetty PK, Murmu NC, Kuila T. Interface engineering for the improvement of mechanical and thermal properties of covalent functionalized graphene/epoxy composites. Journal of Applied Polymer Science. 2017:46124. DOI: 10.1002/app.46124

[90] Jung Y, Moon HG, Lim C, Choi K, Song HS, Bae S, et al. Advanced Functional Materials. 2017;**27**:1700068

[91] Concha BN, Zachary PL, Laker AJM, Neil R, Wilson JPR. Non-covalent functionalization of graphene with a hydrophilic self-limiting monolayer for macro-molecule immobilization. FlatChem. 2017;**1**:52-56. DOI: 10.1016/j.flatc.2016.11.001

[92] Gan L, Shang S, Yuen CWM, Jiang SX. Covalently functionalized graphene with d-glucose and its reinforcement to poly (vinyl alcohol) and poly (methyl methacrylate). RSC Advances. 2015;**5**:15954-15961. DOI: 10.1039/C5RA00038F

[93] Bian J, Wang G, Lin HL, Zhou X, Wang ZJ, Xiao WQ, et al. HDPE composites strengthened–toughened synergistically by l-aspartic acid functionalized graphene/carbon nanotubes hybrid nanomaterials. Journal of Applied Polymer Science. 2017;**134**:45055. DOI: 10.1002/app.45055

[94] Gui D, Xiong W, Tan G, Li S, Cai X, Liu J. Improved thermal and mechanical properties of silicone resin composites by liquid crystal functionalized graphene nanoplatelets. Journal of Materials Science: Materials in Electronics. 2016;**27**:2120-2127. DOI: 10.1007/s10854-015-4000-5

[95] Gouvêa RF, Del Aguila EM, Paschoalin VMF, Andrade CT. Extruded hybrids based on poly(3-hydroxybutyrate-co-3-hydroxyvalerate) and reduced graphene oxide composite for active food packaging. Food Packaging and Shelf Life. 2018;**16**:77-85

[96] Gui D, Yu S, Xiong W, Cai X, Liu C, Liu J. Liquid crystal functionalization of graphene nanoplatelets for improved thermal and mechanical properties of

silicone resin composites. RSC Advances. 2016;**6**:35210-35215. DOI: 10.1039/C6RA01858K

[97] Ou B, Chen M, Huang R, Zhou H. Preparation and application of novel biodegradable polyurethane copolymer. RSC Advances. 2016;**6**:47138-47144. DOI: 10.1039/C6RA03064E

[98] Torres, Castillo CS, Fuentes Agustín JE, García Reyes EM, Zamudio Aguilar MAM, Morales Zamudio L, Lozano T, et al. Effect of non-functionalized and functionalized graphene oxide with a silane agent on the thermal and rheological properties of nylon 6,6. Iranian Polymer Journal. 2022;**32**(2). DOI: 10.1007/s13726-022-01110-3

[99] Chen W, Liu P, Min L, Zhou Y, Liu Y, Wang Q, et al. Non-covalently functionalized graphene oxide-based coating to enhance thermal stability and flame retardancy of PVA film. Nano-Micro Letters. 2018;**10**(3):39. DOI: 10.1007/s40820-018-0190-8

[100] Caselis JLV, Juárez JDS, Aguilar JAG, Alcalá MFD, Rosas ER, Hernández JCM, et al. Effect of addition of non-functionalized graphene oxide in a commercial epoxy resin used as coating. Materials Science - Poland. 2021;**39**(4):467-477. DOI: 10.2478/msp-2021-0039

[101] Cao A, Liu Z, Chu S, Wu M, Ye Z, Cai Z, et al. A facile one-step method to produce graphene–CdS quantum dot nanocomposites as promising optoelectronic materials. Advanced Materials. 2010;**22**(1):103-106. DOI: 10.1002/adma.200901920

[102] Li Y, Wang H, Xie L, Liang Y, Hong G, Dai H. MoS_2 nanoparticles grown on graphene: An advanced catalyst for the hydrogen evolution reaction. Journal of the American Chemical Society. 2011;**133**(19):7296-7299. DOI: 10.1021/ja201269b

[103] Wang H, Casalongue HS, Liang Y, Dai H. Ni $(OH)_2$ nanoplates grown on graphene as advanced electrochemical pseudocapacitor materials. Journal of the American Chemical Society. 2010;**132**(21):7472-7477. DOI: 10.1021/ja102267j

[104] Liang Y, Li Y, Wang H, Zhou J, Wang J, Regier T, et al. Co_3O_4 nanocrystals on graphene as a synergistic catalyst for oxygen reduction reaction. Nature Materials. 2011;**10**(10):780-786. DOI: 10.1038/nmat3087

[105] Shahid MM, Rameshkumar P, Basirunc WJ, Wijayantha U, Chiu WS, Khiew PS, et al. An electrochemical sensing platform of cobalt oxide@ gold nanocubes interleaved reduced graphene oxide for the selective determination of hydrazine. Electrochimica Acta. 2018;**259**:606-616. DOI: 10.1016/j.electacta.2017.10.157

[106] Khalil I, Julkapli NM, Yehye WA, Basirun WJ, Bhargava SK. Graphene–Gold nanoparticles hybrid—Synthesis, functionalization, and application in a electrochemical and surface-enhanced Raman scattering biosensor. Materials. 2016;**9**:406. DOI: 10.3390/ma9060406

[107] Zhao X, Dai P, Xu D, Li Z, Guo Q. Ultrafine palladium nanoparticles anchored on NH_2-functionalized reduced graphene oxide as efficient catalyst towards formic acid dehydrogenation. International Journal of Hydrogen Energy. 2020;**45**(55):30396-30403. DOI: 10.1016/j.ijhydene.2020.08.025

[108] Son J, Lee S, Kim SJ, Park BC, Li HK, Kim S, et al. Hydrogenated monolayer graphene with reversible and tunable wide band gap and its field-effect

transistor. Nature Communications. 2016;**7**:13261. DOI: 10.1038/ncomms13261

[109] Cha J, Choi H, Hong J. Damage-free hydrogenation of graphene via ion energy control in plasma. Applied Physics Express. 2021;**15**(1):5002. DOI: 10.35848/1882-0786/ac4204

[110] Zhao F, Raitses Y, Yang X, Tan A, Tully CG. High hydrogen coverage on graphene via low temperature plasma with applied magnetic field. Carbon. 2021;**177**:244-251. DOI: 10.1016/j.carbon.2021.02.084

[111] Elias DC, Nair RR, Mohiuddin TM, Morozov SV, Blake P, Halsall MP, et al. Control of graphene's properties by reversible hydrogenation: Evidence for graphane. Science. 2009;**323**(5914):610-613. DOI: 10.1126/science.1167130

[112] Matis BR, Burgess JS, Bulat FA, Friedman AL, Houston BH, Baldwin JW. Surface doping and band gap tunability in hydrogenated graphene. ACS Nano. 2012;**6**:17-22. DOI: 10.1021/nn2034555

[113] Li H, Kang Z, Liu Y, Lee ST. Carbon nanodots: Synthesis, properties and applications. Journal of Materials Chemistry. 2012;**22**:24230-24253. DOI: 10.1039/C2JM34690G

[114] Ugeda MM, Brihuega I, Guinea G, Gómez-Rodríguez JM. Missing atom as a source of carbon magnetism. Physical Review Letters. 2010;**104**:096804. DOI: 10.1103/PhysRevLett.104.096804

[115] Yang L, Jiang S, Zhao Y, Zhu L, Chen S, Wang X, et al. Boron-doped carbon nanotubes as metal-free electrocatalysts for the oxygen reduction reaction. Angewandte Chemie. 2011;**123**:7270-7273. DOI: 10.1002/anie.201101287

[116] Chang CK, Kataria S, Kuo CC, Ganguly A, Wang BY, Hwang JY, et al. Band gap engineering of chemical vapor deposited graphene by in situ BN doping. ACS Nano. 2013;**7**:1333-1341. DOI: 10.1021/nn3049158

[117] Luo Z, Lim S, Tian Z, Shang J, Lai L, Macdonald B, et al. Pyridinic N doped graphene: Synthesis, electronic structure, and electrocatalytic property. Journal of Materials Chemistry. 2011;**21**(22):8038-8044

[118] Wang H, Maiyalagan T, Wang X. Review on recent progress in nitrogen-doped graphene: Synthesis, characterization, and its potential applications. ACS Catalysis. 2012;**2**(5):781-794. DOI: 10.1021/cs200652y

Chapter 4

The Graphene Surface Chemistry and Adsorption Science

Enos W. Wambu

Abstract

Graphene (G) has attracted immense attention due to its exceptional physicochemical and electronic properties, and quite a large amount of literature has accumulated on this subject over the last few decades. The current work, based on a systematic review of the relevant literature, was designed to provide an overview of G surface chemistry with respect to its adsorption science. The aim was to improve knowledge of the graphene surface chemistry while informing new strategies for designing and implementing new G materials for emerging applications. The key G surface reactions include: molecular adsorption of gases, bandgap tuning, gas detection tests; alkali metal storage for battery technology; G vacancy engineering; environmental amelioration of pollutants; and sensors and biosensors technology. GO (graphene oxide) or G has been surface-modified using nonmetals, metals, metal oxides, or organics. In general, GO and related functionalized materials have high affinity and adsorption efficacy for ionic adsorbates, whereas pristine G, and reduced graphene oxide (rGO), exhibits enhanced hydrophobic surfaces with propensity to strong π-π interactions. The metals' adsorption and doping can impart G magnetic and metallic character, whereas molecular intercalations tend to induce a G bandgap for nano-electronic and nanophotonic uses among other interactions.

Keywords: adsorption, bandgap tuning, graphene defect engineering, DNA, energy storage, environmental remediation, graphene, biosensors

1. Introduction

Graphene (G), a new layered carbon material, is one of the most studied materials over the last few decades. Investigations about G have centered on new fabrication methods [1]; improving understanding of its properties [2], designing and development of new derivatives [3], and exploring prospects for the new G materials [4]. In examining G properties, most studies focused on specific surface reactions, especially the adsorption of organic contaminants [5] and simple molecular gases [6], as well as heavy metal doping [7], and interaction with pharmaceutical agents [8]. The adsorption performance of G and its derivatives, primarily graphene oxide (GO), reduced graphene oxide (rGO), nonmetal- and transition metal-doped graphene (TM-Graphene), has been assessed extensively for their adsorption potential towards various adsorbates in terms of the reaction conditions; equilibrium, kinetic, and thermodynamic analyses; and the electronic characterization of adsorption energies, geometries, DOS, dipole moments, the reaction work functions, etc. The Langmuir [9], Freundlich [10], Temkin, Polanyi [11], and the Dubinin–Radushkevich [12] isotherms are widely used for evaluating the adsorption capacities, reaction mechanisms and thermodynamic of the adsorption processes. The adsorption kinetics, on the other hand, are often evaluated using the Lagergren first-order [13] and pseudo-second order [14] kinetics equations. Pristine graphene (G) surface presents basically three adsorption sites: the hollow site (H), at the center of a carbon hexagon; the bridge site (B), in the middle of a C−C bond; and the top site (T), on the top of a carbon atom. Molecular orientations are used to determine the most stable adsorption configurations and the adsorption energy, E_{ad}, is defined according to Hess' law as:

$$E_{ad} = E_{molecule} + E_{graphene} - E_{molecule@graphene} \tag{1}$$

where, $E_{molecule}$, $E_{graphene}$, and $E_{molecule@graphene}$ are the energies of the isolated gas molecule, graphene, and molecule–graphene adsorption complex, respectively. The first-principles electronic calculations are used basing on the spin-polarized density-functional theory (DFT) *via* the Vienna *ab initio* simulation package (VASP) [15] with the exchange and correlation energy included through a generalized gradient approximation in PBE format [16]. Similarly projected augmented-wave (PAW) potentials [17] are also used to describe the ion-electron interactions.

However, comprehensive overview of the surface chemistry of G is lacking in the literature and the current work was, therefore, designed to provide an inclusive update of the surface adsorption of G and the related materials. The main goal was to improve the understanding of the graphene chemistry and, thus, contribute to information of emerging strategies for designing new G materials and their applications. GO is characterized by several reactive surface groups with strong acidity and high adsorption capacities for basic and cationic species. G on the other hand, exhibits a hydrophobic surface presenting high adsorption capacities for chemicals that show strong π-π interactions. Suitable modifications of GO or G with metal oxides and/or organic groups can produce nanocomposites with more enhanced tailored adsorption capacities and separation efficiencies for specific diverse applications, including fuel gas adsorption and storage, battery technologies, G vacancy electronic and photonic engineering, pollution ameliorating, and sensors and biosensors fabrications. The current work aims to consolidate relevant literature and provide new insights for understanding G chemistry and spur new strategies, and application of G and its derivatives.

2. Application of graphene in fuel gas storage

The search for sustainable clean energy remains one of the fundamental socio-economic development goal of the 21st century. Actions toward realization of this goal include fuel gas storage techniques using efficient adsorbents and graphene (G) materials have attracted particularly intense attention due to their unique properties that lend them to diverse versatile modifications and applications. The large specific surface area of G has attracted researchers to explore it for diverse adsorptive applications. Ma et al. [18], for example, evaluated hydrogen adsorption using pristine single-layer G samples with a BET-specific surface area of 156 m^2/g and found that only 0.4 wt.% hydrogen could adsorb on G at 77 K under 100 kPa pressure.

Furthermore, even lower hydrogen adsorption rates of <0.2 wt.% were obtained at ambient temperature under elevated 6 MPa-pressure. The small gravimetric hydrogen uptake by G samples were attributed to low specific surface area of the materials and to the weak hydrogen binding into the G surfaces. Elsewhere, while using high pressures of about 40 bars, and Szczęśniak et al. [19], obtained optimum H_2 adsorption of 5.5–7 wt.% and 4.4–4.8 wt.% for activated rGO, and for transition metal (TM) nanoparticles doped-rGO, respectively.

Meanwhile, while investigating different hydrogen adsorption sites of G, Arellano et al. [20], observed that the most stable configuration was obtained when H_2 molecules physisorbed over the hollow (H) site, although the barriers for classical diffusion were, particularly, small. Nevertheless, in an *ab initio* study of atomic hydrogen adsorption on G, Ivanovskayaet al. [21], focused their investigations on the adsorption characteristics of the hydrogenation surface coverage of G. They observed that at high surface coverage, the resulting strain from substrate relaxation controlled H chemisorption leading to localized surface curvature. The chemisorption energy barrier was caused by the relaxation and by the adsorbent carbon atom rehybridization. This showed independence from the optimization technique and the system size. The H desorption barrier was, however, very sensitive to the correct structural relaxation and it was controlled by the degree of system hydrogenation.

According to Miura et al. [22], the carbon atoms reconstructions played an leading role in G interactions with H_2 molecules. The activation barrier for H_2 dissociation from an unrelaxed G was ~4.3 eV for a T–H–T geometry, and ~4.7 eV for a T–B–T geometry. The center of mass position of H_2 was, therefore, at the hollow site, and the two H atoms of the molecule were directed toward the top sites on the G structure. When the carbons relaxed, the activation barrier decreased to 3.3 eV for the T–H–T geometry and 3.9 eV for the T–B–T geometry, in which case, the two carbon atoms near the H atoms moved 0.33 Å toward the gas phase for the T–H–T geometry and 0.26 Å for the T–B–T geometry. Elsewhere, while investigating H atoms adsorption on G using first-principles while employing plane-wave based periodic density functional theory (DFT), Casolo et al. [23] selected a 5 × 5 surface unitcell to study H atoms' single- and multiple-adsorptions on G surface. They computed the binding and barrier energies for sequential sticking of several H atoms' configurations on top of G carbon. The authors recorded binding energies of 0.8–1.9 eV per atom with barriers to sticking of 0.0–0.15 eV. Magnetic structures formed in which spin density localized on a 3 × 3R30° sublattice and the binding (barrier) energies for sequential adsorption increased/decreased with the site-integrated magnetization. These results could be explained on the basis of the valence-bond resonance theory of planar π-conjugated systems. It suggested that preferential sticking due to barrierless adsorption was limited to the formation of hydrogen pairs.

According to Kim et al. [24], however, presence of certain metals at G's vacancy defects increased the binding energy and hydrogen adsorption of the metals tested; however, the Ca-vacancy complexes showed most favorable binding energy and overall hydrogen adsorption capacities. So, Ataca et al. [25], were able to demonstrate the capacity of Ca-adsorbed G has a recyclable hydrogen adsorption media. They found that Ca chemisorbed onto G by donating part of its 4 s charge to the empty π^* G band imparting it with a positive charge and the metallic character, which increased its H_2 adsorption, in turn. The H_2 uptake capacity of Ca-modified G could be improved further by adsorbing Ca on both sides of the G nanosheets and utterly by saturating the Ca surface coverage as because Ca does not agglomerate on G surfaces at high surface concentration. This due to the high coulombic repulsions between the adsorbed Ca atoms.

Besides hydrogen, Gao et al. [26] reviewed G adsorbents for CH_4. They observed weak interactions and charge transfer from intrinsic G to CH_4. The authors noted that the affinity of methane for graphene surfaces could, however, be promoted by doping the Ni atom, setting a single vacancy defect, and/or adding oxygen-containing functional groups into the G surfaces.

3. Application of graphene in electronic technology: bandgap engineering

The unique electronic properties of G means that G could continue to attract attention in potential electronic and photonic applications. The charge carriers in G behave, such as massless Dirac fermions, and G shows ballistic charge transport ideal for circuit fabrication. However, the lack of appreciable bandgap around the Fermi level, which is the key concept for semiconductors critical for controlling electronic conductivity, restricts its applications in nano-electronic and nanophotonic devices. Recent interests in G were, therefore, inspired by its promise in fashioning two-dimensional semiconductor materials with tunable bandgap lending itself to diverse uses. For this, several bandgap engineering methods have been proposed and Berashevich & Chakraborty [27], for example, described adsorption of water and other gases on G and showed that the molecules behaved like defects on nanoscale G surface facilitating bandgap tunability and permitting magnetic ordering of localized states at the G edges. This showed that the molecules pushed the wave functions corresponding to α -spin (up) and β -spin (down) G states to the opposite (zigzag) edges breaking the sublattice and molecular point group symmetry of the material. The wave-function displacement was controlled by the adsorbed molecule lending itself to tunable bandgap opening.

Accordingly, Balog et al. [28] demonstrated a G bandgap tunability using atomic hydrogen adsorption on the Moiré superlattice positions of G supported on an Ir(111) substrate. Then Takahashi et al. [29], while performing high-resolution angle-resolved photoemission spectroscopy of oxygen-adsorbed monolayer G on 6H-SiC(0 0 0 1), found that the energy gap between the π and π-bands increased with oxygen adsorption. This led to a systematic shrinking of π-electron Fermi surface and highlighted the potential of monolayer G oxidization in its bandgap inducement and tunability. On their part, Yavari et al. [30] engineered a tunable G bandgap of ≈ 0.2 eV by reversible H_2O adsorption and showed that the energy gap decreased to ≈ 0.029 eV when humidity was reduced to zero.

According to Şahin & Ciraci [31], bonding of a single Cl atom in G occurs by ionic interaction through charge transfer from G to Cl. This results in a stable direct bandgap semiconducting structure with tunable bandgap controlled by applying uniform strain. Then, by examining structural and electronic properties of F_2 adsorption between the G bilayers, Shayeganfar [32] observed that charge transfer between F_2 and G, and the presence of sp^2 and sp^3 orbitals of C−C and C−F bonds disrupted the G layers symmetry inducing an energy gap that depended on F orientation on the G surfaces. The adsorption of F_2 between the G bilayers also led to linear behavior between dipole moment and the energy gap and its electronic properties. This suggested existence of a tunable bandgap controlled by surface functionalization of G. Also, Tayyab et al. [33] predicted that Br-doping of G could induce sufficient bandgap width for optoelectronic applications.

On the other hand, Quhe et al. [34] demonstrated a tunable bandgap opening attributed to G metal adsorption. They illustrated a single-gated field effect transistor (FET) based on Cu-adsorbed ABC-stacked trilayer G. There was a clear transmission gap comparable to a bandgap showing the promise of metal-adsorbed G as a channel in single-gated FET device. Furthermore, while examining the metal dopants' effect on the structural, electronic, and charge transfer mechanism of G using DFT calculations, Tayyab et al. [33] observed that substituting Al atoms into G lattice induced a bandgap of 0.40 eV, and further predicted values of up to 0.82 eV for Al- and Be-doped G.

Other workers have assessed band gap tuning effects of organic agents intercalated into G surfaces. Chang et al. [35], for instance, while investigating the effect of borazine ($B_3N_3H_6$), triazine ($C_3N_3H_3$), and benzene (C_6H_6) on G's electronic structure, found that the molecular adsorptions were accompanied by bandgap opening of up to 62.9 meV under local density approximation. The band opening scale was controlled by the adsorption site type, for $C_3N_3H_3$ and the heterocyclic molecules were more effective in inducing bandgap opening than the monocyclic ones. Also, most stable configurations led to the largest bandgap opening at the particular adsorption sites, and the charge redistribution patterns controlled the gap on-and-off bandgap switching, which opened whenever the charge redistributed to the bridge site position. This suggested that ionic ability of the dissimilar atoms in the heterocyclic molecules controlled the charge redistribution and the gap tuning efficacy in G. Elsewhere, Hildebrand et al. [36] proposed, a model of self-assembly of halogenated carbene layers on G. They predicted a tunable bandgap opening controlled by adsorbates' self-assembly and surface coverage, which provided for a mechanism for modulated engineering of G electronic structure and its application to electronic technologies.

Thus, graphene is a gapless semiconductor that cannot be used in optoelectronic applications, such as solar cells in its pristine form. This necessitates for its bandgap tuning, which can be achieved by adsorptive doping with different species atoms including simple molecules, metal atoms or organic molecules etc.

4. Application of graphene in water adsorption

The laminar G structure confers unique physicochemical attributes usable in a wide range of applications to G, which are relating to material hygroscopy [37]. According to Liu et al. [38], the oxygen content of GO materials controls its H2O adsorption capacity, which diminishes with increasing levels of GO reduction. Therefore Lian et al. [37] have reported high water uptake capacity of 0.58 g/g for GO. Also, they found that the adsorption-desorption kinetics of H2O uptake by GO was 5-times higher due to high capillary pressure in the GO laminates and micro-sized tunnel-like wrinkles in the GO surfaces. As already reported elsewhere [38], the reduction of GO diminishes its oxygen content to form reduced graphene oxide, rGO [39]. Water adsorption onto rGO is, however, enhanced at high humidity due to water vapor multilayer formation at the rGO surface [38]. Thus, GO is a loosely defined G material containing variable oxygen content, which is dependent on conditions of its production, and it is generally characterized by high capacity to sorb and store H_2O molecules.

So, Leenaerts et al. [40], while investigating H_2O adsorption on pristine G using first-principles calculations, indicated that there were four possible orientations of H_2O molecule on G surface. Water can adsorb, starting from the O atom, with the

H−O bonds pointing up (u), down (d), or parallel to the G surface (n) or based on another orientation (v) with one O−H bond parallel to the surface while the other pointing to the surface. The energy of H_2O adsorption depends on its orientation in the G surfaces and on the site of adsorption. The energy difference between the various configurations is 5–6 meV between the orientations, and about 1–2 meV between the positions. It was noted that H_2O acts as an acceptor on G and the acceptor character (C, v) was the most energetically favored. So, Wehling et al. [41] while investigating the G electronic properties after water adsorption examined the effects of SiO_2 substrate and found that perfect suspended G was unresponsive to H_2O adsorption because doping required highly oriented H_2O clusters. However, adsorbed H_2O molecules shifted the SiO_2 substrate's impurity bands and changed their hybridization with the G bands. So, the H_2O molecules led to doping of G supported on SiO_2 for lower concentrations than those for the free suspended G. This effect was, however, dependent on microscopic substrate properties.

Sanyal et al. [42] found that certain molecules interacted with divacancy in G layers, but, for H_2O molecules, the large DOS in the vicinity of the Fermi level was absent. Accordingly the authors observed a pseudo-gap at the Fermi level, which was comparable to that for pure G. The DOS of the H_2O-adsorbed system was intriguing and the peaks in the C and O atom projected DOS occurred around 4 eV above the Fermi level. Ma et al. [43] determined the H_2O adsorption energy on G using the quantum Monte Carlo and random-phase approximation. Elsewhere, Hamada [44] while using the van der Waals density functional (vdW-DF) to study water interaction with G highlighted the promise of G in water adsorption applications and indicated that the adsorption potency of the materials was controlled by their the intrinsic properties, mode of preparation, and surface modifications employed.

5. Application of graphene in atmospheric detection and amelioration

5.1 Adsorption of nitrogen species

Emmisions of nitrogen compounds, especially the oxides, NO_x, present one of the major environmental concerns today. NOx gases are responsible for various significant environmental problems and they contribute to the poor of human health by resulting in lung dysfunction and respiratory complications. Application of environmentally friendly, mild, and low-cost methods in mitigating presence of NOx in the environment is desirable both for a healthier and more eco-friendly environment [45]. The affinity of G surfaces for N-species has attracted the attention of a significant number of researchers. Sanyal et al. [42] studied G having divacancy defects on its surface, as an eco-friendly, mild, and inexpensive adsorbent for N_2 and found that N_2 molecule chemisorbed onto G in plane with the G surface with the two N atoms substituting in the usual positions of the G hexagonal network as substitutional impurity with chemisorption energies, E^{ch}, of −4.53 eV. The N_2 inclusion in the G plane resulted in a perfect honeycomb lattice with electrons doped into the G's conduction.

Elsewhere, Leenaerts et al. [40] proposed two orientations for NH_3 adsorption on G. In one orientation, NH_3 adsorbs with the H atoms pointing away from the G surface (*u*) and in the other with the H atoms pointing to the surface (*d*). This indicated that both the adsorption site and the orientation influenced the adsorption energies of NH_3 but charge transfer was determined only by the surface orientation of the NH_3 molecule. The authors found that there was a small charge transfer of 0.03 eV from NH_3 to the G surface in the *u* orientation but almost no charge transfer in the *d* orientation. The *u* orientation, which was consistent with the donor character observed elsewhere [46], was therefore more energetically favored. So, Zhou et al. [47] then compared NH_3 adsorption on pristine G and on transition metal (TM) doped G (TM–graphene) and confirmed that the lowest-energy configurations for NH_3 adsorption on pristine and TM–graphene were through the N atom. However, for TM–graphene, NH_3 bonds *via* the N atom to the metal atom, by coordinating N lone-pair orbital with the metal valence bands.

It has been stated that adsorbates with a magnetic moment result in a larger doping [48]. So, Leenaerts et al. [40], while examining NO_2 adsorption on G and proposed that adsorbed NO_2 molecule can adopt three orientations on G. Thus, starting from the N atom, the NO_2 molecule can adsorb on G with the N−O bonds pointing up (*u*), down (*d*), or parallel to the G surface (*n*). The authors found that the total G and NO_2 magnetic moment in the (*B*, *d*) orientation was 0.862 μ_B, and the corresponding charge transfer from G ($M = 0\ \mu_B$) to NO_2 ($M = 1\ \mu_B$) was 0.099 e. The orbital mixing, therefore, led to a charge transfer of ±0.039e from NO_2 to G but, when compared with NO, the charge transfers for the latter were an order of magnitude smaller. Accordingly, Zhou et al. [47] have stated that the NO_2 molecule can bind to G surfaces through the N atom (nitro configuration), through an O atom (nitrite configuration), or through both O atoms (cycloaddition configuration). Thus, for pristine G, NO_2 bonds by the cycloaddition configuration at the bridge site position with adsorption energy (E_{ad}) of ~0.18 eV and charge transfer of 0.2e to NO_2. However, for TM–graphene, the cycloaddition configuration is preferred with enhanced $E_{ad} \sim 2$ eV, followed by the nitro configuration with $E_{ad} \sim 1.8$ eV. This is by a large charge transfer of ~0.6e from G to NO_2. However, Tang and Cao [49] found that the adsorption of NOx on GO was stronger than that on G leading to enhanced binding energies and charge transfers from NOx to GO through chemisorption interaction and with transition of the doping properties of NO_2 and NO from acceptor to donor character. Therefore, the interaction of NOx with GO forms H-bonds OHO (N) between -OH and the NOx besides other covalent bonds, including C–N and C−O. This is also accompanied by H abstraction to form nitrous acid- and nitric acid-like moieties. The spin-polarized DOS revealed hybridization of frontier orbitals of NO_2 and NO_3 with the electronic states around the Fermi level of GO resulting in strong acceptor doping and remarkable charge transfer characteristics from the molecules to GO.

Elsewhere, Dai et al. [50] evaluated NO and NO_2 adsorption on G modified with B, N, Al, and S dopants using DFT. They found that both B and N atoms retained their planar configuration on G surface but both Al and S atoms protruded out of the G plane. Nevertheless, both NO and NO_2 molecules were bonded to B-doped G but only NO_2 was adsorbed on S-doped G surfaces. The Al-doped G was, however, most reactive toward the gases although B- and S-doped G presented the most plausible prospects for practical NO and NO_2 gas adsorption and detection. So, the most relevant work to environment detection and remediation has been dedicated to NOx molecules mobilization on G, TM-graphene, and GO, but GO. Nonetheless, both TM-graphene and GO displayed enhanced NOx adsorption compared to pristine G adsorbents.

5.2 Adsorption of carbon oxides

Controlling toxic gas emissions is the foremost goal of the twenty-first century environmentalists. The adsorption technique offers a credible approach to dealing with these challenges and many carbon nanostructures, due to their unique surface morphologies and divergent potential for modification, have shown promise in fashioning plausible CO_2 sequestrators [19]. So, Mishra & Ramaprabhu, [51] reported high CO_2 adsorption capacity of 21.6 mmol/g for G at an 11-bar pressure and 25°C, based on a physisorption process. According to Szczęśniak et al. [19] polymers- and/or metal species-modified G nanomaterials possess high specific-surface areas and tailored surface attributes, which confers effective adsorbent properties for CO_2 and other gases.

Several scientists have focused their research on the mechanism of molecular adsorption onto G surfaces. While investigating adsorption of several molecules on G, Leenaerts et al. [40] showed that CO molecule assumed three different orientations on G, two of which involve the CO molecule in perpendicular position to the G surface, with the O atom toward the C atom (u) or away from the carbon (d), and one orientation with the CO molecule parallel to the G surface (n). In this case, CO acts as a donor molecule with charge transfer governed by the orientation assumed by the molecule on the G plane. Nevertheless, minor variations in charge transfer results due to differences in orbital overlap between the HOMO of the CO molecule and G. So, the C and O atoms remain bonded to each other, while CO is bonded to two pairs of G's carbon atoms. However, the CO molecule does not align perfectly in the plane of G nanosheets so that the C atom of the CO molecule stays in the G plane, making the usual hexagonal planar structure while the O atom stays out of plane. So, Zhou et al. [47] found that the most stable adsorption sites for CO on pristine G involved physisorption interactions at the hollow site, and the CO molecule preferentially assumes a parallel orientation to the G surface at an adsorption distance of ~3.6 Å and adsorption energy of ~0.017 eV. For TM–graphene, however, CO bonds to the C atom bonding the metal atom with enhanced adsorption energy and charge transfer accompanied with shortened adsorption distance. Intriguingly, the magnetic moment of TM-graphene gets enhanced from 1 μB (without adsorption) to ~3 μB.

According to Lee and Kim [52], CO_2 chemisorption energies on graphene-**C40** at high pressure are 71.2–72.1 kcal/mol for the lactone systems are controlled by C−O orientations at the UCAM-B3LYP level of theory. Nonetheless, the physisorption energies of CO_2 on G are only 2.1 and 3.3 kcal/mol at the single-point UMP2/6-31G** level of theory for the perpendicular and parallel orientations, respectively.

5.3 Dioxygen and ozone molecules

Ozone and other reactive oxygen species play an important role in atmospheric photochemistry of pollutants. Detection and removal of reactive oxygen species is, therefore, of essence. Dai et al. [50] analyzed the adsorption of several gases on G and showed that O_2 molecules adsorbed well onto B- and S-doped G samples. On the other hand, Sanyal et al. [42], have suggested that O_2 adsorbs on G by dissociating the two O atoms, which bond to two pairs of C atoms around divacancies in the G structure. The O atoms get oriented in opposite directions out of plane with the G surface. They, however, retain their bond length as in the isolated dimolecular state. In the absence of divacancy, however, O_2 adsorbs on pristine G *via* the hollow site with low adsorption energy of <0.1 eV and with the O−O bond oriented perpendicularly to the G surface at an adsorption length 3.7 Å [47]. The length of the O−O bond remains almost the same as in the free dioxygen molecule (1.24 Å). The Bader charge population analysis showed that the adsorption involved weak physisorption interaction. For metal-embedded G, however, O_2 bonds with the O−O bond parallel to the G plane with increased adsorption energies (> 1 eV) and decreased adsorption lengths (~2 Å). This is accompanied by charge transfer for O_2 adsorbed on TM–graphene samples and a large expansion in the O−O bond of 1.44 Å.

Lee et al. [53], who investigated O_3 adsorption on G using the *ab initio* DFT, found that O_3 molecule adsorbed on G basal plane with binding energies of 0.25 eV *via* physisorption interaction before it undergoes G-surface epoxidation with release of an O_2 molecule. The activation energy barrier for the physisorption to chemisorption reaction is 0.72 eV, and the binding energy of the chemisorbed state is about 0.33 eV. Ozone adsorption on G is, thus, a gentle reversible reaction relevant to covalent functionalization of the G basal plane.

5.4 Adsorption of sulfur species

Volatile sulphur compounds represent another class of species of relevance to environmental hygiene. Gao et al. [26] while using first principles based on DFT method , discussed in depth the stable configuration, adsorption energy, DOS, and charge transfer of H2S adsorption on intrinsic G, GO, Ni-doped G, and vacancy defect G samples. They observed weak adsorption and charge transfer for intrinsic G, which was enhanced by introducing oxygen-containing functional groups, doping with Ni atom, and setting its single vacancy defect. The single vacancy defect, in particular, promoted G interaction with H_2S molecule imparting excellent adsorption performance for H_2S molecule relevant to studying G-based sensors for the gas.

In another study, Shao et al. [54] investigated SO_2 adsorption on intrinsic G and on heteroatom-doped G samples containing B, N, Al, Si, Cr, Mn, Ag, Au, and Pt atoms using first principles based on DFT. They reported that the structural and electronic properties of the adsorption adducts of the molecules depended on the dopant type. SO_2 adsorbed weakly on intrinsic G, on the G samples doped with non-metals B-, and N. However rapid and strong SO_2 chemisorption occurred on metal and semimetal Al-, Si-, Cr-, Mn-, Ag-, Au-, and Pt-doped G. The analyses of the adsorption mechanisms showed the sensitivity of G-based SO_2 sensors was enhanced by introducing appropriate dopants with both Cr and Mn giving particularly promising results.

5.5 Halogens and halides

In Section 3 of this chapter, we discussed the role of Cl_2 and F_2 molecules in tuning the G bandgap. Şahin &and Ciraci [31] found that a single Cl atom bonding in G occurs by ionic interaction *via* charge transfer from G to Cl with local distortion in the underlying G layer. Meanwhile, a single Cl adatom migration on perfect G surfaces occurs unhindered but Cl adatom accumulation on the G surface to produce various conformations becomes unsustainable because strong Cl−Cl interactions occur leading to desorption and formation of Cl_2 molecules. Fully chlorinated G with single Cl atoms bonded alternately to each carbon atom from opposite sides of the G sheets with sp^3-type covalent bonds is buckled. So, Shayeganfar [32] discovered that F_2 adsorbs on G *via* mixed mechanisms involving both in-plane and out-of-plane molecular orientation on the G surface.

Consequently, Chen et al. [55] presented a detailed Br_2 adsorption and charge transfer study of G by combining *in situ* Raman spectroscopy and DFT techniques. They observed that when G is encapsulated in hexagonal boron nitride (h-BN) layers on either sides, the G surface is protected from Br_2 doping, but when G is supported on only one side by h-BN layer, it undergoes strong hole doping by adsorbing Br_2. The authors obtained molecular adsorption isotherm by plotting surface coverage versus pressure using a combination of Raman spectra and DFT calculations. They saw that the adsorption data fitted the Fowler-Guggenheim model with an adsorption equilibrium constant of ~0.31 per Torr. The repulsive lateral interaction between adsorbed Br_2 molecules was ~20 meV, whereas the binding energy for the Br_2 molecule was ~0.35 eV. At the monolayer coverage, each Br_2 molecule accepts 0.09 e^- of charge from single-layer G, but when the adsorbent was supported on SiO_2 instead of h-BN, a threshold pressure was observed for which the diffusion of Br_2 along the SiO_2/G interface resulted in Br_2 adsorption on both sides of the SiO_2-doped G.

On the other hand, Sun et al. [56], while studying HF molecules' adsorption on intrinsic and on Al-doped G by first-principles calculations, found that HF adsorption mechanisms were different for the adsorbents. The Al-doped G depicted greater adsorption energy and stronger interactions with HF molecule than the intrinsic G. The calculated net electron transfers, electronic density difference images, and DOS provided evidence that HF adsorption on Al-doped G was a chemisorption process, while the molecule's uptake by intrinsic G was physisorption.

6. Adsorption of alkali metals

Atomic-level investigations of metal adsorption and migration on G surface offer a plausible way of dealing with the challenges of developing novel alkali metal battery technologies [57]. Yang [58] pioneered work in which they reported that Li atoms bonded on a G layer interchangeably on both sides of the nanosheets by distorting the relative positions of the C atoms in the G honeycomb plane. They compared the results with graphane obtained by hydrogenation in which each C is pulled out of the plane by H and found that for Li-doped G, the carbon atom is pushed off by the attached Li instead. This resulted in a conducting counter-intuitive structure. Xue et al. [59], then, studied the adsorption of single Li and the formation of Li clusters on G using first-principles including van der Waals interactions. They found that while Li can exist on the surface of defect-free G under favorable conditions, the bonding was weaker and the surface concentration lower than in the metal adsorption on graphitic surfaces. At low concentrations, Li ions spread out on G surfaces due to coulombic repulsions between the adsorbing metal ions. However, as the Li content on the G surfaces increased, small Li clusters formed so that even though G has higher ultimate Li adsorption capacity than graphite, the nanoclusters nucleate Li dendrites, leading to failure. So, while studying Li adsorption using first-principles DFT calculations and surface diffusion on pristine and defect G structures, Fan et al. [60] observed that Li/C ratios lower than 1/6 for the single-layer G were energetically favorable. Further more, the existence of vacancy defects increased the Li/C ratio of the material. For double-vacancy and higher-order defects, however, Li ions diffused more freely in the direction perpendicular to the G sheets boosting the diffusion energetics in turn.

Indeed, Jin et al. [61] compared alkali metals; adsorption on single-layer G using first principles and observed a common trend in binding distance, charge transfer, and work function relating to increasing metal adsorption proportion ρ (adatom/C atom) on the G surface. There was a dip in the properties at $\rho \approx 0.04$ for all metals except for Li, for which the dip occurred at $\rho \approx 0.08$. This represented the transition of the adsorbed metals from individual atoms to two-dimensional metallic sheets exerting a depolarizing effect. Thus, G exhibited asymmetric function showing dependence on ρ with a dip on the adatom layer side and saturation on the G side, which was different from the case of bulk graphite.

Therefore, in order to appreciate the influence of G point defects on its Li adsorption, Zhou et al. [62] studied the uptake and diffusion of Li on G with divacancy and Stone–Wales defect using first-principles calculations. They found that in the presence of divacancy, Li adatom adsorbed on the hollow site above the center of an octagonal ring rather than on the top site of carbon atoms next to vacancy site. For the Stone–Wales defects, however, the Li atom adsorbed on the top site of the carbon atom in a pentagonal ring shared by two hexagonal rings. This resulted in buckling of the G sheet. In the case of both the divacancy and Stone–Wales defects, the interactions with Li adatom were attractive and the analysis of both the difference charge density and the Bader charge showed a significant charge transfer from Li adatom to the adjacent carbon atoms. Yang et al. [63], then, investigated sodium adsorption and intercalation into the G bilayers using DFT calculations. They systematically assessed the specific capacity, voltage, and migration energy barriers for Na storage in pristine and mono-vacancy defective bilayer G using DFT. The authors found that with an appropriate voltage (>0.5 V), the mono vacancy defects improved the specific capacity from 123.97 to 382.54 mAh/g. Thus, Na^+ ions transported from the defect-free regions (low energy barriers, 0.15–0.32 eV) to defect regions (large energy barriers, 0.56–0.59 eV) and decelerated. It demonstrated that a defect bilayer G was a promising material for making negative electrodes of the Na-ion batteries. Nonetheless, according to Olsson et al. [57], Li and K adsorption on pristine G was more favorable than Na adsorption. According to these workers, N- and O-containing defects dominated on the G surfaces, acting as metal trapping sites, and hindering the metal diffusion and migration on the G surfaces. This diminished the battery cycling performance.

Kim et al. [24] while studying the impact of intercalating G surfaces with alkali metal on its hydrogen uptake, however, found that G surface defects enhanced its metal binding energy and its dispersion and a considerable increase in binding energy was observed for alkaline earth metals. Furthermore, additional alkali metal adsorption of G was studied by Ataca et al. [25]. These studies could therefore give consequence to use of alkali-G material in metal storage devices.

7. The graphene-transition metal doping

Exploring new ways to diversify the basic properties of G continues to attract immense research interest. Transition metal-doping has also gained attention in the recent past, paving way to developing improved G materials with improved magnetism, and other fascinating properties with promise for diverse applications. Chan et al. [64], for instance, explored the adsorption of 12 metal adatoms on G using first-principles and computed the adsorption energy, geometry, DOS, dipole moment, and work function of each adatom-G system. They found that the adsorption groups I-III elements on G aligned with ionic bonding and it was characterized by large charge transfer and minimal change in the G electronic states. However, for the TMs, noble metals, and group IV metal, the adatoms interacted covalently with G surfaces leading to strong hybridization between adatom and G electronic states. Dipole moments across individual adatoms but the work-function shift correlated to the induced interfacial dipole of the G-adatom system and to the ionization potential of the isolated atom. According to Hu et al. [65], chemisorption of *3d* and noble metals adatoms on G occurs at the bridge site and at the top site with hybridization between the adatom and the G electronic states, which resulted in distorted G layer. The half-filled *3d* shell metal atoms as well as Zn, Ag, and Au atoms depicted low adsorption energies with a general decrease in the magnetic moment compared to the corresponding free adatoms. This was attributed to charge transfer and electron shift between the states of the adatom.

Elsewhere, Sevinçli et al. [66] while assessing the electronic and magnetic properties of *3d* TM atoms' adsorption on G and on G-nanoribbons, reported metal-dependent binding energies of 0.10–1.95 eV. The metals' adsorption imparted magnetic and metallic properties of G with armchair edge shapes (AGNR's) of adsorbed TM yielding minimum energy states. The resulting nonmagnetic semiconductor AGNR assumed metal/semiconductor and ferromagnetic/antiferromagnetic spin alignments. So, Cao et al. [67], while studying the geometries, electronic states, and magnetic properties of TM adatom and dimer adsorption on G, found that, except for Cu, chemisorption interactions yielded the most stable adsorption and dimerization on G surfaces controlled by exchange-correlation.

According to Valencia et al. [68], who modeled G and (8,0) single-walled carbon nanotubes (SWCNTs) functionalized with *3d* TM atoms, the 4 s occupation with Pauli repulsion was responsible for the Cr, Mn, and Cu physisorption behavior of G. Using a new physical model involving coulomb interaction, $3d^n4s^x \rightarrow 3d^{n+x}$ electronic promotion energy and occupation of the $1e_2(\delta)$, $2e_1(\pi)$, and $2a_1(\sigma)$ metal orbitals, the authors found that Sc, Ti, Fe, and Co metals were present in the G surface as isolated individual atoms but all other *3d* TM atoms diffused with clustering. In another work, Amft et al. [69] conducted density functional investigations on Cu, Ag, and Au adsorption on pristine G. While accounting for van der Waals (vdW) interactions using the vdW-DF and PBE + D2 methods, the authors analyzed the favorable adsorption sites, the adsorption-induced distortions in G sheets, and adatom diffusion paths and found from the vdW schemes that the three metal atoms adsorbed on the G sheets with buckling of the G layer. Only the results for Ag qualitatively differed from those obtained from generalized gradient approximation, which gives no binding for it. Otherwise, the results for the rest of the metals showed quantitative variations for the vdW-DF and PBE + D2 models.

Thus, by applying DFT calculations for elements of atomic number of 1–83, Nakada and Ishii [70] focused their work on the adsorbed adatoms' migration on G and showed that, adsorption favored the H6-site for the metals and the B-site for the nonmetals. Nonetheless, the migration energy was particularly high for the *3d* TMs and for certain nonmetals. In the same way, Liu et al. [71], while applying first principles, noted that the H6 was the favored site for the rare earth adatoms of Nd, Gd, Eu, and Yb adsorption. The adsorption energies and the diffusion barriers of Nd and Gd were, however, larger than those of Eu and Yb, and all the adatoms induced significant electric dipole and magnetic moments in the adsorption complex. Eu formed flat islands on G attributed to its low diffusion barrier and large adsorption energy compared to the bulk cohesive energy. Nonetheless, the adsorption of Nd and Gd adatoms led to in-plane G lattice distortions.

Vacancies or defects in G nanostructures present sites of altered chemical reactivity and open possibilities for tuning G properties by defect engineering. The understanding of the chemical reactivity of G defects is, thus, critical in implementing carbon materials in several advanced technologies. Pašti et al. [72] investigated atomic adsorption on G surfaces with single vacancy (SV) using DFT analyses of elements. They based their calculations on PBE, long-range dispersion interaction-corrected PBE (PBE + D2 and PBE + D3), and nonlocal vdW-DF2 functionals. They found that most elements, except groups 11 and 12 elements and the noble gases for which the contribution of dispersion interaction was most significant, bonded to the vacancy sites with interaction strengths that correlated to the cohesive energy of the elements in their stable phases. As most atoms could be trapped at the SV site, the calculated dissolution potentials showed that the adsorbed metals became more "noble" than they were in their respective stable phases. Then, Malola et al. [73] reported that Au adatoms adsorbed in-plane at G double-vacancies with diffusion barriers >4 eV and < 2 eV at the larger vacancies in line with the results earlier reported by Gan et al. [74].

8. Graphene adsorption of organic molecule

G materials present attractive pore volume, high conductivity, rich surface chemistry, and exceptionally large aspect ratio, among other promising properties, relevant to adsorptive catalytic applications in organic chemistry [5].

8.1 Pristine graphene

According to Lazar et al. [75] acetone, acetonitrile, dichloromethane, ethanol, ethyl acetate, hexane, and toluene adsorb on G surfaces with adsorption energies that range from 5.9 to 13.5 kcal/mol. The interaction strength of the organic molecules with G surface sites and the corresponding adsorption enthalpies were controlled by London dispersive forces, which dominated (~60%) the interactions. Basing on DFT calculations, Chi & Zhao [76] investigated formaldehyde (H_2CO) adsorption on intrinsic and Al-doped G. They found that the process was characterized by high binding energies and short connecting distances controlled by chemisorption interactions between the adsorbing molecule and G surfaces. The DOS showed the presence of orbital hybridization between H_2CO and Al-doped G, but there was no evidence of hybridization between H_2CO molecule and intrinsic G surfaces. So, Cortés-Arriagada [77] analyzed 1,4-dioxane adsorption on Al-, Ti-, Mn-, and Fe-doped G and found that the metal-doped G displayed enhanced interaction with 1,4-dioxane molecules compared to intrinsic G. The adsorption energies (1.2–1.6 eV), in this case, were accompanied by changes in the electronic structure of the substrates, especially for the Mn and Fe dopants. The *ab initio* dynamics simulations showed stable adsorbent–adsorbate interactions of highest aqueous stability of interaction for Al- and Fe- followed by the Mn- and Ti-doped G, respectively.

While applying 50 mg/L initial adsorbate concentration at pH of 6.3 and 285 K reaction temperature, Li et al. [78], on their part, reported G adsorption capacity of 28.26 mg/g for phenol. The phenol adsorption equilibrium simultaneously fitted the Langmuir and the Freundlich isotherms while the adsorption kinetics could be described by pseudo-second order kinetics model. Elsewhere, Xu et al. [79], reported that aqueous bisphenol A (BPA) adsorption on G is also spontaneous and exothermic. According to the authors, the reaction was the Langmuir isotherm and the pseudo-second order kinetics with a high BPA adsorption capacity of 182 mg/g reported at 302.15 K. They suggested that both π-π interactions and H-bonding contributed to BPA immobilization into G. Notwithstanding, Pei et al. [80] investigated 1,2,4-trichlorobenzene (TCB), 2,4,6-trichlorophenol (TCP), 2-naphthol and naphthalene (NAPH) adsorption on both G and GO by applying batch equilibrium technique and micro-Fourier transform infrared spectroscopy. The results showed that the adsorption isotherms for the four aromatic compounds were nonlinear, which indicated that both general hydrophobic interactions and specific interactions played part and the four aromatic compounds depicted similar adsorption efficiencies at pH 5.0. In the case of G, however, greater adsorption efficacy of 2-naphthol displayed under alkaline conditions due to π-π interactions resulting from the higher π-electron density of anionic 2-naphthol than that of neutral 2-naphthol. For GO, however, the affinity for the four compounds was in the order of: NAPH < TCB < TCP < 2-naphthol, and the FTIR spectroscopy revealed dominant π-π interaction for TCB, TCP, and 2-naphthol adsorption on G.

8.2 Graphene oxide

According to Pei et al. [80], high GO adsorption for both TCP and 2-naphthol was attributed to existence of H-bonding between the adsorbates and O-containing groups in GO. However, Chen and Chen [81], while studying *m*-dinitrobenzene, nitrobenzene, and *p*-nitrotoluene adsorption on GO, rGO, and G-nanosheets using IR spectroscopy, observed that the hydrophilic GO displayed inferior adsorption capacity for the three nitroaromatic compounds (NACs). This was ascribed to the greater hydrophobic π-conjugation of the active sites in the adsorbents. Even so, greater adsorptions observed for rGO over those of pristine G were linked to the π-π electron donor-acceptor interactions between the NACs phenyls and the π-electron-enriched G lattice, and to charge electrostatics and the polar interactions between G defects and the NACs nitro groups groups. Accordingly, while investigating bisphenol A, nitrobenzene, phenol, benzoic acid, and salicylic acid adsorption on GO, Tang et al. [82], also opined that the aromatic compounds' π-stacking ability controlled the adsorption processes of the compounds. So, Wang & Chen, [83] compared the adsorption and co-adsorption of naphthalene, 1-naphthol and Cd^{2+} on graphene oxide (GO), chemically-reduced graphene (CRG) and annealing-reduced graphene (ARG). They found that CRG had superior adsorption capacity the adsorbates than either ARG and GO. This, together with observation that the affinity of 1-naphthol for the adsorbents was greater than that of naphthalene was attributed to additional n-π electron-donor-acceptor (EDA) interactions between the 1-naphthol -OH groups and the electron-depleted G sites besides the π-π interactions. Furthermore, 1-naphthol uptake, by both the CRG and the ARG, had a maximum near the respective pKa values, which was consistent with the n-π EDA interaction mechanism. However, GO with greater surface functional groups density than both CRG and ARG, displayed greater affinity for Cd^{2+}. This somehow enhanced naphthalene and 1-naphthol co-adsorption on GO and CRG through surface-bridging cation-π interactions. Nonetheless, somewhat suppressed naphthalene co-adsorption with Cd^{2+} on ARG was attributed to the sieving effect of hydrated Cd^{2+} on the micropore edges of ARG.

8.3 Reduced graphene oxide

Meanwhile, Yu et al. [84] synthesized rGO and utilized it in benzene and toluene adsorption under dynamic conditions. The synthesized rGO samples displayed maximum adsorption of 276.4 and 304.4 mg/g for benzene and toluene, respectively. In a comparative study, greater adsorption capacities and breakthrough times were observed for rGO than GO, and the spent rGO was readily regenerated by heating at 150°C. So, Wang et al. [85], while investigating effect of change in concentration of oxygen-containing groups in rGO observed that higher degrees of reduction of GO enhanced the interactions between the π system of G and the π unit of the phenolic molecules. Thus, the adsorption, which was an exothermic and spontaneous process, was promoted by enhancing GO reduction and introducing electron-donating/withdrawing functional groups on the rGO benzene ring.

9. DNA adsorption and bio-sensing tests

Interfacing DNA oligonucleotides with G materials has paved the way to the production of diverse new sensor devices. GO is an outstanding fluorescence quencher and fluorescently-labeled DNA molecules get quenched when they adsorb on GO. However, when a complementary DNA (cDNA) is introduced, it desorbs the probe DNA molecule from GO, and the fluorescence is restored. This permits for devising GO-based DNA optical sensors [86]. This also provides a fascinating topic for bio-interface science. DNA adsorb on GO *via* π–π stacking and H-bonding while overcoming electrostatic repulsion. The mechanism by which cDNA induces probe DNA desorption from G is, however, still a subject of intriguing discussion [87].

Whereas the analytical aspects of this phenomenon have been demonstrated extensively, the fundamental understanding of the binding mechanisms has lagged behind. Wu et al. [86] therefore studied 12-, 18-, 24-, and 36-mer single-stranded DNA adsorption on GO as a function of DNA length, pH, ionic strength, solvents, temperature, and cDNAs. They found that faster more robust DNA adsorption on G was experienced using shorter DNAs or cDNAs, at low pH, high ionic strength, appropriate solvents, and optimum temperatures. At the same time Lu et al. [39] compared DNA adsorption of GO and rGO using fluorescently labeled DNA and showed that, under similar conditions, DNA adsorbed with a 2.6-fold higher capacity on rGO than on GO. However, the corresponding GO systems showed a higher absolute rise in sensing and signaling kinetics.

10. Application of graphene materials in studies on remediation of water pollution

Environmental pollution is one of the leading global challenges of the twenty-first century and the removal of various pollutants from the environment is of essence [88]. Adsorption methods are effective in the remediation of water and air pollution, but their efficiencies depend on intrinsic porosity, surface characteristics of the adsorbent, and on the applied conditions. Graphene lends itself to tailored surface modification suited for wide range of application in water and atmospheric remediation and pollution prevention.

10.1 Heavy metals removal from water

10.1.1 Pristine graphene

Many studies on environmental water amelioration have focused on individual [89] and competitive [90] adsorption of hazardous metals. Hao et al. [91] obtained SiO_2/G composite for selective removal of Pb(II) ions from water with 113.6 mg/g adsorption efficiency Cortés-Arriagada & Toro-Labbé [92], on the other hand, while investigating adsorption capacity of Al- and Fe-doped G for trivalent and pentavalent methylated arsenic using quantum chemistry computations, observed that trivalent methyl arsenicals adsorption was reached under neutral pH conditions with adsorption energies of 1.5–1.7 eV whereas for the pentavalent methyl arsenicals' the adsorption was achieved more optimally in acidic to neutral media with adsorption energies of 1.2–2.4 eV and 3.3–4.2 eV, respectively. The interacting σAs [sbnd] O bond weakening in the pollutant structure influenced the capacity and the stability of the adsorption complex.

10.1.2 Graphene oxide

Sitko et al. [93] while studying Cu(II), Zn(II), Cd(II) and Pb(II) adsorption on GO reported high adsorption capacities of 294, 345, 530, and 1119 mg/g for the metals, respectively. The metal affinities for GO surfaces, which was based on chemisorption reactions, was in the order: Pb(II) > Cu(II) ⋙ Cd(II) > Zn(II). Meanwhile, Wang & Chen [83] reported strong affinity of GO for Cd^{2+} ions, while studying Zn(II) adsorption on GO, Wang et al. [94] reported a high Zn(II) adsorption capacity of 246 mg/g, which was achieved in 20 min under adsorption conditions of neutral pH of 7.0, 2 mg adsorbent dosage, and 20°C temperature. The process had a Langmuir equilibrium constant was 5.7 L/g based on pseudo second-order kinetics [14]. Elsewhere, Lingamdinne et al. [95] elucidated Co(II) adsorption properties of GO and obtained a maximum adsorption of 21.28 mg/g while using medium pH of 5.0–8.0 and GO adsorbent dosage of 1.0 g/L. It was found that the adsorption process followed pseudo-second order [14] kinetics based on mixed reaction mechanism involving the π−π bonds electrons interaction Langmuir, which was described by Langmur, Freundlich, and Temkin isotherms.

10.1.3 Functionalized graphene oxide

While investigating EDTA-functionalized GO (EDTA-GO) for aqueous Pb(II) removal, Madadrang et al. [96], reported high adsorption capacity of 479 ± 46 mg/g achieved at pH 6 in 20 min. They showed that experimental data fitted the Langmuir adsorption model and the EDTA-GO adsorbent was regenerated by washing in HCl. Yet, in another study, White [97] while evaluating aqueous Cu(II) adsorption on GO and on carboxylated graphene oxide (GO–COOH) under conditions of ambient temperature, pH of 6 and 60 min equilibration, Cu^{2+} removal reached 97–99.4%, with 277.77 and 357.14 mg/g maximum adsorption capacities for GO and GO–COOH, respectively. The process, which was spontaneous and exothermic, was also described by the Langmuir-type adsorption equilibrium. Kumar & Jiang [98], at the time proposed a chitosan-functionalized GO adsorbent for aqueous arsenic adsorption and reported that arsenic oxyanions adsorption onto the novel adsorbent was facilitated via cation-π interactions, based on RNH_3^+-aromatic π moieties; electrostatic interactions involving $H_2AsO_4^-$, $HAsO_4^{2-}$–$^+NH_3R$; intra- and intermolecular H–bonding; and on anion-π interaction based on the R-COO^--aromatic π moieties with the oxygen-containing groups on GO surfaces.

10.2 Adsorption of nutrient ions from water

On their part, Vasudevan & Lakshmi [99] evaluated aqueous phosphate adsorption on G at varying pH, ionic strength, and temperatures conditions and determined the G adsorption capacity for phosphate to be 89.37 mg/g using 100 mg/L initial phosphate concentration and 303 K temperatures. The time-dependent phosphate adsorption fitted the second order kinetics [14], while the adsorption equilibrium was a spontaneous, endothermic process governed by Langmuir adsorption mechanism.

10.3 Dyes removal from aqueous solutions

While investigating aqueous methylene blue (MB) adsorption on G, Liu et al. [100] found that the adsorption data followed the Langmuir isotherm with maximum adsorption capacity of 153.85 mg/g at 293 K. The corresponding kinetics fitted the pseudo-second order kinetics model [14], and the process was both spontaneous and endothermic. At the time, Yao et al. [101], testing the potential of magnetic Fe_3O_4@Graphene composite for aqueous MB and Congo red (CR) removal, reported respective adsorption capacities of 45.27 and 33.66 mg/g for the dyes.

Wu et al. [102], then, utilized rGO and also evaluated MB adsorption alongside that of acrylonitrile (AN), p-toluenesulfonic acid (p-TA), and 1-naphthalenesulfonic acid (1-NA) and found that the larger molecules with greater number of benzene rings showed greater propensity to adsorb on GO than the smaller compounds. Accordingly, MB uptake by G was a π-π stacking process, which allowed for up to five adsorbent reuse cycles. The recorded MB, p-TA, and 1-NA adsorption capacities were ~1.52 ~ 1.43, and~1.46 g/g, respectively. At the same time, Sun et al. [103] studied the enhanced GO for aqueous adsorption of acridine orange based on *in situ* GO reduction by sodium hydrosulfite and observed an improved GO adsorption capacity from 1.4 g/g to 3.3 g/g, which was based on sodium hydrosulfite conversion of carbonyl groups on GO surfaces into hydroxyl groups forming the principal dye adsorption sites in rGO.

Yan et al. [104], then, assessed aqueous MB adsorption on oxidized G samples with varying oxidation oxygen levels and found that GO samples displayed a fast and efficient pH independent adsorption process, which increased with the increasing oxidation of G surface. The MB adsorption behavior was transformed from Freundlich-type at low G oxidation levels to Langmuir-type at high oxidation levels. The binding features of MB-adsorbed GO also changed from parallel stacking of MB molecules on the graphitic plane through hydrophobic π-π interaction to a vertical standing interaction *via* groups forming MB surface adsorption. The adsorption efficiency of the regenerated GO displayed little loss up in efficacy to four cycles of reuse. Later, Molla et al. [105] when studying MB, rhodamine B, and methyl orange adsorption on GO samples, observed that, unlike negative dyes, which did not adsorb, the positive dyes (methylene blue and rhodamine B) adsorbed on GO surfaces through electrostatic interactions between their positive dipoles and the GO negative dipole of the oxygen surface groups. This process was fast and reached 97% and 88% in just 15 min, for MB and for rhodamine B, respectively. Accordingly, the *ab initio* molecular dynamics showed that favorable adsorption configuration was at 2298 fs for methyl orange and 2290 fs for MB but MB was more strongly (−2.25 eV/molecule) adsorbed than methyl orange (−1.45 eV/molecule).

At the time, Nguyen-Phan et al. [106] studied the role of GO in photocatalytic removal of MB by titanium dioxide/GO composites reporting enhanced adsorption-photocatalysis for TiO_2/GO composites compared to those of pure TiO_2. Both the removal efficacy and the corresponding MB photodegradation increased with the GO proportion in the composite up to 10 wt%. The enhancement of these properties was attributed to the components' synergy in increasing the composite specific surface area, π-π conjugation between the dye and the aromatic rings, and in the ionic interactions between MB and oxygen-containing functional groups in GO.

10.4 Pesticides and other organic pollutants

Cationic surfactants present a considerable water pollution problems. They are difficult to degrade, leading to persistence in water sources. Chen et al. [107] evaluated aqueous adsorption of a cationic dodecyl amine hydrochloride (DACl) surfactant on GO by analyzing its zeta potential and applying FTIR and X-ray photoelectron spectroscopy (XPS). They found that the adsorption equilibrium was consistent with the Freundlich isotherm while the adsorption kinetics followed pseudo-second order model [14]. Accordingly, they reported an adsorption process that was endothermic and consistent with electrostatic interaction and H-bonding between DACl and GO.

Then, Wuest & Rochefort [108] while studying amino triazines adsorption on G using DFT calculations reported strong adsorbate affinity for G, which they attributed to specific attractive interactions of -NR_2 groups within the underlying surfaces as the driving force for the process. Afterwards, the authors [109] reported unprecedented aqueous GO and rGO adsorption of capacities of ~1200, 1100, and 800 mg/g for chlorpyrifos (CP), endosulfan (ES), and malathion (ML) respectively. They noted that the process was mediated through water because direct interactions between G and the pesticides were weak and unlikely. However, the adsorption of the compounds was both pH- and background ions-independent.

10.5 Pharmaceuticals and other emerging pollutants

Pharmaceutical agents have been become common persistent pollutants in in many water bodies around the world [7]. The occurrence of such emerging contaminants in water bodies poses unique threats to living things. Conventional techniques of pharmaceuticals removal from water are complex, expensive and generate secondary hazardous residues. Due to inefficient wastewater management systems, new treatment technologies are needed to deal with new products for the safety of the environment [110]. The adsorption technique has gained popularity but the engineering of new adsorbents for emerging trace pollutants is desired. G materials have lent themselves to application in various fields due to their robust physicochemical attributes and several researchers are already exploring them in environmental remediation. Gao et al., for instance, investigated removal of three tetracycline antibiotics, including tetracycline, oxytetracycline, and doxycycline [111] from water using GO. They reported that the tetracycline antibiotic deposited on GO surfaces *via* the π-π and cation-π interactions. The adsorption data could be fitted the Langmuir and Temkin isotherms with Langmuir adsorption capacity of 313, 212, and 398 mg/g for three antibiotics, respectively. The kinetics adsorption data fitted the pseudo-second order kinetics model [14] with a sorption rate constant (k) of 0.065 g/mg/h. Nonetheless, the adsorption capacities of GO for tetracycline decreased with increasing solution pH and increasing Na^+ concentration.

In another study, Nam et al. [112], while studying the adsorption behaviors of diclofenac (DCF) and sulfamethoxazole (SMX), evaluated the effect of GO dosage, contact time, pH, and sonication. They predicted binding energies between the drugs and GO surface groups and found that GO adsorption of the drugs was controlled by the oxygen-containing functional groups in GO, which exhibit negative surface charge over a wide range of pH values of 3–11. DCF showed favorable binding energy of −18.8 kcal/mol compared to that of SMX, which was −15.9 kcal/mol. The removal efficiencies for the two agents reached 35 and 12% within the initial 6 h respectively. however they were strongly enhanced to 75 and 30% under sonication, respectively. This is because sonication facilitates dispersion of exfoliated GO particles diminishing their surface density of oxygen-containing functional groups. It was found that the equilibrium data for both adsorbates fitted the Freundlich model.

Zhu et al. [110], on the other hand, while testing GO for metformin removal from water, observed an initial rapid and efficient metformin uptake, which was also strongly temperature-, pH-, ionic strength-, and the background electrolyte-controlled. The GO optimum adsorption of metformin was achieved at pH 6.0 and 288 K. The authors then found that the adsorption process was both spontaneous and exothermic and suggested that both π–π interactions and H-bonds played the leading role. The adsorption kinetics showed that 80% metformin removal was achieved within 20 min with high-rate constants of k_1 (0.232 per min) and k_2 (0.007 g/mg/min) for Lagergren first order [13] and pseudo-second order kinetic [14], respectively.

In the same way, Pavagadhi et al. [113] employed GO the removal of microcystin-LR (MC-LR) and microcystin-RR (MC-RR) –two common algal toxins in natural water. The adsorption experiments employed typical aqueous matrix containing environmental water anions and cations. The authors reported 1700 and 1878 μg/g GO adsorption capacities for MC-LR and MC-RR, respectively. The adsorption kinetics for the process were achieved within 5 min and the GO samples could be recycled up to ten cycles without significant loss in their adsorption potency.

11. Conclusions

G exhibits hydrophobic surfaces presenting high adsorption to chemicals that exhibit strong π-π interactions. GO, on the other hand, has a number of reactive functional groups with strong acidity suited for high adsorption for basic and cationic adsorbates. We find that the key application for G materials in the last decades have included hydrogen and other fuel gas adsorption and storage; alkali metal storage and battery technology; design and development of nano-electronic and nanophotonic devices; vacancy engineering for modifying the G surface reactivity for various application; G environmental application in water and atmospheric amelioration of both organic and inorganic pollutants; and in fashioning sensors and biosensors for most diverse applications. Suitable modifications of GO or G with nonmetals, metals, metal oxides, or organics can produce nanocomposites with enhanced adsorption capacities and separation efficiencies toward various groups of adsorbate species. The metals' adsorption, for example, imparts magnetic and metallic character to G. A number of molecular intercalations have also been found useful in inducing sufficient bandgap opening into G, which is critical for controlling electronic conductivity and opening up the materials to application in nano-electronic and nanophotonic devices among other applications.

Author details

Enos W. Wambu
Department of Chemistry and Biochemistry, University of Eldoret, Eldoret, Kenya

*Address all correspondence to: ewambu@uoeld.ac.ke

References

[1] Adetayo A, Runsewe D. Synthesis and fabrication of graphene and graphene oxide: A review. Open Journal of Composite Materials. 2019;**9**:207-229

[2] Farjadian F, Abbaspour S, Sadatlu MAA, Mirkiani S, Ghasemi A, Hoseini-Ghahfarokhi M, et al. Recent developments in graphene and graphene oxide: Properties, synthesis, and modifications: A review. Chemistry Select. 2020;**5**:10200-10219

[3] Yu W, Sisi L, Haiyana Y, Jie L. Progress in the functional modi fi cation of graphene/graphene oxide: A review. RSC Advances. 2020;**10**:15328-15345

[4] Fang B, Chang D, Xu Z, Gao C. A review on graphene fibers: Expectations, advances, and prospects. Advanced Materials. 2020;**1902664**:1-29

[5] Thakur K, Kandasubramanian B. Graphene and graphene oxide-based composites for removal of organic pollutants: A review. Journal of Chemical & Engineering Data. 2019;**64**(3):833-867

[6] Kong L, Enders A, Rahman TS, Dowben PA. Molecular adsorption on graphene. Journal of Physics. Condensed Matter. 2014;**26**(44):443001

[7] Xu J, Cao Z, Zhang Y, Yuan Z, Lou Z, Xu X, et al. A review of functionalized carbon nanotubes and graphene for heavy metal adsorption from water: Preparation, application, and mechanism. Chemosphere. 2018;**195**:351-364

[8] Carmalin Sophia A, Lima EC, Allaudeen N, Rajan S. Application of graphene based materials for adsorption of pharmaceutical traces from water and wastewater- a review. Desalination and Water Treatment [Internet]. 2016;**57**(57):27573-27586. DOI: 10.1080/19443994.2016.1172989

[9] Langmuir I. The adsorption of gases on plane surfaces of glass, mica, and platinum. Journal of the American Chemical Society. 1918;**40**:1361

[10] Freundlich HMF. Über die Adsorption in Lösungen. Zeitschrift für Physikalische Chemie. 1906;**57U**:385-471

[11] Polanyi M. Adsorption from the point of view of the third law of thermodynamics. Verhandlungen der Deutschen Physikalischen Gesellschaft. 1914;**16**:1012-1016

[12] Dubinin MM, Radushkevich LV. The equation of the characteristic curve of the activated charcoal. Proceedings of the Academy of Sciences of the USSR. Physical chemistry section. 1947;**55**:327-331

[13] Ho YS, Mckay G. Sorption of dye from aqueous solution by peat. Chemical Engineering Journal. 1998;**70**:115-124

[14] Ho YS, Mckay G. Pseudo-second order model for sorption processes. Process Biochemistry. 1999;**34**:451-465

[15] Kresse G, Hafner J. *Ab initio* molecular dynamics for liquid metals. Physical Review B. 1993;**47**(1):558-561

[16] Perdew J, Burke K, Ernzerhof M. Generalized gradient approximation made simple. Physical Review Letters [Internet]. 1996;**77**(18):3865-3868. Available from: http://www.ncbi.nlm.nih.gov/pubmed/10062328

[17] Kresse G, Joubert D. From ultrasoft pseudopotentials to the projector

augmented-wave method. Physical Review B. 1999;**59**(3):11-19

[18] Ma LP, Wu ZS, Li J, Wu ED, Ren WC, Cheng HM. Hydrogen adsorption behavior of graphene above critical temperature. International Journal of Hydrogen Energy [Internet]. 2009;**34**(5):2329-2332. DOI: 10.1016/j.ijhydene.2008.12.079

[19] Szczęśniak B, Choma J, Jaroniec M. Gas adsorption properties of graphene-based materials. Advances in Colloid and Interface Science [Internet]. 2017;**243**:46-59. DOI: 10.1016/j.cis.2017.03.007

[20] Arellano JS, Molina LM, Rubio A, Alonso JA. Density functional study of adsorption of molecular hydrogen on graphene layers. The Journal of Chemical Physics. 2000;**112**(18):8114-8119

[21] Ivanovskaya VV, Zobelli A, Teillet-Billy D, Rougeau N, Sidis V, Briddon PR. Hydrogen adsorption on graphene: A first principles study. European Physical Journal B. 2010;**76**(3):481-486

[22] Miura Y, Kasai H, Diño W, Nakanishi H, Sugimoto T. First principles studies for the dissociative adsorption of H_2 on graphene. Journal of Applied Physics. 2003;**93**(6):3395-3400

[23] Casolo S, Løvvik OM, Martinazzo R, Tantardini GF. Understanding adsorption of hydrogen atoms on graphene. The Journal of Chemical Physics. 2009;**130**(5):1-11

[24] Kim G, Jhi SH, Lim S, Park N. Effect of vacancy defects in graphene on metal anchoring and hydrogen adsorption. Applied Physics Letters. 2009;**94**(17):2012-2015

[25] Ataca C, Aktürk E, Ciraci S. Hydrogen storage of calcium atoms adsorbed on graphene: First-principles plane wave calculations. Physical Review B: Condensed Matter and Materials Physics. 2009;**79**(4):1-4

[26] Gao X, Zhou Q, Wang J, Xu L, Zeng W. Performance of intrinsic and modified graphene for the adsorption of H_2S and CH_4: A DFT study. Nanomaterials. 2020;**10**(299):1-15

[27] Berashevich J, Chakraborty T. Tunable band gap and magnetic ordering by adsorption of molecules on graphene. Physical Review B: Condensed Matter and Materials Physics. 2009;**80**(3):1-4

[28] Balog R, Jørgensen B, Nilsson L, Andersen M, Rienks E, Bianchi M, et al. Bandgap opening in graphene induced by patterned hydrogen adsorption. Nature Materials [Internet]. 2010;**9**(4):315-319. DOI: 10.1038/nmat2710

[29] Takahashi T, Sugawara K, Noguchi E, Sato T, Takahashi T. Band-gap tuning of monolayer graphene by oxygen adsorption. Carbon NY [Internet]. 2014;**73**:141-145. DOI: 10.1016/j.carbon.2014.02.049

[30] Yavari F, Kritzinger C, Gaire C, Song L, Gulapalli H, Borca-Tasciuc T, et al. Tunable bandgap in graphene by the controlled adsorption of water molecules. Small. 2010;**6**(22):2535-2538

[31] Şahin H, Ciraci S. Chlorine adsorption on graphene: Chlorographene. Journal of Physical Chemistry C. 2012;**116**(45):24075-24083

[32] Shayeganfar F. Energy gap tuning of graphene layers with single molecular F2 adsorption. Journal of Physical Chemistry C. 2015;**119**(22):12681-12689

[33] Tayyab M, Hussain A, Adil W, Nabi S, Asif Q, ul A. Band-gap engineering of graphene by Al doping

and adsorption of Be and Br on impurity: A computational study. Computational Condensed Matter [Internet]. 2020;**23**:e00463. DOI: 10.1016/j.cocom.2020.e00463

[34] Quhe R, Ma J, Zeng Z, Tang K, Zheng J, Wang Y, et al. Tunable band gap in few-layer graphene by surface adsorption. Scientific Reports. 2013;**3**:1-8

[35] Chang CH, Fan X, Li LJ, Kuo JL. Band gap tuning of graphene by adsorption of aromatic molecules. Journal of Physical Chemistry C. 2012;**116**(25):13788-13794

[36] Hildebrand M, Abualnaja F, Makwana Z, Harrison NM. Strain engineering of adsorbate self-assembly on graphene for band gap tuning. Journal of Physical Chemistry C. 2019;**123**(7):4475-4482

[37] Lian B, De Luca S, You Y, Alwarappan S, Yoshimura M, Sahajwalla V, et al. Extraordinary water adsorption characteristics of graphene oxide. Chemical Science. 2018;**9**(22):5106-5111

[38] Liu R, Gong T, Zhang K, Lee C. Graphene oxide papers with high water adsorption capacity for air dehumidification. Scientific Reports [Internet]. 2017;**7**(1):1-9. DOI: 10.1038/s41598-017-09777-y

[39] Lu C, Huang PJJ, Liu B, Ying Y, Liu J. Comparison of graphene oxide and reduced graphene oxide for DNA adsorption and sensing. Langmuir. 2016;**32**(41):10776-10783

[40] Leenaerts O, Partoens B, Peeters FM. Adsorption of H_2O, NH_3, CO, NO_2, and NO on graphene: A first-principles study. Physical Review B: Condensed Matter and Materials Physics. 2008;**77**(12):1-6

[41] Wehling TO, Lichtenstein AI, Katsnelson MI. First-principles studies of water adsorption on graphene: The role of the substrate. Applied Physics Letters. 2008;**93**(20):1-4

[42] Sanyal B, Eriksson O, Jansson U, Grennberg H. Molecular adsorption in graphene with divacancy defects. Physical Review B. 2009;**79**(11):1-4

[43] Ma J, Michaelides A, Alfè D, Schimka L, Kresse G, Wang E. Adsorption and diffusion of water on graphene from first principles. Physical Review B: Condensed Matter and Materials Physics. 2011;**84**(3):1-4

[44] Hamada I. Adsorption of water on graphene: A van der Waals density functional study. Physical Review B: Condensed Matter and Materials Physics. 2012;**86**(19):1-6

[45] Nikokavoura A, Trapalis C. Applied surface science graphene and g-C 3 N 4 based photocatalysts for NO x removal: A review. Applied Surface Science [Internet]. 2018;**430**(x):18-52. DOI: 10.1016/j.apsusc.2017.08.192

[46] Schedin F, Geim AK, Morozov SV, Hill EW, Blake P, Katsnelson MI, et al. Detection of individual gas molecules adsorbed on graphene. Nature Materials. 2007;**6**:652-655

[47] Zhou M, Lu YH, Cai YQ, Zhang C, Feng YP. Adsorption of gas molecules on transition metal embedded graphene: A search for high-performance graphene-based catalysts and gas sensors. Nanotechnology. 2011;**22**(38):1-9

[48] Wehling TO, Novoselov KS, Morozov SV, Vdovin EE, Katsnelson MI, Geim AK, et al. Molecular doping of graphene. Nano Letters. 2018;**8**(1):173-177

[49] Tang S, Cao Z. Adsorption of nitrogen oxides on graphene and

graphene oxides: Insights from density functional calculations. The Journal of Chemical Physics. 2011;**134**(4):044710

[50] Dai J, Yuan J, Giannozzi P. Gas adsorption on graphene doped with B, N, Al, and S: A theoretical study. Applied Physics Letters. 2009;**95**(23):10-12

[51] Mishra AK, Ramaprabhu S. Carbon dioxide adsorption in graphene sheets. AIP Advances. 2011;**1**(3):10-16

[52] Lee KJ, Kim SJ. Theoretical investigation of CO_2 adsorption on graphene. Bulletin of the Korean Chemical Society. 2013;**34**(10):3022-3026

[53] Lee G, Lee B, Kim J, Cho K. Supporting information ozone adsorption on graphene: *Ab initio* study and experimental validation. The Journal of Physical Chemistry C [Internet]. 2009;**113**(32):14225-14229. Available from: https://arxiv.org/ftp/arxiv/papers/0906/0906.2243.pdf

[54] Shao L, Chen G, Ye H, Wu Y, Qiao Z, Zhu Y, et al. Sulfur dioxide adsorbed on graphene and heteroatom-doped graphene: A first-principles study. European Physical Journal B. 2013;**86**(2):2-6

[55] Chen Z, Darancet P, Wang L, Crowther AC, Gao Y, Dean CR, et al. Physical adsorption and charge transfer of molecular Br_2 on graphene. ACS Nano. 2014;**8**(3):2943-2950

[56] Sun Y, Chen L, Zhang F, Li D, Pan H, Ye J. First-principles studies of HF molecule adsorption on intrinsic graphene and Al-doped graphene. Solid State Communications [Internet]. 2010;**150**(39-40):1906-1910. DOI: 10.1016/j.ssc.2010.07.037

[57] Olsson E, Chai G, Dove M, Cai Q. Adsorption and migration of alkali metals (Li, Na, and K) on pristine and defective graphene surfaces. Nanoscale. 2019;**11**:5274-5284

[58] Yang CK. A metallic graphene layer adsorbed with lithium. Applied Physics Letters. 2009;**94**(16):3-6

[59] Xue T, Cui X, Chen J, Liu C, Wang Q, Wang H, et al. Adsorption of single Li and the formation of small Li clusters on graphene for the anode of lithium-ion batteries A switch of the oxidation state of graphene oxide on a surface plasmon resonance chip. ACS Applied Materials & Interfaces. 2013;**5**(16):7793-7797

[60] Fan X, Zheng WT, Kuo JL. Adsorption and diffusion of Li on pristine and defective graphene. ACS Applied Materials & Interfaces. 2012;**4**(5):2432-2438

[61] Jin KH, Choi SM, Jhi SH. Crossover in the adsorption properties of alkali metals on graphene. Physical Review B: Condensed Matter and Materials Physics. 2010;**82**(3):4-7

[62] Zhou LJ, Hou ZF, Wu LM. First-principles study of sulfur immobilizing and interacting with lithium on graphene with point defects. American Chemical Society. 2012;**2012**:21780-21787

[63] Yang S, Li S, Tang S, Dong W, Sun W, Shen D, et al. Sodium adsorption and intercalation in bilayer graphene from density functional theory calculations. Theoretical Chemistry Accounts. 2016;**135**(7):1

[64] Chan KT, Neaton JB, Cohen ML. First-principles study of metal adatom adsorption on graphene. Physical Review B: Condensed Matter and Materials Physics. 2008;**77**(23):1-12

[65] Hu L, Hu X, Wu X, Du C, Dai Y, Deng J. Density functional calculation of

transition metal adatom adsorption on graphene. Physica B: Condensed Matter [Internet]. 2010;**405**(16):3337-3341. DOI: 10.1016/j.physb.2010.05.001

[66] Sevinçli H, Topsakal M, Durgun E, Ciraci S. Electronic and magnetic properties of 3d transition-metal atom adsorbed graphene and graphene nanoribbons. Physical Review B: Condensed Matter and Materials Physics. 2008;**77**(19):1-7

[67] Cao C, Wu M, Jiang J, Cheng HP. Transition metal adatom and dimer adsorbed on graphene: Induced magnetization and electronic structures. Physical Review B: Condensed Matter and Materials Physics. 2010;**81**(20):1-9

[68] Valencia H, Gil A, Frapper G. Trends in the adsorption of 3d transition metal atoms onto graphene and nanotube surfaces: A DFT study and molecular orbital analysis. Journal of Physical Chemistry C. 2010;**114**(33):14141-14153

[69] Amft M, Lebègue S, Eriksson O, Skorodumova NV. Adsorption of Cu, Ag, and Au atoms on graphene including van der Waals interactions. Journal of Physics. Condensed Matter. 2011;**23**(39):1-9

[70] Nakada K, Ishii A. Migration of adatom adsorption on graphene using DFT calculation. Solid State Communications [Internet]. 2011;**151**(1):13-16. DOI: 10.1016/j.ssc.2010.10.036

[71] Liu X, Wang CZ, Hupalo M, Yao YX, Tringides MC, Lu WC, et al. Adsorption and growth morphology of rare-earth metals on graphene studied by *ab initio* calculations and scanning tunneling microscopy. Physical Review B: Condensed Matter and Materials Physics. 2010;**82**(24):1-7

[72] Pašti IA, Jovanović A, Dobrota AS, Mentus SV, Johansson B, Skorodumova NV. Atomic adsorption on graphene with a single vacancy: Systematic DFT study through the periodic table of elements. Physical Chemistry Chemical Physics. 2018;**20**(2):858-865

[73] Malola S, Häkkinen H, Koskinen P. Gold in graphene: In-plane adsorption and diffusion. Applied Physics Letters. 2009;**94**(4):6-8

[74] Gan Y, Sun L, Banhart F. One- and two-dimensional diffusion of metal atoms in graphene**. Small. 2008;**4**(5):587-591

[75] Lazar P, Karlický F, Jurecka P, Kocman M, Otyepková E, Šafářová K, et al. Adsorption of small organic molecules on graphene. Journal of the American Chemical Society. 2013;**135**(16):6372-6377

[76] Chi M, Zhao YP. Adsorption of formaldehyde molecule on the intrinsic and Al-doped graphene: A first principle study. Computational Materials Science [Internet]. 2009;**46**(4):1085-1090. DOI: 10.1016/j.commatsci.2009.05.017

[77] Cortés-Arriagada D. Expanding the environmental applications of metal (Al, Ti, Mn, Fe) doped graphene: Adsorption and removal of 1,4-dioxane. Physical Chemistry Chemical Physics. 2016;**18**(47):32281-32292

[78] Li Y, Du Q, Liu T, Sun J, Jiao Y, Xia Y, et al. Equilibrium, kinetic and thermodynamic studies on the adsorption of phenol onto graphene. Materials Research Bulletin [Internet]. 2012;**47**(8):1898-1904. DOI: 10.1016/j.materresbull.2012.04.021

[79] Xu J, Wang L, Zhu Y. Decontamination of bisphenol A from

aqueous solution by graphene adsorption. Langmuir. 2012;**28**(22):8418-8425

[80] Pei Z, Li L, Sun L, Zhang S, Shan XQ, Yang S, et al. Adsorption characteristics of 1,2,4-trichlorobenzene, 2,4,6-trichlorophenol, 2-naphthol and naphthalene on graphene and graphene oxide. Carbon N Y [Internet]. 2013;**51**(1):156-163. DOI: 10.1016/j.carbon.2012.08.024

[81] Chen X, Chen B. Macroscopic and spectroscopic investigations of the adsorption of nitroaromatic compounds on graphene oxide, reduced graphene oxide, and graphene nanosheets. Environmental Science & Technology. 2015;**49**(10):6181-6189

[82] Tang H, Zhao Y, Shan S, Yang X, Liu D, Cui F, et al. Theoretical insight into the adsorption of aromatic compounds on graphene oxide. Environmental Science. Nano. 2018;**5**(10):2357-2367

[83] Wang J, Chen B. Adsorption and coadsorption of organic pollutants and a heavy metal by graphene oxide and reduced graphene materials. Chemical Engineering Journal [Internet]. 2015;**281**:379-388. DOI: 10.1016/j.cej.2015.06.102

[84] Yu L, Wang L, Xu W, Chen L, Fu M, Wu J, et al. Adsorption of VOCs on reduced graphene oxide. International Journal of Environmental Science and Technology (China) [Internet]. 2018;**67**:171-178. DOI: 10.1016/j.jes.2017.08.022

[85] Wang X, Huang S, Zhu L, Tian X, Li S, Tang H. Correlation between the adsorption ability and reduction degree of graphene oxide and tuning of adsorption of phenolic compounds. Carbon N Y [Internet]. 2014;**69**:101-112. DOI: 10.1016/j.carbon.2013.11.070

[86] Wu M, Kempaiah R, Huang PJJ, Maheshwari V, Liu J. Adsorption and desorption of DNA on graphene oxide studied by fluorescently labeled oligonucleotides. Langmuir. 2011;**27**(6):2731-2738

[87] Liu B, Salgado S, Maheshwari V, Liu J. DNA adsorbed on graphene and graphene oxide: Fundamental interactions, desorption and applications. Current Opinion in Colloid & Interface Sciencei [Internet]. 2016;**26**:41-49. DOI: 10.1016/j.cocis.2016.09.001

[88] Wang S, Ang HM, Tadé MO. Adsorptive remediation of environmental pollutants using novel graphene-based nanomaterials. Chemical Engineering Journal [Internet]. 2013;**2014**(226):336-347. DOI: 10.1016/j.cej.2013.04.070

[89] Wambu EW, Muthakia GK, wa-Thiong'o JK, Shiundu PM. Regeneration of spent bleaching earth and its adsorption of copper (II) ions from aqueous solutions. Applied Clay Science. 2009;**46**(2):176-180

[90] Areco MM, Dos Santos AM. Copper, zinc, cadmium and lead biosorption by Gymnogongrus torulosus. Thermodynamics and kinetics studies. Colloids and Surfaces. B, Biointerfaces. 2010;**81**(2):620-628

[91] Hao L, Song H, Zhang L, Wan X, Tang Y, Lv Y. SiO_2/graphene composite for highly selective adsorption of Pb(II) ion. Journal of Colloid and Interface Science. 2012;**369**(1):381-387

[92] Cortés-Arriagada D, Toro-Labbé A. Aluminum and iron doped graphene for adsorption of methylated arsenic pollutants. Applied Surface Science. 2016;**386**(June):84-95

[93] Sitko R, Turek E, Zawisza B, Malicka E, Talik E, Heimann J, et al.

Adsorption of divalent metal ions from aqueous solutions using graphene oxide Rafal. Dalton Transactions. 2013;**42**(16):5682-5689

[94] Wang H, Yuan X, Wu Y, Huang H, Zeng G, Liu Y, et al. Adsorption characteristics and behaviors of graphene oxide for Zn(II) removal from aqueous solution. Applied Surface Science. 2013;**279**:432-440

[95] Lingamdinne LP, Koduru JR, Roh H, Choi YL, Chang YY, Yang JK. Adsorption removal of Co(II) from waste-water using graphene oxide. Hydrometallurgy [Internet]. 2016;**165**:90-96. DOI: 10.1016/j.hydromet.2015.10.021

[96] Madadrang CJ, Kim HY, Gao G, Wang N, Zhu J, Feng H, et al. Adsorption behavior of EDTA-graphene oxide for Pb (II) removal. ACS Applied Materials & Interfaces. 2012;**4**(3):1186-1193

[97] White RL, White CM, Turgut H, Massoud A, Tian ZR. Comparative studies on copper adsorption by graphene oxide and functionalized graphene oxide nanoparticles. Journal of the Taiwan Institute of Chemical Engineers [Internet]. 2018;**85**:18-28. DOI: 10.1016/j.jtice.2018.01.036

[98] Kumar ASK, Jiang SJ. Chitosan-functionalized graphene oxide: A novel adsorbent an efficient adsorption of arsenic from aqueous solution. Journal of Environmental Chemical Engineering [Internet]. 2016;**4**(2):1698-1713. DOI: 10.1016/j.jece.2016.02.035

[99] Vasudevan S, Lakshmi J. The adsorption of phosphate by graphene from aqueous solution. RSC Advances. 2012;**2**(12):5234-5242

[100] Liu T, Li Y, Du Q, Sun J, Jiao Y, Yang G, et al. Adsorption of methylene blue from aqueous solution by graphene. Colloids and Surfaces B: Biointerfaces [Internet]. 2012;**90**(1):197-203. DOI: 10.1016/j.colsurfb.2011.10.019

[101] Yao Y, Miao S, Liu S, Ma LP, Sun H, Wang S. Synthesis, characterization, and adsorption properties of magnetic Fe_3O_4@graphene nanocomposite. Chemical Engineering Journal [Internet]. 2012;**184**:326-332. DOI: 10.1016/j.cej.2011.12.017

[102] Wu T, Cai X, Tan S, Li H, Liu J, Yang W. Adsorption characteristics of acrylonitrile, p-toluenesulfonic acid, 1-naphthalenesulfonic acid and methyl blue on graphene in aqueous solutions. Chemical Engineering Journal [Internet]. 2011;**173**(1):144-149. DOI: 10.1016/j.cej.2011.07.050

[103] Sun L, Yu H, Fugetsu B. Graphene oxide adsorption enhanced by in situ reduction with sodium hydrosulfite to remove acridine orange from aqueous solution. Journal of Hazardous Materials. 2012;**203-204**:101-110

[104] Yan H, Tao X, Yang Z, Li K, Yang H, Li A, et al. Effects of the oxidation degree of graphene oxide on the adsorption of methylene blue. Journal of Hazardous Materials [Internet]. 2014;**268**:191-198. DOI: 10.1016/j.jhazmat.2014.01.015

[105] Molla A, Li Y, Mandal B, Kang SG, Hur SH, Chung JS. Selective adsorption of organic dyes on graphene oxide: Theoretical and experimental analysis. Applied Surface Science [Internet]. 2019;**464**:170-177. DOI: 10.1016/j.apsusc.2018.09.056

[106] Nguyen-Phan TD, Pham VH, Shin EW, Pham HD, Kim S, Chung JS, et al. The role of graphene oxide content on the adsorption-enhanced photocatalysis of titanium dioxide/graphene oxide composites. Chemical Engineering Journal [Internet].

2011;**170**(1):226-232. DOI: 10.1016/j.cej.2011.03.060

[107] Chen P, Li H, Song S, Weng X, He D, Zhao Y. Adsorption of dodecylamine hydrochloride on graphene oxide in water. Results in Physics [Internet]. 2017;**7**:2281-2288. DOI: 10.1016/j.rinp.2017.06.054

[108] Wuest JD, Rochefort A. Strong adsorption of aminotriazines on graphene. Chemical Communications. 2010;**46**(17):2923-2925

[109] Maliyekkal SM, Sreeprasad TS, Krishnan D, Kouser S, Mishra AK, Waghmare UV, et al. Graphene: A reusable substrate for unprecedented adsorption of pesticides. Small. 2013;**9**(2):273-283

[110] Zhu S, Guo LY, Bo LS, Ming ZG, Hua JL, Fei TX, et al. Adsorption of emerging contaminant metformin using graphene oxide. Chemosphere [Internet]. 2017;**179**:20-28. DOI: 10.1016/j.chemosphere.2017.03.071

[111] Gao Y, Li Y, Zhang L, Huang H, Hu J, Shah SM, et al. Adsorption and removal of tetracycline antibiotics from aqueous solution by graphene oxide. Journal of Colloid and Interface Science [Internet]. 2012;**368**(1):540-546. DOI: 10.1016/j.jcis.2011.11.015

[112] Nam SW, Jung C, Li H, Yu M, Flora JRV, Boateng LK, et al. Adsorption characteristics of diclofenac and sulfamethoxazole to graphene oxide in aqueous solution. Chemosphere [Internet]. 2015;**136**:20-26. DOI: 10.1016/j.chemosphere.2015.03.061

[113] Pavagadhi S, Tang ALL, Sathishkumar M, Loh KP, Balasubramanian R. Removal of microcystin-LR and microcystin-RR by graphene oxide: Adsorption and kinetic experiments. Water Research [Internet]. 2013;**47**(13):4621-4629. DOI: 10.1016/j.watres.2013.04.033

Section 3

Applications of Graphene

Chapter 5

Mechanical Performance of Carbon-Fiber-Reinforced Composite Textile Laminates Integrated with Graphene Nanosheets

Vishwas Jadhav and Ajit D. Kelkar

Abstract

This chapter discusses the fabrication and mechanical characterization of nano-engineered composite laminates fabricated using variable-thickness graphene sheets incorporated in non-crimp carbon fiber prepregs. The effect of graphene sheet thickness on interlaminar strength (Mode I fracture toughness) of the carbon fiber composites was evaluated. The graphene lattice structure used in the present research had linear and square grids. Linear grids were arranged parallel and perpendicular to the 0° fibers in the composite laminates and labeled as vertical and horizontal grid patterns, respectively. Mechanical characterization involved the study of the effects of sheet thickness and grid pattern with and without nanoengineered enhanced laminates at the midplane. The composite laminates fabricated using a lattice graphene structure had better interlaminar strength than those fabricated with straight graphene sheets. Nanoengineered sheets with minimal thickness showed better interlaminar strength than the thicker sheets. The polymer used to manufacture the graphene sheet could not bond with the epoxy used in the composite laminate. In the literature, when the graphene nanoparticles are dispersed in the epoxy, the challenge is a uniform distribution of the nanoparticles. To overcome this dispersion problem, sheets made using nanomaterials can be used to enhance the mechanical properties of the composite laminates.

Keywords: carbon-fiber-reinforcement, composite textile laminates, doble cantilever beam, fabric characterization, graphene nanoplatelets, graphene sheet, non-crimp fibers

1. Introduction

Due to superior specific properties over traditional materials, the use of textile composites for aerospace, automotive, and marine applications has increased dramatically. Metal failure depends on the yield strength of the material, whereas the applied stresses influence the failure of composite material, the interaction between the two phases, the stacking sequence, and voids or defects in the composite laminate. Composite laminates are bonded together by a thin layer of resin between them. The function of this interface layer is to transfer the displacement and the force from one layer to another. As this matrix is a weaker element in the composites, the matrix fails at the beginning and then the fibers. When these layers get damaged or weakened, they form a crack between adjacent plies, separating the adjacent layers. This results in a failure of the lamination, which is known as delamination. The weak interlaminar region of each ply is the primary reason for the interlaminar-induced delamination of composite laminates. Researchers have developed various methods from different perspectives in diverse fields to overcome this challenge and enhance interlaminar strengths. The next-generation reinforcements, non-crimp fibers (NCF), are being explored for various structural applications. Non-crimp fabric provides excellent laminate strength, and the fabrication cost is lower than that of traditional composite manufacturing due to its drape ability and lack of fiber crimp in the woven fabric.

Various techniques such as three-dimensional weaving, stitching, Z-pinning, edge notching, braiding, fiber hybridization, and edge capping have been explored. These techniques suppress delamination but degrade in-plane laminate properties and incur additional manufacturing costs due to the process modifications. Degradation of in-plane properties and an increase in the weight of the composite laminate also lead to macroscale defects in the composite laminate, which may act as a crack initiation point for failure.

Tsai and coworkers studied the effect of delamination and tensile strength in the thick and thin ply composite laminate. To achieve a thickness of less than one-third of conventional plies (thin ply), a constant airflow through a sagged fiber filament technique was used to spread thick tows without damaging the fibers. Smaller ply thickness provides more choice in optimizing the laminate composite structure. The authors found that the tensile strength was higher for the thin-plies than the thicker-plies composite laminates due to the less resin-rich area used, but manufacturing thinner plies is more challenging [1]. With the advanced technology and concept of Prof. Stephen W Tsai's theory, known as the CHOMARAT, which was founded in 1998, four industrial sites in France, Tunisia, the USA, and China, manufactures non-crimp fabric (NCF) [2].

Nanomaterials, such as carbon nanotubes (CNTs)/graphene nanoplatelets (GNPs) incorporated polymer composites, are becoming popular due to their excellent mechanical, thermal, and electrical properties. Adding a small portion of nanomaterials improves the mechanical properties of the composite laminates, such as their strength, stiffness, impact toughness, and electrical and thermal conductivity by several orders of magnitudes, and allows them to work with inner plies However, due to strong intermolecular interactions based on dipole interactions and van der Waals forces, non-functionalized nanomaterials form bundles or agglomerates. Agglomeration causes the collection of bundles of these nanomaterials together, distributing them unevenly in the matrix.

Nanomaterials can be aligned and dispersed in the polymer composite by applying external force. Various techniques, such as mechanical stretching, electrical fields, spinning processes, and magnetic fields, are used to overcome the agglomeration issue. Mechanical methods, such as ultrasonication (bath or probe), three-roll milling (calendaring) ball milling techniques, or a combination of these methods in series or parallel, are used to achieve dispersion. Each method has its pros and cons. High energy created during the sonication or milling process often damages the nanomaterials. So, surface modification techniques became famous for dealing with agglomeration problems. In these methods, the nanomaterials are treated with chemicals such as nitric, phenolic, and carboxylic groups to improve the interaction with the solvent, which leads to uniform dispersion and avoid agglomeration.

Various tactics are used to integrate carbon nanotubes or graphene into fiber-reinforced plastic structural composites to improve the mechanical properties of the composite laminate. Micro/nanoscale materials incorporation into the matrix resin to enhance the interlaminar properties have become popular compared to mechanical insertion due to both in-plane and out-of-plane enhancement of the properties per unit mass. Furthermore, even though mechanical insertion improves the interlaminar strength, it degrades in-plane properties due to cutting of the fibers during the insertion and adding mass to the composite laminate. To overcome the agglomeration problem, researchers have used nanoengineered interleaving sheets typically inserted at various locations through the thickness of the laminate [3].

In the present research work, graphene sheets of various thicknesses were incorporated into composite laminate and the interlaminar strength investigated by performing a double cantilever beam test. Graphene sheet-embedded composite laminates showed degraded interlaminar strengths, due to low bonding between the resin and graphene sheets. To overcome this problem, grids were formed on the graphene sheet, which helped to form the thin layer of resin at the midplane by enhancing the interlaminar fracture toughness of the composite laminates. After the test to check the bonding at midplane, the tested specimens were opened at midplane for microscopic analysis.

2. Literature background

Nanomaterials have superior specific (per unit mass) mechanical properties as compared to resins and fibers. A small amount of nanomaterial added to the composites increases the mechanical properties by a few orders of magnitudes. Carbon nanotubes (CNTs)/graphene nanoplatelets (GnPs)-incorporated polymer composites have become popular in composite industries. The carbon nanotubes (CNTs) improve mechanical properties such as strength, stiffness, impact toughness, and electrical and thermal conductivity of the composites.

Toughening of the matrix resin, interleaving with short fibers or micro/nanoscale particles, became popular in recent years to overcome delamination issues by avoiding mechanical inserts (pins). The addition of nanofillers leads to additional toughening mechanisms such as fiber pull-out, peeling/pull-out hackle formation, and stick-slip mechanism, which are controlled by: (i) the method by which the laminate is manufactured, (ii) the even dispersion of the nanofiller, and (iii) the matrix mixing method. **Table 1** presents the techniques used to improve the interlaminar using different nanofillers with various results/laminar improvements reported by researchers in the last few decades.

A key method of incorporating nanomaterials in composite processing is based on the use of vacuum assisted resin transfer molding method (VARTM). However, the use of nanomaterials in composite processing using VARTM presents a significant challenge due to the narrow gaps between fibers as nanoparticles filter out during infusion as depicted in **Figure 1**. To overcome these challenge, researchers utilize advanced three-roll mills, centrifugal mixers, functionalization, sprayers, etc., for uniform dispersion of nanoparticles into the resin to produce nanomodified prepregs [19].

Improvement techniques	Key results/improvements	Reference
The adhesive layer of 0.008″	Load-carrying capacity was increased by 50%, with a loss of in-plane strength.	Lagace et al. [4]
Graphene platelets	Mode I fracture toughness enhanced by ~53% over that of neat epoxy.	Rafiee et al. [5]
Alumina nanoparticles	Mode I fracture toughness was increased by 51% for non-functionalized alumina nanoparticles, while 74% with functionalized silica nanoparticles.	Kelkar et al. [6]
Electrospun nanofibers	Mode I fracture toughness of composites with electrospun nanofiber interleaving increased by 80% as compared to neat composite.	Kelkar et al. [7]
Multi-wall carbon nanotubes (MWCNTs), graphene nanoplatelets, and carbon blanks polymer nanocomposite	The interlaminar shear strength of the carbon fiber reinforced composite increased marginally.	Shrivastava et al. [8]
0.2 wt.% of highly dispersed rGO (reduced graphene oxide)	A 52% improvement in fracture toughness (K_{IC}) of the cured epoxy.	Long-Cheng Tang [9]
0–7 wt.% GO with an increment of 1 wt.% as a function of PVP content	Mode I fracture toughness improved by ~100% for 7 wt.% GO.	Vaidyanathan et al. [10]
POSS-PVP (polyhedral oligomeric silsesquioxanes-polyvinylpyrrolidone) 1, 3, 5, and 10 wt.% to PVP	For 5 wt.% of POSS loading 70% increase in interlaminar fracture toughness.	Mishra et al. [11]
CNT concentration of 0.02% to 5%	0.047 wt.% caused an increase in the fracture toughness by 22%.	Han Zhang et al. [12]
Aligned CNTs in unidirectional carbon tape composite	The Mode I fracture toughness increased by 1.5–2.5 times and Mode II by three times.	Enrique and coworkers [12]
Buckypaper intervals carbon nanofibers unidirectional prepregs $[0]_{12}$ fabricated by hand layup	ILSS increased by 31%, whereas Mode II interlaminar fracture toughness increased by 104%.	Khan et al. [13]
CF/Biaxial NCF/Toray T1700Sc 50C) 0.5 and 1 wt.% CNT premixing, shear mixing/degassing	Mode I fracture toughness increased from 428 ± 27 (pristine) to 462 ± 25 (0.5 wt.%) and 537 ± 34 J/m^2 (1 wt.%).	Quan et al. [14]
Polyphenylene-sulfide (PPS) veils doped with multiwall carbon nanotubes and GNPs	Fracture toughness of MWCNTs veils improved while the incorporation of GNPs on the veils decreased adhesion.	Quan et al. [15]
0.5 wt.% GO-epoxy nanocomposites + use of high pressure & shear rates	Fracture toughness increased up to 220% with adequate process.	Guijosa [13]
Graphene nanoplatelets (GnPs)	Fracture toughness was found optimum at 1 wt.% for higher loading, the properties deteriorate due to agglomeration of GnPs.	Valorosi et al. [16]
Laminates interleaved by multiwall carbon nanotubes loadings of 5, 10, and 15%	Mode I interlaminar fracture toughness increased by 206% as compared to the controlled sample.	Liu et al. [17]

Table 1.
Summary of the techniques used to fabricate nanoengineered composite laminates and their effect on the mechanical properties.

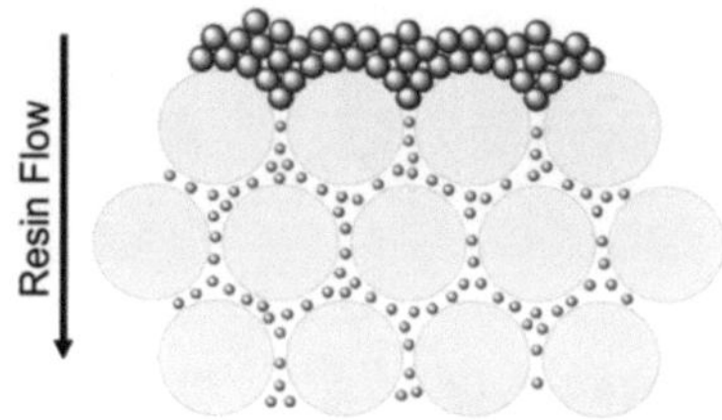

Figure 1.
Filtering of larger particles by fiber array and infiltration of smaller particles between fibers [18].

Furthermore, the use of nanofillers poses additional challenges of nonuniform dispersion resulting from nano-agglomeration, which degrades the properties of the laminate. The advanced method of inserting the carbon nanotube buckypaper was used by various researchers with promising results. Wang et al. [20], for example, studied glass fiber reinforced polymer composite with and without buckypaper. They reported 25% more shear strength due to incorporating carbon nanotubes (CNTs) buckypaper. Li et al. [21], on the other hand, studied the effect of curing pressure on glass fiber reinforced plastic (GFRPs) laminates fabricated by adding CNTs buckypaper to the midplane layer resulting in fracture toughness increments of 174.81% for initiation and 179.97% for propagation relative to baseline composite for 2 MPa curing pressure. Then, Chen et al. [22] investigated fracture toughness by incorporating buckypaper at the midplane of reinforced fiber and found the initiation interlaminar fracture toughness was increased by 29% compared to that of the one without the buckypaper composite laminate. These studies demonstrated that the interlaminar fracture toughness of fiber-reinforced plastics could be significantly enhanced by incorporating CNTs buckypaper interleaf. The researchers used nanomaterials with a small percentage of the composite and the nanoengineered micro thickness sheets were incorporated into composite laminates.

On the other hand, the use of graphene sheet interleaf is not found in the literature. So, in the present study, the graphene sheets with 50, 120, and 240 μm thickness were incorporated at midplane in order to study the initiation and propagation fracture toughness of quasi-isotropic layered non-crimp carbon fibers. Straight graphene sheets of 50, 120, and 240 μm thickness incorporated composite laminates degraded the initiation interlaminar fracture toughness by 75, 92, and 86% compared to the composite laminate without graphene sheet. To overcome this problem, uniform grids were formed on the graphene sheet, aligned parallel or perpendicular to 0° fibers at the midplane, and labeled as either vertical or horizontal grids in the present research work.

3. Materials and the process

3.1 Procurement and characterization of graphene sheets

Graphene sheets of three different thicknesses: 50, 120, and 240 μm (PN40003, PN40008, and PN 40009) are supplied by XG Sciences Inc, Lansing, MI, USA, were used as interleaving material. SEM images of the graphene sheet, as shown in **Figure 2**, depict the two-dimensional graphene nanoplatelets (GnPs).

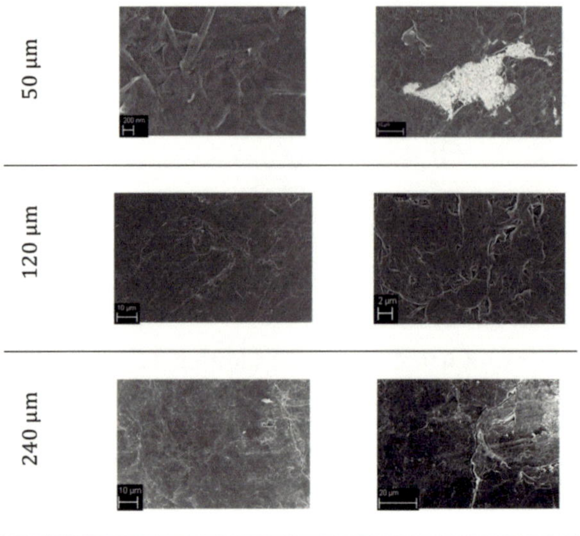

Figure 2.
SEM images of the graphene sheet.

3.2 Panel fabrication from NCF-MTM45-1 Prepreg

Almost all high-tech composite companies manufacturing aerospace and sporting goods use prepreg because of the high quality of fiber alignment and uniform fiber volume fraction. In the present study, NCF-MTM45-1 pre-impregnated prepregs supplied by SHD composites, NC, USA, were used to fabricate nanoengineered composite laminates. The graphene sheets were employed as interleaf at the midplane of the composite laminate to analyze the fracture toughness at the initiation and the propagation point. Other supporting materials, including glass mold, vacuum bag, release film, breather, and sealant tape, were assembled as shown in **Figure 3**.

The NCF composite laminates were fabricated using 48 layers (24 bilayers) of biaxial spread-tow carbon NCF fabric prepreg. Each NCF prepreg layer had [0/−45] orientation. These 24 bilayers were symmetrically stacked about midplane with the orientation of [0/−45/90/45] to achieve a balanced layup in the final cured laminate. Hand-cutting fibers produced the composite laminate with dimension 12″ × 12″, with layup $[(0/-45/90/45)_6/GS/(45/90/-45/0)_6]$ [22, 23]. **Figure 4a** depicts the debulking, using vacuum pressure to remove the entrapped air in the prepregs. Then, the 12 layers of biaxial spread-tow carbon fibers were placed on the mold, and a Teflon sheet of 14″ × 3″ was inserted at one end such that the 14″ side is perpendicular to 90° fibers in the laminate, as shown in **Figure 4b** and **c**.

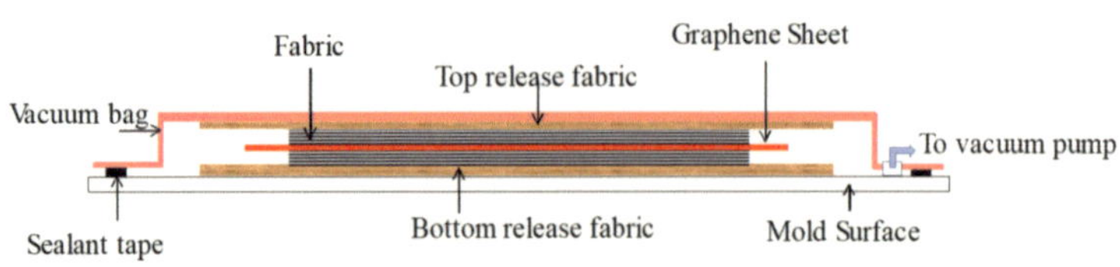

Figure 3.
Fabrication set up for nanoengineered composite laminate.

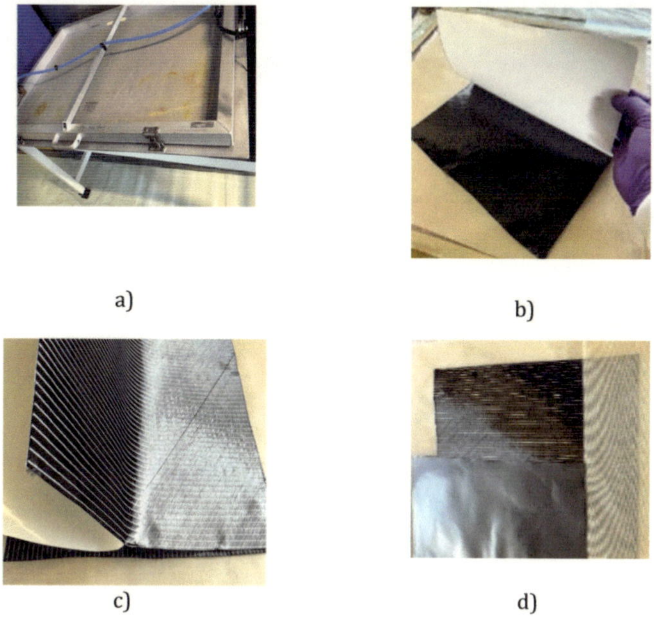

Figure 4.
Fabrication—nanoengineered composite laminate. (a) Debulking of Prepregs. (b) NCF prepreg layer preparation. (c) NCF prepreg layer arrangement. (d) GS at midplane.

The graphene sheet (either 50, 120, or 240 μm) was incorporated next to the Teflon sheet at the midplane to produce nanoengineered composite laminate. The fabrication process included composite laminate, with and without nanoengineered graphene as depicted in **Figure 4d** [2, 23, 24].

Weak bonding between the matrix and graphene resulted in degraded interlaminar resistance causing delamination and separation at the midplane when manual force was applied. This was then corrected by perforating the graphene sheets forming a lattice structure.

3.3 The curing cycle of the process

The curing cycle for the process is shown in **Figure 5**. After oven curing, the coupons were waterjet cut as shown in **Figure 6a**, and tested as per American Society for Testing and Materials (ASTM) D5528 standards [25]. The edges were made smooth and uniform using the grinding machine to achieve uniform width, as depicted in **Figure 6b**.

3.4 Fracture toughness (G_{IC})

Strain energy release rate G, expressed as Mode I interlaminar fracture toughness in ASTM D 5528 standards is given as:

$$G_{IC} = \frac{3P\delta}{2b\left(a + |\Delta|\right)} \tag{1}$$

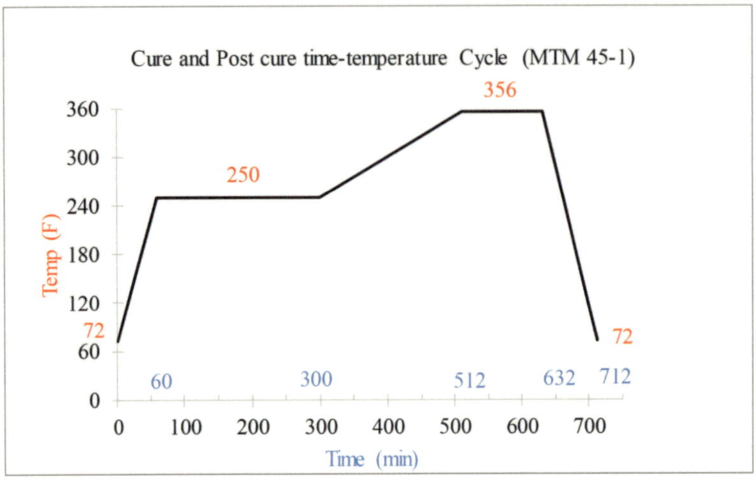

Figure 5.
Cure and post cure time-temperature cycle (MTM 45-1).

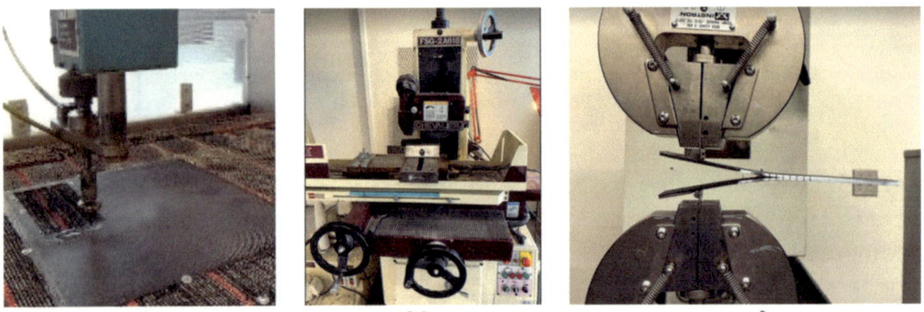

a) b) c)

Figure 6.
(a–c) Preparation and DCB test. (a) Waterjet cutting, (b) edge grinding of coupons and (c) DCB Test.

where P = load, δ = load point displacement, b = specimen width, a = delamination length and Δ is constant, determined experimentally by generating the least-square plot of the cube root of compliance, $C^{1/3}$ as a function of delamination length. Compliance C is the ratio of crack opening displacement to the applied load, δ/P. For more details, readers are referred to appropriate literature sources [25]. Instron electromechanical system is used to record the load and displacement at every 10 s after test analysis was done to calculate fracture toughness and coupons were separated about midplane to observe the midplane bonding using microscopic images.

The present work uses the modified beam method to calculate fracture toughness as per ASTM 5528 standards.

4. Results and discussion

Strain energy release rate (G_{IC}) values for with- and without-graphene sheet reinforced-composite laminate were calculated using Eq. (1) and the results are shown in **Figure** 7. The initiation and propagation values of strain energy release rate

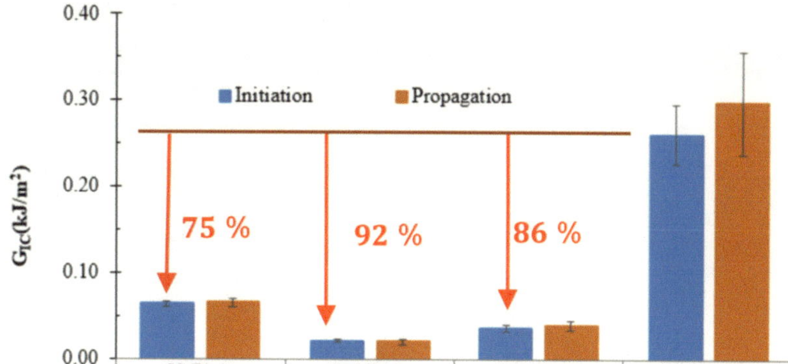

	50-Plain	120-Plain	240-Plain	Without GS
Initiation	0.0640	0.0210	0.0357	0.2605
Propagation	0.0654	0.0199	0.0392	0.2975

Figure 7.
Initiation and propagation of GIC with and without graphene sheet.

(G_{IC}) with brittle graphene-epoxy for graphene sheet reinforced-composite reduced initiation fracture toughness by 75, 92, and 86% for 50, 120, and 240 μm thick graphene sheets, and similar trends observed for propagation fracture toughness [26].

Weak bonding between the matrix and graphene resulted in degraded interlaminar resistance. Because of this, the specimens were easily delaminated and separated at the midplane when manual force was applied, as shown in **Figure 8**. To overcome this challenge, the graphene sheet was perforated by forming a lattice structure.

The lattice grids were developed with 5 mm spacing horizontally, vertically, and in square form, as shown in **Figure 9**. The gap between the graphene sheets helped to form the fiber-epoxy bonding.

Figure 10 represents the delamination resistance curve (R curve) for the horizontal, vertical, and square grid embedded graphene sheet at midplane in the composite laminate. It then compares it with the composite laminate without a graphene sheet. Delamination resistance for composite laminate without nanoengineered graphene sheet was dominant as compared to the plain and lattice structure graphene sheet-embedded composite laminates. The lattice structure formed on the graphene sheet helped to improve the delamination resistance of the composite laminate.

Figure 11 compares the fracture toughness of the 24 bilayers, NCF carbon fiber with and without incorporated lattice structure graphene sheet at the midplane with bare composites. Irrespective of the thickness, the vertical grids (grids parallel to 90° fibers in the laminate) had better performance than horizontal and square grids.

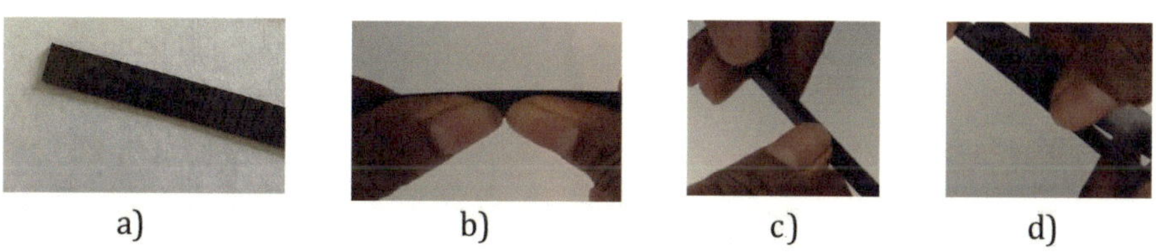

Figure 8.
Testing of coupon manually. (a) Waterjet coupon, (b) manually applied force and (c and d) separation of the laminate at the midplane.

Straight GS	Horizontal Grids (grids arranged parallel to 0° fibers)	Vertical Grids (grids arranged perpendicular to 0° fibers)	Square Grids
			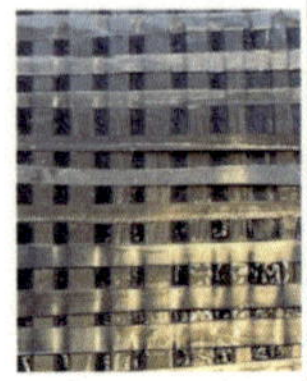

Figure 9.
Lattice structure formed on a graphene sheet.

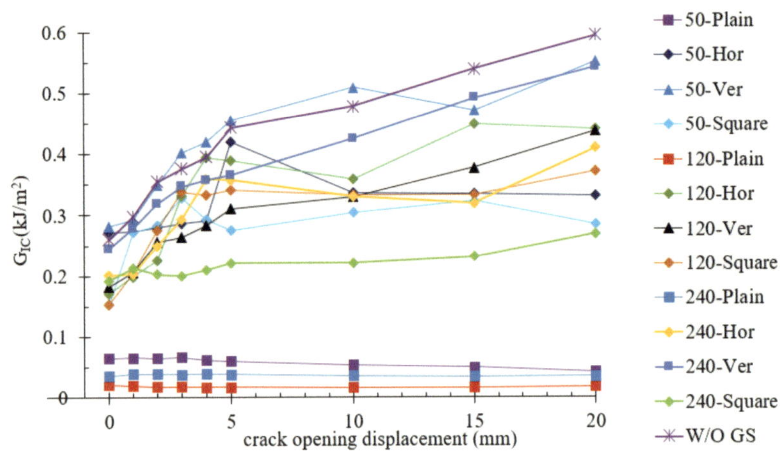

Figure 10.
Delamination resistance curve (R curve) for all GS.

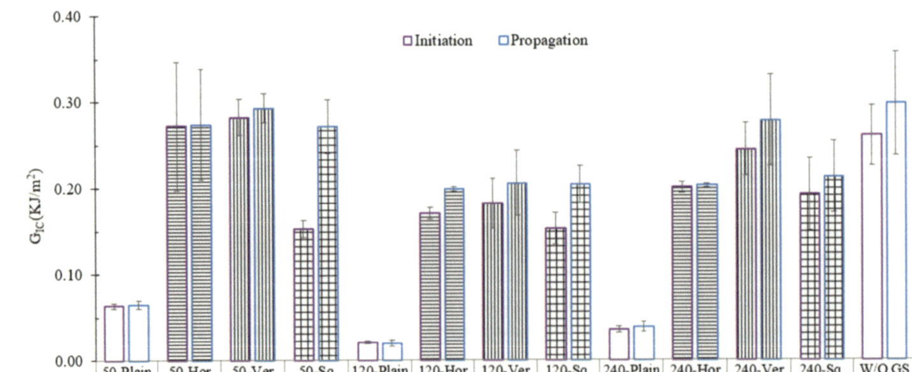

	50-Plain	50-Hor	50-Ver	50-Sq	120-Plain	120-Hor	120-Ver	120-Sq.	240-Plain	240-Hor	240-Ver	240-Sq.	W/O GS
Initiation	0.06	0.27	0.28	0.15	0.02	0.17	0.18	0.15	0.04	0.20	0.24	0.19	0.26
Propagation	0.07	0.27	0.29	0.27	0.02	0.20	0.21	0.20	0.04	0.20	0.28	0.21	0.30

Figure 11.
Fracture toughness for all samples.

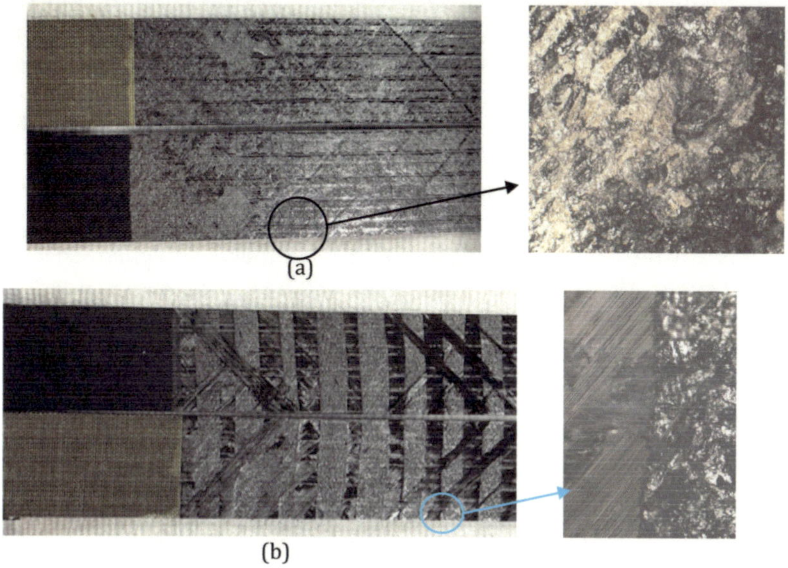

Figure 12.
(a and b) Images of the coupons after the DCB test. (a) 50 μm straight GS and (b) 50 μm horizontal grid.

The graph for the horizontal grids showed a sudden change in values due to graphene strips, because of the alternating resin patch.

The fracture toughness of the vertical grid 50 μm graphene sheet was the highest and was almost the same as the baseline composite laminate. The fracture toughness of the square grids was observed to be less than that of the other grid patterns. The fracture toughness of a straight 120 μm-thick graphene sheet was the lowest compared to those of similar groups.

Sample images for the above configurations are shown in **Figure 12a** and **b**.

For the straight graphene sheet-incorporated panels, it was observed that there was very little binding between the graphene sheet and adjacent plies. This degraded its interlaminar fracture toughness compared to the other configurations. In straight graphene composite laminates, laminates fabricated using 50 μm-thick graphene sheet-embedded showed higher fracture toughness when compared to that of the 120 and 240 μm thick graphene sheet-embedded composite laminate. This could be due to the smaller thickness of the sheet. Also, the vertical lattice structure on 50 μm sheet showed a fracture toughness of 0.28 kJ/m^2 as compared to 0.18 and 0.24 kJ/m^2 when 120 and 240 μm vertical lattice structures graphene sheets were used, respectively, as the matrix formed a thin layer in between the lattice structure, which helped to improve the interlaminar resistance. As vertical grids were arranged parallel to 0° fibers in the composite laminates, which formed the thin alternate layer of epoxy along the length of the DCB test specimen, it helped to increase the interlaminar strength of the composite laminate. The grids are aligned horizontally perpendicular to the 0° fibers in the composite laminate, which forms alternate thin layers of the matrix between two strips of graphene sheet along the length of the DCB test specimen, resulted in instability that was observed for delamination resistance (in the delamination curve) of horizontal grid pattern of graphene sheet embedded composite laminates.

To maintain uniformity in the fabrication process, horizontal-grid-formed sheets were arranged in a manner that a graphene sheet follows the same gap after the Teflon at midplane. Experimentally, maximum fracture toughness was observed for vertical grids. Furthermore, all the vertical grid samples exhibited fiber pullout, which showed bonding between the top and bottom layers at midplane, and this was not observed in the specimens with other grid patterns. Khan and Kim [13] used buckypaper intervals and reported a 31% increase in the Mode II interlaminar shear strength while reporting doubled fracture toughness, the cost of making CNT buckypaper is way too expensive compared to the commercial graphene sheets used in the present study. Also, Kelkar et al. [6] reported an 80% increase in interlaminar strength as compared to neat composite using electrospun fibers.

5. Conclusion

Graphene sheets-embedded composite laminate was used to incorporate the nanoparticles and avoid the nanoparticles' uniform dispersion problem in the epoxy resin. Mode I fracture toughness (strain energy release rate) studies indicated that when graphene sheets were inserted in the midplane of the carbon fiber composite laminate, the results for fracture toughness were poorer as there was no bonding between the graphene sheet and adjacent prepregs. Therefore, the graphene sheets were converted into three different lattice configurations, including horizontal, vertical, and square structures, so as to alleviate this problem. Grids formed on the graphene sheet showed enhanced properties and vertical grids showed peak mechanical properties as compared to horizontal and square grids. The addition of grids helped to form a tiny epoxy layer and to improve the fracture toughness of the 50 μm lattice structure sheet, more than that of the composite without a graphene sheet. Vertical grids formed on 50, 120, and 240 μm thick graphene sheets showed improvement in the fracture toughness by 4.5, 10, and 7 times compared to the straight graphene sheet, but the values were always less compared to composite without graphene sheet. Overall use of two-dimensional nanoengineered graphene sheet showed degraded interlaminar strength as compared to the composite laminate without a graphene sheet.

Author details

Vishwas Jadhav[1*] and Ajit D. Kelkar[2]

1 Joint School of Nanoscience and Nanoengineering, North Carolina A&T State University, Greensboro, NC, USA

2 Department of Mechanical Engineering, North Carolina A&T State University, Greensboro, NC, USA

*Address all correspondence to: vsjadhav@aggies.ncat.edu; vishwas.ncat@gmail.com

References

[1] Sihn S, Kim RY, Kawabe K, Tsai SW. Experimental studies of thin-ply laminated composites. Composites Science and Technology. 2007;**67**(6):996-1008

[2] Jadhav VS, Kelkar AD. Fabrication, processing and characterization of carbon fibre reinforced laminated composite embedded with graphene lattice sheets. In: Proc. ASME 2021 Int. Mech. Eng. Congr. Expo. 2021. pp. 1-9

[3] Stahl JJ, Bogdanovich AE, Bradford PD. Carbon nanotube shear-pressed sheet interleaves for Mode i interlaminar fracture toughness enhancement. Composites Part A: Applied Science and Manufacturing. 2016;**80**(January):127-137

[4] Lagace P. Suppression of delamination in a gradient stress field in graphite/epoxy laminates. ICCM/9. Composites: Properties and Applications. 1993;**6**:705-713

[5] Graphene L et al. Enhanced mechanical properties of. ACS Nano. 2009;**3**(12):3884-3890

[6] Kelkar AD, Mohan R, Bolick R, Shendokar S. Effect of nanoparticles and nanofibers on Mode I fracture toughness of fiber glass reinforced polymeric matrix composites. Materials Science and Engineering B. 2010;**168**(1-3):85-89

[7] Shendokar S, Kelkar A, Bolick R. Application of electrospun nanofibers for the improvements in mode-I fracture toughness of fiberglass composites. In: Proc. Int. Conf. Nanosci. Eng. Technol. ICONSET 2011. 2011. pp. 5-10

[8] Shrivastava R, Singh KK. Interlaminar fracture toughness characterization of laminated composites: A review. Polymer Reviews. 2020;**60**(3):542-593

[9] Tang LC et al. The effect of graphene dispersion on the mechanical properties of graphene/epoxy composites. Carbon N. Y. Aug. 2013;**60**:16-27

[10] Mishra K, et al. Effect of graphene oxide on the interlaminar fracture toughness of carbon fiber/epoxy composites. Polymer Engineering and Science. 2019;**59**(6):1199-1208

[11] Mishra K, Babu LK, Vaidyanathan R. Influence of POSS-PVP in modification of CFRP interlaminar fracture toughness. Journal of Applied Mechanical Engineering. 2019;**8**:1-6

[12] Zhang H, Liu Y, Kuwata M, Bilotti E, Peijs T. Improved fracture toughness and integrated damage sensing capability by spray coated CNTs on carbon fibre prepreg. Composites Part A: Applied Science and Manufacturing. 2015;**70**:102-110

[13] Khan SU, Kim J-K. Improved interlaminar shear properties of multiscale carbon fiber composites with bucky paper interleaves made from carbon nanofibers. Carbon N. Y. 2012;**50**(14):5265-5277

[14] Quan D, Urdániz JL, Ivanković A. Enhancing mode-I and mode-II fracture toughness of epoxy and carbon fibre reinforced epoxy composites using multi-walled carbon nanotubes. Materials and Design. 2018;**143**:81-92

[15] Quan D, Mischo C, Binsfeld L, Ivankovic A, Murphy N. Fracture behaviour of carbon fibre/epoxy composites interleaved by MWCNT- and graphene nanoplatelet-doped

thermoplastic veils. Composite Structures. 2020;**235**(November 2019):111767

[16] Valorosi F et al. Graphene and related materials in hierarchical fiber composites: Production techniques and key industrial benefits. Composite Science and Technology. 2020;**185**(October 2019):107848

[17] Liu J, Li Y, Xiang D, Zhao C, Wang B, Li H. Enhanced electrical conductivity and interlaminar fracture toughness of CF/EP composites via interleaving conductive thermoplastic films. Applied Composite Materials. 2021;**28**(1):17-37

[18] Islam AI, Kelkar A. An assessment of cure kinetics of Epon 862-W epoxy resin system in the presence of electrospun carbon nanofibers during the fabrication of nanoengineered Prepregs. In: SAMPE Spring Conference. 2017

[19] Islam ABMI, Kelkar AD. Prospects and challenges of nanomaterial engineered prepregs for improving interlaminar properties of laminated composites: A review. MRS Communications. 2017;**7**(2):102-108

[20] Wang X et al. Helical milling response of glass fiber-reinforced polymer composite with carbon nanotube buckypaper interlayer. Polymers and Polymer Composites. 2020;**28**(6):378-387

[21] Li N et al. Interlaminar properties of GFRP laminates toughened by CNTs buckypaper interlayer. Composites Structure. 2019;**208**(August 2018):13-22

[22] Chen C, Li Y, Yu T. Interlaminar toughening in flax fiber-reinforced composites interleaved with carbon nanotube buckypaper. Journal of Reinforced Plastics and Composites. 2014;**33**(20):1859-1868

[23] Jadhav VS, Kelkar AD. Mechanical characterization of graphene nanoplatelets integrated non-crimp carbon fiber laminated composite laminates. In: International SAMPE Technical Conference. Vol. 2022-May. 2022. pp. 1-14

[24] Jadhav VS. Process development and characterization of carbon fiber composites cured using nanoengineered graphene sheets [PhD thesies]. North Carolina A&T State Univ.; 2022. pp. 1-199

[25] ASTM D5528-01. Standard test method for mode I interlaminar fracture toughness of unidirectional fiber-reinforced polymer matrix composites. American Standard Testing Methods. 2014;**3**(Reapproved 2007):1-12

[26] Kelkar AD, Jadhav VS. Carbon Fiber Reinforced Composite with Graphene Sheets. U.S. Serial No.: 63/274,384. 1 Nov 2021

Chapter 6

Advanced Graphene-Based Materials for Electrochemical Biomarkers and Protein Detection

Carmen Ioana Fort, Liviu Cosmin Cotet, Lucian Cristian Pop, Monica Baia and Lucian Baia

Abstract

In this chapter, recent advances in the field of graphene materials-based (bio)sensors that are used for biomarker and protein electrochemical detection are presented. Approaches related to the synthesis of electrode material for (bio)sensors construction as well as to their morphological and structural characterization, are highlighted, pointing out the advantages of using graphene-based materials for (bio)sensors applications. This chapter focuses on obtaining graphene-based electrodes, detecting biomarkers and proteins, and evaluating the performance of the sensors. Different methods for designing sensors for a large variety of biomolecules are described and comparatively discussed. In order to increase their electroanalytical performances, such as sensitivity, selectivity, detection limit, linear range, and stability, the research performed in the last years was focused on different types of graphene structures including graphene oxide, reduced graphene oxide, graphene nanofoams, graphene nanocomposites, different functionalized graphene, etc. The detection of analytes including neurotransmitters and neurochemicals (dopamine, ascorbic acid, uric acid, serotonin, epinephrine, etc.), hydrogen peroxide, and proteins, have been discussed. The studies related to electrochemical (bio)sensors are presented in three subchapters, and the key results—electroanalytical performances—of the sensors are summarized. The final chapter provides the conclusions derived from the comparative analyses of different approaches related to these types of (bio)sensors based on graphene materials.

Keywords: biosensors, electrochemical sensors, graphene materials, graphene oxide, reduced graphene oxide, graphene-based composites, graphene nanomaterials

1. Introduction

In the last decade, the necessity for rapid, simple hand-held testing devices in medicine has prompted the development of biosensors for clinical purposes. Biological sensors were developed as optical, electrical, piezoelectrical devices or systems, which consist of biological and electronic components and are able to discover and detect biological compounds like nucleic acids, proteins, enzymes, and genes. Nowadays, biosensors are known as devices that are employed for qualitative and quantitative detection of biological analytes [1]. The biomolecules, the biological structures, or the microorganisms can play the role of biological analytes. As it is well-known, an electrochemical (bio)sensor involves three components, which are responsible for (i) the recognition of the analyte and the signal formation, (ii) the transducer of the received signal, and (iii) the reader device [2]. Owing to an interdisciplinary combination of approaches from physics, chemistry, biology, nanotechnology, and medical science, the achievements in the biosensor field are impressive. Consequently, they are becoming essential devices developed for diagnosis of life-threatening syndromes [3, 4].

Electrochemical sensors are of high interests in different applications because of their great sensitivity, selectivity, inexpensive and simplistic production, and facile miniaturization [5, 6]. They depend on the use of sensitive biological molecules immobilized on the surface of solid electrodes that are able to capture target molecules by specific recognition [2, 3]. This process, at the electrode surface, occurs with a reaction signal. The modified electrode transforms the produced chemical signal into a measurable electrical signal, such as current, voltage, conductivity, impedance, etc. This way, the technique enables both the qualitative and quantitative analysis of target species [7, 8].

Electrochemical biosensors have some advantages as compared to other biosensor categories. First of all, the theory behind it is well-developed, and it involves a facile design based on simple structures for easy measurements [5, 9]. As compared to other analytical methods including fluorescence [10, 11] colorimetric [12], and chemiluminescence methods [13, 14], electrochemical biosensors can be used in many important fields such as biomedicine, pharmaceutical industry, food industry, and environmental analysis [3, 4, 15]. In order to increase the sensor performances with respect to detection limits, sensitivity, selectivity, time stability, and linear detection range and to lessen the time of response, recent research was focused on developing a new preparation method for biosensor electrodes using different electrode materials.

Thus, due to their low cost, compared to other materials such as Pt and Au metals, and because of their good conductivity, a variety of modified electrodes incorporating carbonaceous materials [16], such as graphite [17, 18], carbon aerogel [19–22], glassy carbon [23–28], graphene [25, 29–33], carbon fiber [34], and screen-printed electrode [35], have been developed for sensing of numerous biological compounds. Also, various materials including metals nanoparticles Au [25, 36–39], Ag [40]; phosphates such as zirconium phosphates [28, 41] and titanium phosphates [27]; mediators [42], polymers (PANI [37, 43], chitosan [44–46]), enzymes (HRP) [17, 23, 47], plasmodium falciparum lactate dehydrogenase (PfLDH) [48], cellobiose dehydrogenase (CDH) [20]; oxides (α-Fe_2O_3 [49], Al_2O_3 [50], ZnO [36, 51], Cu_2O [52]), and non-metals (N [53]) have been used widely to modify different carbonaceous electrode materials.

Among the carbonaceous modified electrodes, those based on graphene matrix, i.e., graphene (G), graphene oxide (GO), and reduced graphene oxide (rGO) [54, 55], exhibit important advantages over the conventional ones. This is because the latter classes of materials, present significant drawbacks that range from limited active surface areas easy inactivation and high cost of production. G structure is composed only of carbon atoms that are all sp^2-hybridized and organized in a single atomic layer. Every carbon atom belonging to G is bonded to three other neighboring carbon atoms using three of its valency electrons, and the fourth electron is delocalized, thus facilitating the electrical conductivity of G [15]. The use of G for sensors' development is attributed to its additional unique features that include its ease of production; good chemical, morpho-structural, and mechanical characteristics, joined with other atypical properties including the high surface area to volume ratio, exceptional optical characteristics, outstanding carrier mobility, and extraordinary electrical and thermal properties as compared to those exhibited by other carbon allotropes [56]. Besides its exceptional properties, G can be functionalized easily, and many of its biomedical applications can benefit from non-toxic, biocompatible, and water-dispersible G layers obtained by chemical functionalization with different ligands [57].

The grafting of functional groups on the G surface structure is extremely reliable for binding molecules so as to further examine their communications with well-defined targets. In this respect, oxidized G presents one of the best solutions to obtain surface charges by anchoring on its surface oxygen-containing substituents like hydroxyl, carbonyl, carboxyl, and epoxide groups, which make it easier to understand the future specific interactions that take place on the surface of the G structure. Thus, G and its oxidized derivatives (GO or rGO) combine well-known properties of the carbonaceous structures, i.e., the great electric and thermal conductivities, mechanical strength, and impermeability to gases, with those related to the enhancement of their surface response after interaction with specific molecules. Accordingly, G-based materials are desired because of their essential functional groups such as those found in GO or N-G, very beneficial and effective in aligning with biomedical and related fields [58]. This is the reason why, in the last years, investigations were focused on different types of G structures including GO, rGO, G nanofoams, G nanocomposites, and different functionalized G derivatives [59], and the biosensor applications are now focused on biomarkers and protein detection.

In this chapter, the latest developments in the domain of biosensors, based on G materials, are described. Details about the synthesis and their morpho-structural investigations are also provided. Different approaches for designing sensors for a great variety of biomolecules are described and discussed. The detected analytes contain neurotransmitters and neurochemicals (dopamine, ascorbic acid, uric acid, serotonin, epinephrine, etc.), hydrogen peroxide, proteins, DNA, glucose, and others. The aim was to highlight biosensor performances for glucose, hydrogen peroxide, neurotransmitters and neurochemicals, and cancer and disease biomarkers detection. A special importance was accorded to dopamine, ascorbic acid, uric acid, serotonin, and epinephrine detection. The conclusions resulting from the presented research on the biosensors are summarized in the last subchapter.

2. Synthesis, structure, and morphology of graphene materials for sensing applications

G materials have been widely promoted due to their suitable properties and versatility in biomedical applications as like biosensors, diagnosis-imaging, and drug delivery [58]. G consists of a one-atom-thick carbon structure that forms a 2D hexagonal honeycomb nanomaterial (**Figure 1a**), but besides the pristine G materials, the functionalized G structures (containing hydroxyl, carbonyl, carboxyl, and epoxide groups on the edge or plane of the nanosheets) such as GO (**Figure 1b**) and rGO (**Figure 1c**) are tremendously involved and tested in the same biomedical application fields. As a convention, the limits of 2D G are accepted to have the thickness of up to 10 atomic layers (i.e., a few nanometers) and the lateral size larger than 10 nm up to more than 20 μm [61]. Those having less than 10 atomic layers and a few thousand

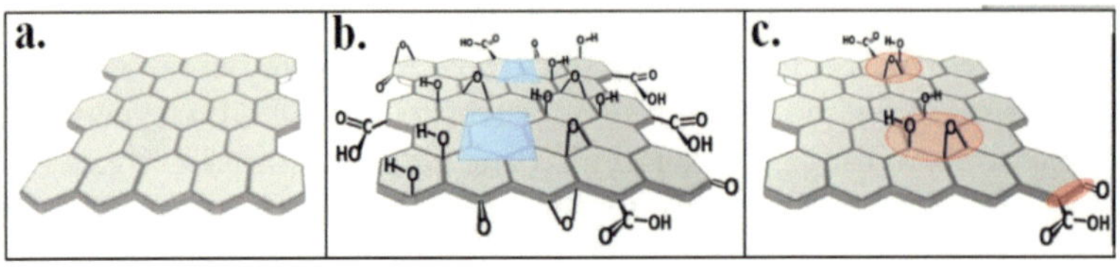

Figure 1.
Suggested structural models for monoatomic layered nanosheet of G (a), GO (b), and rGO (c); the blue square (b) depicted the electrical conductor oasis of GO. The red ovoid (c) depicted electrical insulator oasis of rGO. Reproduced from Ref. [60] with permission from the Royal Society of Chemistry.

nanometers in size will be termed as nanosheets. The properties that make G materials suitable for various purposes can be adjusted by selecting the appropriate synthesis pathway. If the target is represented by the sensing applications, then the obtained structures are desired to recognize entities having similar physical properties and to provide information about them [58].

One very important parameter of G for electrochemical sensing applications is its semiconductor behavior with zero band gap energy, which shows that G has electrical conductivity much greater than that of copper, resulting from its extended π-π conjugation of sp^2 hybridized carbon atoms. In the same way, rGO is considered an electrical conductor that contains electrically insulating oases (**Figure 1c**), whereas GO is an electrical insulator with electrically conductive oases (**Figure 1b**). The electrical insulation behavior of GO results from the presence of sp^3 hybridized carbon atoms that are involved in anchoring oxygen-containing groups on the edges or planes of GO/rGO nanosheets. Among them, the carboxyl and epoxide groups are very commonly involved in biomolecular immobilization for biosensor construction [58].

Generally, the synthesis of G materials can be grouped into two main classes: bottom-up and top-down methods [62]. The representative pathway for the first class is carbon vapor deposition (CVD) that consists of a G growth on catalytic support (e.g., nickel or copper) using a carbon-source gas (e.g., CH_4, CH_4/H_2 mixture, C_2H_4) at a high temperature (around 1000°C). Free-standing monoatomic layer G as a crystal sheet with the size of catalytic metallic support is released after support etching [62].

Among the top-down methods is the one proposed by Paton et al. [63], which enables the obtaining of G nanosheets in a large-scale production method by physical exfoliation based on shear mixing of graphite. Even though, the obtained G has a high yield, good conductivity, and strong mechanical properties, its poor dispersion in common solvents (e.g., water and ethanol) is the primary disadvantage [64]. The most efficient ways to overcome this drawback is to use a dispersing agent (e.g., amphiphilic polymers, alkylamines, and molecules with hydrophilic carboxyl groups) or to generate organic functional groups (e.g., hydroxyl, carbonyl, carboxyl, and epoxide) on the planes and the edges of G sheets. The most known of such methods are Hummer [65] and its derivates [66], which are based on chemical oxidative exfoliations of graphite using concentrated sulfuric acid (H_2SO_4) and sodium nitrate ($NaNO_3$) as the reaction medium combined with potassium permanganate ($KMnO_4$) as oxidant. But toxic and hazardous gases (e.g., NO_2 and N_2O_4) are generated if the reaction temperature overcomes certain limits. Marcano et al. [67] proposed an improved method of Hummers that consists in an oxidative exfoliation, in which phosphoric acid (H_3PO_4) was involved. Better efficiency in GO production and no hazardous gas generation were claimed. By using sonication and a lower reaction temperature, a more controlled sono-oxidative-exfoliation process is obtained [60, 68]. A new GO preparation route is proposed by Peng et al. [69], who used the potassium ferrate (K_2FeO_4) as the oxidant and concentrated H_2SO_4, as acidic medium for the reaction. Besides the higher quantity of GO that can be obtained in comparison with the other reported pathways, the approach presented a simple way of preparation with the additional benefit brought about by the recycling of H_2SO_4. In another pathway version promoted by Yu et al. [70], an oxidant mixture of K_2FeO_4 and $KMnO_4$ was employed with boric acid (H_3BO_3) as a stabilizer in H_2SO_4. By involving such liquid phase exfoliation methods, GO nanosheets that are proper to form stable aqueous suspensions can be obtained. For sensor construction, the obtained 2D GO nanosheets have to be capable to form stable individual 3D structures as nanobricks or to form (nano)composites/modified structures by combination with other components.

Generally, the 3D structures of G materials can be grouped into three main classes: films (i.e., supported membranes), membranes (i.e., unsupported films), and porous structures (e.g., foams, aerogels/xerogels, frameworks, etc.).

In the case of G obtained by using the CVD technique and flat metallic support, a large-area continuous 2D film or membrane can be obtained after etching the support by using acid solutions. But, if the template is a porous 3D metallic structure (e.g., nickel foam, copper powder, magnesium, or aluminum oxide nanoparticles), a 3D G, as a coating film of metallic framework support, stabile hollow or macroporous scaffolds without metallic part can be achieved after the metallic support is etched [71]. Due to the lack of promoter-link groups on the surface of G it is difficult to obtain stable 3D structures if non-functionalized G nanosheet/nanoflake suspensions are involved. Besides, although the individual G nanosheets exhibit a very high electrical conductivity, because of the poor connection between the resulted 3D materials, a lower conductivity is measured. To overcome this drawback, the G nanosheets should be functionalized. The most known and used functionalized G material is GO. In this case, the most common methods for obtaining films are vacuum filtration, dip or drop-casting, spraying, screen printing, and the Langmuir–Blodgett method [62]. Due to the strong connection between functional groups of neighboring GO nanosheets, such as hydroxyl, carboxyl, carbonyl, and epoxy (see **Figure 1b**), very stable 3D laminar structures are formed as 3D GO films on the support materials after their drying. By removing (generally by a simple exfoliation) the forming support, stable free-standing GO membranes (i.e., unsupported films) can be obtained [68].

Besides this, Cotet et al. [60] proposed an elegant method that enables the harvesting of a stable membrane formed as a self-assembled pellicle at air-liquid interface after performing an isopycnic separation of a certain GO fraction. A proposed schematic distribution of functional groups present in the obtained dried GO membrane is illustrated in **Figure 2**.

For a deep understanding of the interaction processes between GO nanosheets that build up such a GO membrane, an assessment of the zeta potential was performed for initial GO suspension [72]. A variation in time of the zeta-potential values with high negative values (i.e., −63 and −67 mV) at about 3 weeks from the GO synthesis, followed by a stabilization (at about −45 mV) after 4–6 weeks, was observed. These high values of negative charge indicated a high density of functional groups present both on the plans and edges of GO nanosheets [72]. The presence of high number of functional groups is beneficial for functionalization due to the potential of attaching

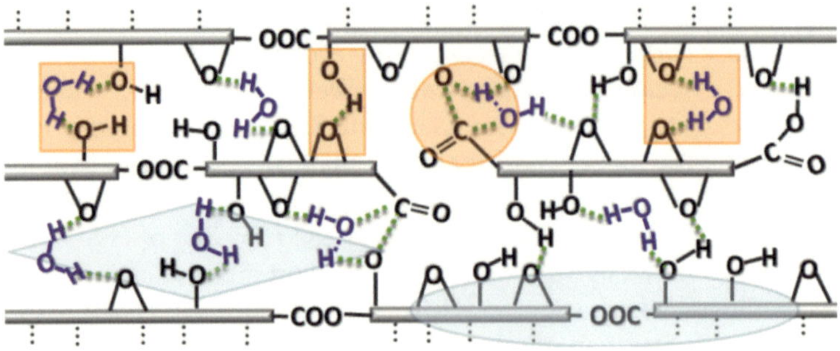

Figure 2.
Suggested 3D laminar GO structure with intraplanar (ovoid) or interplanar (rhombic) interactions, hydrogen bonding involving H_2O (square) or not (rectangular), and ester linking (circular). Reproduced from Ref. [60] with permission from the Royal Society of Chemistry.

specific groups or structures including sensitive biological molecules that can be involved in targeted sensing activity.

Porous 3D structures from the class of aerogel/xerogel/cryogel are obtained by drying in supercritical conditions of liquid CO_2, in ambient conditions or by freezing-sublimation of wet hydrogels. A simple way to obtain hydrogels with GO nanosheets in their structure was promoted by Worsley et al. [73]. A GO suspension was involved as reaction medium for the polycondensation reaction of resorcinol with formaldehyde catalyzed by sodium carbonate (Na_2CO_3). In another approach based on a hydrothermal route [71], the added carbohydrates (e.g., glucose, cyclodextrin, and chitosan) played both the role of morphology-oriented agents and of reduction agents, which define the structural, physical, and electrochemical properties of the obtained 3D interconnected-pore aerogel.

A cheaper method for 3D GO porous structure synthesis without the necessity of complex equipment and supporting substrates is the room-temperature vacuum centrifugal evaporation of a properly concentrated GO suspension, which is described in the literature [71].

During the research and testing of GO materials for sensing applications, a very important step is represented by the reduction process that partially restores the structure and electrical properties of G. In this way, a conversion from insulating GO into electrically conductive rGO occurred with the increase of the C:O molar ratio. These reduction processes can be thermal, chemical, electrochemical, irradiation-induced, etc. [62, 68]. Among these approaches, the thermal and chemical reduction methods are identified as the most used techniques, and we will therefore refer to these methods later in this chapter.

In the case of GO materials modified with bio-structures, particular sensitivity (e.g., pH and temperature) of this kind of class of substances have to be taken into account. Moreover, if the modification will be carried out with more sensitive bio-structures, the reduction process, which is harsh, should be performed prior to this process. Additionally, it is known that the higher electrochemical properties of G materials are due to their intrinsic favorable properties represented by the π-π stacking interactions. These can enhance electrochemical signals and the irreversible anchoring of catalytic sites onto G materials. Thus, it is important that the new 3D structures formed present these structural advantages [68]. A method used to improve these properties is the doping of G material with nitrogen. Shao et al. [74] obtained nitrogen-doped G by exposing G materials to nitrogen plasma. The

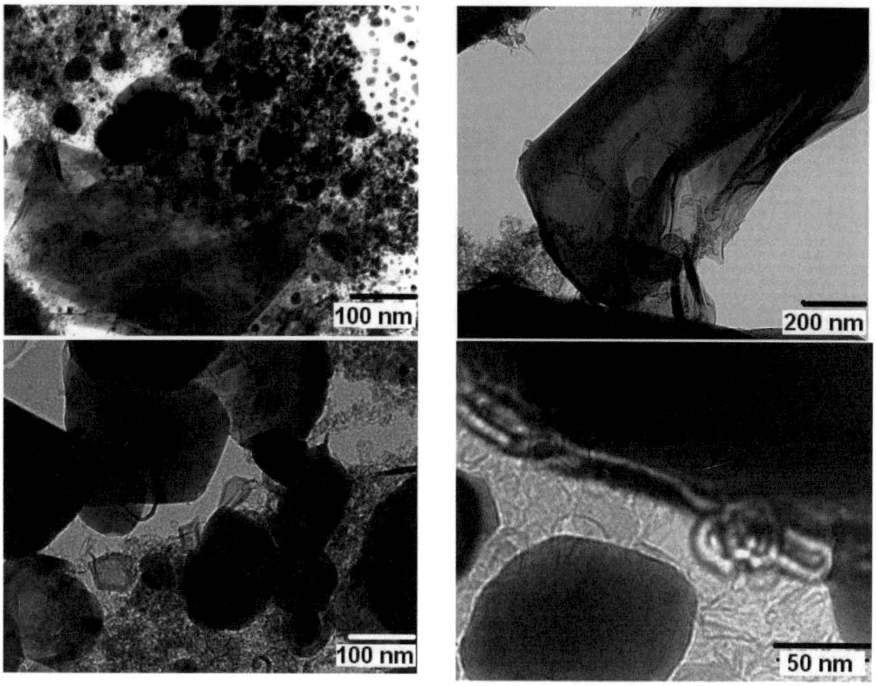

Figure 3.
TEM (transmission electron microscopy) images of Fe-doped carbon aerogels showing G-like nanosheets (a and b); carbon nanoribbons, which are isolated (c) or around the Fe nanoparticles (a and c); and carbon nanotubes (d). Reproduced from Ref. [21] with permission from the Elsevier.

as-obtained N-doped G materials produce the best result in the detection of hydrogen peroxide (H_2O_2) due to nitrogen and oxygen-containing groups. Interesting results were obtained by Fort et al. [21], who observed that the presence of Fe in the carbon aerogel matrix led to the formation of G-like structure as single-layer nanosheets (**Figure 3**), which increased the conductivity of the electrode materials and the electroanalytical performance of the obtained modified electrode for H_2O_2 detection [21].

The graphitic nanostructure formation in the presence of Fe and Bi was observed by Rusu et al. [75]. The graphitic structure obtained (**Figure 4**) is relevant in charge transport and thermo-chemical processes.

del Pino et al. showed that the use of the ultraviolet laser irradiation of flexible free-standing GO membranes in the ammonia-rich atmosphere and liquid environments generates significant integration of nitrogen groups, especially amines, in a partially reduced GO structure [68]. This method allowed for the obtaining of films with controlled geometry of reduced areas up to hundreds of square centimeters of N-doped rGO materials with high potential for (bio)sensing applications. So, flexible electrodes prepared on flexible polymeric supports with 10 μm thickness and low resistivity (6×10^{-4} Ω m) were obtained by an innovative laser reduction protocol [76].

There are various chemical agents that are used in GO reduction: hydrazine (N_2H_4), alcohols, sodium borohydride ($NaBH_4$), hydriodic acid in acetic acid (HI/CH_3COH), sodium/potassium hydroxide (NaOH, KOH), iron/aluminum powder (Fe/Al), ammonia (NH_3), hexylamine (CH_3-$(CH_2)_5$-NH_2), sulfur-containing compounds, hydroxylamine hydrochloride ($HONH_2$·HCl), urea (CH_4N_2O), manganese(II) oxide

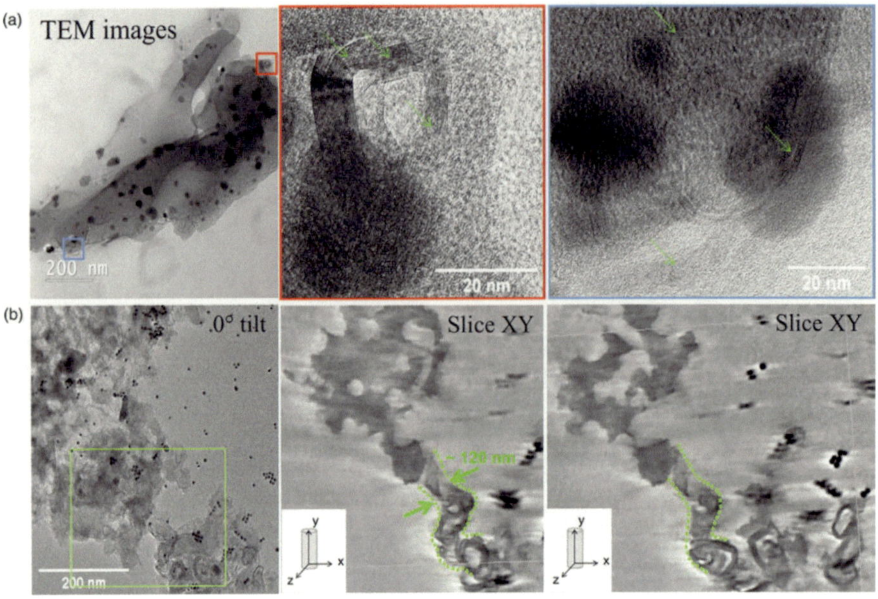

Figure 4.
Graphitic nanostructures evidenced in carbon matrix that contains bismuth-iron nanoparticles: (a) TEM and HR-TEM (high resolution-Transmission Electron microscopy) images after heating at 900°C with two highlighted regions during the in situ TEM experiment, (b) electron tomography results on a grain representative for the graphitized sample. On the left, a 2D projection extracted from the tilt series was used to reconstruct the grain volume; on the middle and right, two orthoslices took at different depths within the reconstructed volume evidencing a tubular morphology of the graphitic structure having a length of roughly 120 nm [75]. Reproduced from Ref. [75] with permission from Cambridge University Press.

(MnO), enzymes, vitamin C, bacteria, etc. [77]. However, the most used reducing agent is hydrazine, which allows for the obtaining of rGO materials with a C:O of about 12.5:1 and a conductivity of 99.6 S cm^{-1} [78]. But, because of its toxicity, green alternatives are encouraged to be developed. From these, reduction with alcohols [79], hydriodic acid [80], and vitamin C [81] are reported [82].

Thermal reduction of GO materials consists of organic (i.e., hydroxyl, epoxy, carbonyl, carboxyl, etc.) group removal as O_2, CO, CO_2, and H_2O by the aid of high temperatures (>120°C) [83]. Because hazardous reagents are not involved, this reduction method is considered to be safer than certain chemical reduction approaches. The degree of reduction can be adjusted by controlling the reaction temperature and the process time. A high temperature (of about 400–1050°C) and an oven equipped with an inert atmosphere (i.e., Ar or N_2) or in hydrogen flow are required to obtain rGO materials with high electrical conductivity [62]. Nonetheless, even by the use of these methods, the materials obtained show a low surface and a smaller number of insulated oases (**Figure 1c**) due to the remaining functional groups from rGO materials that could be further involved in applications related to the sensing electrode construction.

To the best of our knowledge there has not yet been established a direct relationship between the morphology-structure of a composite and its sensing properties. However, there are several studies reported in which particular correlations between the structure and sensing properties were made by keeping in mind the possibility of improving the sensing performances as a result of understanding the structural particularities.

Besides the excellent electrical conductivity, another remarkable property of G, when used in sensing applications, consists of its high specific surface area (theoretical specific surface area ~2630 m^2/g), which promotes the attachment of (bio) molecules, polymers, or nanoparticles to the G surface by exposing all carbon atoms to sense biomolecules. These results indicate an increased G's sensitivity as biosensor [84]. After synthesizing cobalt oxide (Co_3O_4) nanowires using a hydrothermal method based on 3D G foam grown by CVD, Dong et al. used this composite (G/Co_3O_4) as a monolithic free-standing electrode for free-enzyme electrochemical detection of glucose. The porous structure of the G foam proved to play the role of a scaffold for the sensors fabrication [85]. A double-layered membrane of rGO-and-polyaniline (PANI) nanocomposites (to enhance the sensitivity of sensor) and molecularly imprinted polymers decorated with gold nanoparticles was prepared by Xue's research team in order to obtain an electrochemical serotonin sensing interface. When the morphology of the material was studied by SEM (Scanning Electron Microscopy), a uniform distribution of rGO/PANI nanocomposites spheres with the diameter of the single sphere of approximately 93 nm was observed. A rough and foliated structure embedded with AuNPs was observed. This indicates the achievement of imprinted membrane embedded with AuNPs, which led to an increased conductivity and electrocatalytical activity of the membrane [37]. By employing electrophoretic deposition in a magnesium salt/GO electrolyte, Akhavan et al. prepared GO nanowalls with extremely sharp edges and preferred vertical orientation, and then they deposited them on a graphite electrode. The results revealed a uniformly deposited film of GO nanowalls on the electrode surface. Interesting petal-like G nanosheets with lateral sizes of ~500 nm and very sharp edges (1–15 nm thickness) were observed. Because of the formation of a large fraction of graphitic edge-plane defects that can lead to the obtaining of a higher surface activity than the G nanosheets placed parallel to the substrate, the authors speculate that such vertical nanosheets may exhibit distinctive electrochemical properties [86]. By SEM investigation [60] of free-standing self-assembled GO membranes (**Figure 5**), a continuous wavy feature (**Figure 5a**) and a compact layer-by-layer stacking (**Figure 5b**) were observed at the surface and the cross-section of obtained dried GO membranes.

In the meantime, in the case of this preparation pathway, by changing the time used for self-assembled process from 15 to 120 min, GO membranes with about 2.85 μm and 11.40 μm thickness, respectively, were obtained. This kind of GO membrane type could be properly reduced (e.g., by protected laser irradiation, **Figure 5**, [76]) to

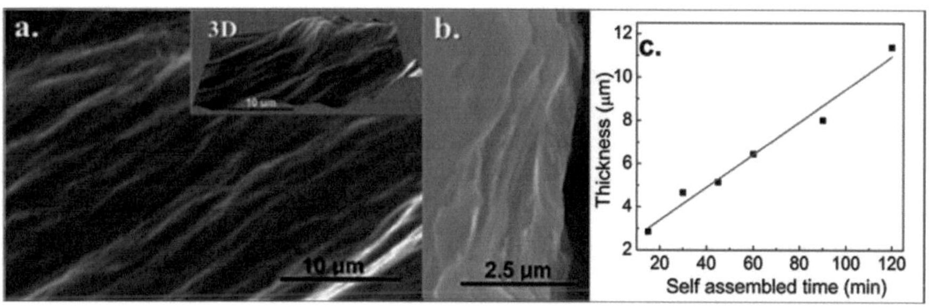

Figure 5.
SEM images of the self-assembled GO membrane surface (a), and its 3D view (see the inset), the cross-section (b) obtained by dispersing a certain GO fraction on an open area framework support and harvesting it after 15 min, and the average membrane thickness evolution against the self-assembly time (c) [60]. Reproduced from Ref. [60] with permission from the Royal Society of Chemistry.

obtain 3D rGO membranes or films with controlled geometry, reduction degree, and morpho-structural properties suited for biosensor construction.

3. Biosensor applications

As already reported, the attractive properties of G-based nanomaterials can be successfully used in biomedical applications as electrochemical-based biosensors with improved analytical performance with respect to low detection limits, selectivity, high sensitivity, low response time, reproducibility, and large linear detection range [54, 56, 59, 71]. In addition, a significant improvement in the sensors is mechanical properties, relating to flexibility. An important work focused on the importance of flexible sensors was reported by Giaretta et al. [87]. They compared the biosensors performances based on the obtained material substrates. It was concluded that the polymeric substrates are the cheapest, but their performances were inferior when compared to those of the other substrates obtained from carbonaceous materials. This is due to the poorer electrical conductivity, reduced permeability and lack of porosity, respectively.

A bioelectrochemical sensor transforms a biological modification occurrence into an electrical signal. Hence, the working principle is based on the transfer of a negative elementary charge (the electron) between the G interfaces and species that present electrical activity. The species can be either the molecule to be analyzed or a species whose electrochemical measured signal can be associated with the existence of the target analyte.

3.1 Biomarkers and protein detection

Real-time quantitative monitoring of biomarkers, which represent a particular class of biological compounds whose presence in serum and saliva shows a relationship with a certain disease, has become essential for early disease detection, leading to adapted treatments and investigation of treatment efficiency. Guo et al. synthesized Ni nanosheet/G composites, which have an exceptional electrocatalytic signal for the detection of L-alanine by using a direct current (DC) arc plasma jet CVD method [88].

The obtaining of 1 to 10-layered G nanosheets, G nanoribbons, and core-shell Ni/G nanosheets by the above-mentioned method was evidenced. These G-based composites were used as sensors to detect proteins. It was revealed that the higher value of the specific surface area of the G, the higher adsorption ability for L-alanine, and the good transfer of the electron between the Ni nanosheet/G composite and the surface of glassy carbon electrodes (GCE) had considerably improved the performances of the obtained sensor (**Table 1**). Furthermore, by means of the obtained sensor, without enzyme presence, the direct electrooxidation process of amino acids was achieved [88].

Zhang and co-workers developed functionalized graphene oxide (FGO), which exhibited high affinity to (His)-tagged acetylcholinesterase (AChE) for paraoxon manufacture, an acetylcholinesterase inhibitor, biosensors [89]. In order to estimate the functionalization of GO and their capacity to be involved as multipurpose enzyme immobilization nanomaterials for the bioelectrochemical sensor design, the AChE was nominated as being a perfect enzyme. The authors evidenced the existence of an optimum amount of Nα,Nα-bis (carboxymethyl)-L-lysine hydrate ($NTA\text{-}NH_2$) and

Electrodes	Analyte	Linear range	Sensitivity	Detection limit	Ref.
Ni nanosheet/G/GCE	L-alanine	1–7 μM 0.1–1 μM	2.15 A M^{-1} 8.34 A M^{-1}	0.1 μM	[88]
(His)-tagged (AChE)	Paraoxon	10 μM–1 mM	2.23 mA M^{-1}	$6.5 \cdot 10^{-10}$ M	[89]

GCE—glassy carbon electrodes, AChE—acetylcholinesterase.

Table 1.
Electrochemical parameters of some G-based biosensors for different analytes detection.

that Ni^{2+} (Ni-NTA) needed to attach on the GO surface. This connection has succeeded to an increased enzyme loading, which lead to an enhanced electrocatalytic activity and sensitivity. This work proved excellent stability, for both short-term and long-term, being the effect of the stable binding between Ni-NTA and His-tagged AChE. The paraoxon concentrations also influence the inhibition response, and a low detection limit value was reached (**Table 1**). The authors showed that the obtained FGO composite can be changed to multipurpose biosensor construction [89].

3.1.1 Glucose

Glucose, a monosaccharide, serves as an energy source and metabolic fuel and is involved in the processes of photosynthesis and respiration in most organisms. The high blood-glucose concentration, recognized as hyperglycemia, or the reduced glucose presence, identified as hypoglycemia, is caused by the effect of insufficiency of insulin in the body, known as Diabetes mellitus, an incurable disease. By monitoring the glucose concentration in blood, as an illness marker, it is possible to extend life expectancy. Thus, people with diabetes, people who have problems with glucose concentration in blood, can manage episodes of hypo- or hyper-glycaemia, hence providing improved control over their conditions and avoiding some of the incapacitating side effects. Moreover, by monitoring the glucose, the patient's treatment strategies can be optimized, and also, the effect of medications, physical exercise, and nutrition on the patient can be controlled.

Xue and co-workers prepared a biocompatible AuNPs/PPy/rGO nanocomposite that demonstrated exceptional electro-catalytical activity toward O_2 reduction [37]. By encapsulating glucose oxidase (GOD) *via* chitosan (N-deacetylated derivative of chitin) cross-linking in the obtained composite, AuNPs/PPy/rGO/GOD/chitosan modified electrode was fabricated. The developed electrochemical sensor revealed a linear range for glucose from 0.2 to 1.2 mM with a good sensitivity value of 123.8 mA M^{-1} cm^{-2} (**Table 2**). The obtained performances were correlated with (i) the high surface area and high electric conductivity of rGO, (ii) good shielding of PPy and fixing capacity on rGO surfaces, (iii) biocompatibility of the dispersed small AuNPs and their electron transfer promotion ability. The authors showed that the enzyme, GOD, has a good connection with AuNPs/PPy/rGO nanocomposite, which was proved by the fast electron transfer between the used enzyme and electrode. Also, between the electrode surface and the enzyme, there seems to be a rather low barrier. Furthermore, the achieved AuNPs/PPy/rGO ternary nanocomposites can be favorable for encapsulating some other biomolecules, indicating, thus, a potential substitute in constructing multipurpose bioelectrochemical sensors and additional devices [37]. Chitosan, due to its uncommon mixture of properties including the exceptional membrane-forming capability, good bond, high mechanical strength

Electrodes	Analyte	Linear range /mM	Sensitivity	Detection limit	Ref.
AuNPs/PPy/rGO/GOD/CS	Glucose	0.2–1.2	123.8 mA M^{-1} cm^{-2}	—	[37]
GCE/rGO/PTZ-O/GDH	Glucose	0.5–12	42 mA M^{-1} cm^{-2}	—	[90]
rGO/PAMAM/Ag	Glucose	0.032–1.89	75.72 mA M^{-1} cm^{-2}	4.5 μM	[40]
3D G/Co_3O_4	Glucose	Up to 0.08	3.39 A M^{-1} cm^{-2}	25 nM	[85]
Pt-CuO/rGO	Glucose	Up to 12	3577 mA M^{-1} cm^{-2}	10 nM	[35]
Cu_2O/GNs	Glucose	0.3–3.3	0.28 A M^{-1} cm^{-2}	3.3 μM	[52]
	H_2O_2	0.3–7.8	—	20.8 μM	
Au nanocubes/G	Glucose	0.1–0.8	221.0 mA M^{-1} cm^{-2}	—	[91]

GCE—glassy carbon electrodes, CS—chitosan, His—histidine, GOD—encapsulated glucose oxidase, AuNPs—gold nanoparticles, PPy—polypyrrole, GDH—glucose dehydrogenase, PTZ-O—phenothiazone, PAMAM—poly(amidoamine), GNs—G nanosheets.

Table 2.
Electrochemical parameters of some G-based modified electrodes for glucose and H_2O_2 detection.

and water permeability, and extraordinary biocompatibility, is one of the best-suited biopolymers for the surface-deposition of nanocomposite for the electrode fabrication [45, 46, 92, 93].

With the aim of encapsulating redox enzymes, rGO film with adsorbed phenothiazine represents another composite that was synthetized by Ravenna et al. [90]. The prepared composite was very efficient for the electron transfer process between flavin adenine dinucleotide (FAD)-dependent glucose dehydrogenase and the obtained modified sensor. The study showed that for the glucose oxidation process, the determined redox potential was lower than 0 V vs. Ag/AgCl reference electrode. Furthermore, the obtained rGO-based biosensor presented an increased value of sensitivity and a large linear range for glucose detection (**Table 2**). Also, the obtained biosensor achieved reasonable reproducibility and stability, a great selectivity for different interfering compounds. The results demonstrated a promising sensor for several bioelectrochemical applications [90].

Luo and co-workers showed that by combining the rGO with poly(amidoamine) and silver (rGO–PAMAM–Ag), the newly developed nanocomposite offers an exceptional microenvironment to assure the direct electron transfer process of GOD enzyme fixed on the modified surface of GCE [40]. Besides, it was confirmed that the developed arrangement can preserve a great electrocatalytic activities of the enzyme. By using GO self-assembled with PAMAM-G3.5, as a growth pattern, and microwave irradiation, rGO-PAMAM-Ag new nanocomposite electrode material was prepared. Then, based on this type of sensitive nanocomposite, a biosensor for glucose detection was manufactured. A high value of sensitivity, a low value of the detection limit, and a wide linear range, were the analytical performances of the obtained biosensor (**Table 2**). The authors also showed that the interference between the signals originating from uric acid (UC) and ascorbic acid (AA), which are regularly detected in blood fluids together with glucose, is insignificant compared to the signal attained by the glucose biosensor. The achieved characteristics recommended the obtained rGO-PAMAM-Ag nanocomposite as an innovative exceptional electrode material for the construction of glucose biosensors, involving a direct electron transfer process [40].

An interesting study is related to the involvement of an easy synthesis path for 3D G/Co_3O_4 composite production, which was further used as a sensor for glucose [85]. Thus, the synthesis process was based on the (i) Co_3O_4 nanowires achievement by *in situ* hydrothermal synthesis, and (ii) CVD growth of G foam. Through this study, it was shown that Co_3O_4 nanowires possess high crystallinity and constant diameter and form a compact nanomesh covering the 3D G matrix, which can function as a free-standing electrode due to the higher mechanical strength of G. 3D G/Co_3O_4 obtained composite was employed as electrode material and the acquired results proved significant in performance as sensor for glucose and as a supercapacitor. These results were based on the synergistic incorporation of the two, G and Co_3O_4, nanomaterials and represent an important step for developing enzyme-free ultrasensitive sensors for glucose. Co_3O_4 nanowires proved excellent electrochemical and electrocatalytic properties (**Table 2**). The authors showed that the 3D multiplexed and extremely conductive matrix provided by the defect-free G foam depicted fast electron transfer and conduction, and offered a high available active surface area. In addition, the open pores of 3D G/Co_3O_4 composite were found to be advantageous to ion diffusion and transport kinetics. Thus, the G foams showed unique morphological properties, and these could serve as 3D supports for embedding great capacity to homogeneously bind metal oxides with well-defined properties (size, shapes, and crystallinity), as demonstrated by Dong et al. [85]. Agglomeration, which represents a regular fact of metallic oxide synthesis, is no longer a problem. The multifunctional character and the improved performance of composites based on G and metal oxides were achieved thanks to the synergies between components, representing a positive aspect for developing novel applications [85].

In their work, Dhara et al. presented an easy and low-cost fabrication process with no enzyme-limited screen-printed electrodes [35]. Thus, it was possible to synthesize Pt-CuO/rGO nanocomposite electrocatalyst by a one-step chemical reduction. The electronic microscopy measurements of the composite revealed the nanocubes structure for Pt, and nanoflowers structure for CuO. The obtained sensor for glucose oxidation process exhibited exceptional sensitivity (**Table 2**) and good selectivity. The acquired linear response for glucose and the detection limit value are presented in **Table 2**. When the newly developed sensor measured the glucose quantity in blood sample, the results could be considered as being satisfactory [35].

Another study that is worth mentioning was performed by Liu et al. and was focused on glucose detection. They demonstrated that by the use of a non-enzymatic sensor composed of G nanosheet-covered Cu_2O nanocubes (Cu_2O/GNs), good sensor's characteristics for glucose oxidation could be obtained (**Table 2**) [52]. They also showed that Cu_2O/GNs hybrids can be produced by an easy technique at low temperatures, and additionally, an outstanding electrocatalytic activity for electro-reduction process of H_2O_2 can be obtained (**Table 2**). Therefore, electroanalytical detection of glucose and H_2O_2 was achieved, and the developed sensor revealed low limits of detection, good selectivity, good linear response, and a large detection range due to the enlarged electroactive surface area and the enhanced capacity for electron transfer of the fabricated G-hybrid nanostructure. Furthermore, the authors affirmed that due to the surface covered by G nanosheets, the Cu_2O/GNs composite revealed enhanced electrochemical stability related to that acquired for Cu_2O nanocubes only. The innovative Cu_2O/GNs nanocomposite led to new possible applications in different categories of biosensors, such as sensitive electrode materials, bioelectronic devices, and (electro)catalysts [52].

By a simple electrodeposition method without any template, Chu et al. synthesized *in situ* an originally designed Au nanocubes/G composite film [91]. The obtained composite film possessed exceptional performance for glucose detection under a low potential value of −0.4 V *vs.* Ag/AgCl (**Table 2**). A possible explanation for this behavior was the synergistic consequence of regular morphology and uniform isolated distribution. The authors stated that Au nanocubes/G composite film presents the possibility of being used in applications for trace analysis for new physiological activators by changing the corresponding proteins. The high catalytic activity and conductive capacities of the new composite material represent important properties that were demonstrated not only in biosensor construction but also offer the possibility of being successfully used in other studies closely related to electrocatalysis [91].

3.1.2 Hydrogen peroxide

The production of reactive oxygen species represents a process that can be associated with early signs of cancer or neurodegenerative diseases such as Alzheimer's, Parkinson's, and multiple sclerosis. Precisely, reactive oxygen species appear as a consequence of changed cellular metabolism rising from a given disease state. These species are extremely reactive and induce redox reactions of cell structures leading to activation of immune response and apoptosis. Between these, hydrogen peroxide (H_2O_2) was intensively investigated as analyte to describe disease conditions. Easier H_2O_2 detection is as well a challenge in various domains like medicine, manufacturing, food production, and pharmaceuticals. Among the variety of possible applied techniques, electrochemistry consists of the simplest way of getting rapid and accurate results while demanding only a simple device, and it is realized by analyte oxidation or reduction (**Figure 6**) [21, 94–96].

By electrochemical reduction of GO-horseradish peroxidase (GO-HRP) to rGO-horseradish peroxidase (rGO-HRP) a new biocomposite was obtained by Selvakumar et al. [97]. The enzyme immobilization process was facile, and the obtained composite material was employed for the preparation of rGO-HRP modified screen-printed carbon electrode (SPE) in order to detect H_2O_2. The obtained sensor characteristics were a wide linear range, high selectivity, and good stability, which recommended its extended practical applications. The sensitivity value toward H_2O_2 reduction is

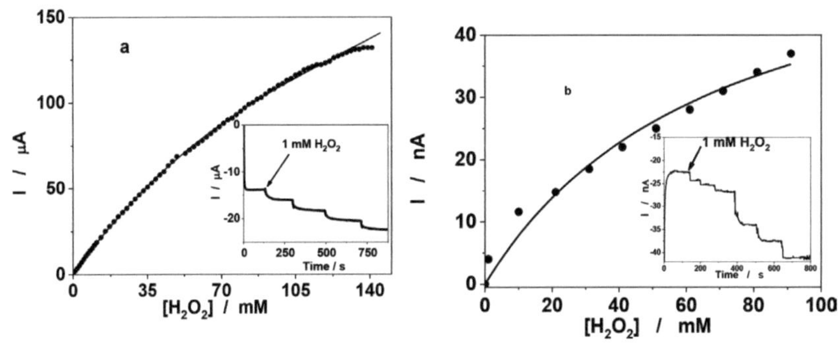

Figure 6.
I vs. [H_2O_2] calibration curve recorded at (Fe-CA)-CPE (a) and undoped CA-CPE (b). Inset: amperograms recorded at an applied potential of 0.3 V vs Ag/AgCl, KCl_{sat}, and a rotating speed of 500 rpm, in 0.1 M phosphate buffer (pH 7) at (Fe-CA)-CPE (a) and CA-CPE (b) for consecutive addition of 1 mM H_2O_2. Reproduced from Ref. [21] with permission from the Elsevier.

presented in **Table 3**. The authors suggested that the fabrication process used for obtaining of rGO-HRP biocomposites can be utilized for the construction of G-based composite materials of a large range of electrochemically essential molecules with redox properties [97].

Elsewhere, an innovative biosensor based on GCE, which was modified with the AuNPs and hemin-G nanosheets (H-GNs), was successfully fabricated by Song et al. [98]. Their work demonstrated that the newly fabricated H-GNs/AuNPs/GCE bioelectrochemical sensor exhibited an improved electrocatalytic activity for H_2O_2 reduction process, when compared with the AuNPs/GCE, H-GNs/GCE, and bare GCE. The authors proved that the improved electrocatalytic signal was the result of two properties: (i) the improved specific surface area of the electrode surface, and (ii) the great loading of the H-GNs on the modified electrode surface. Therefore, an enhanced synergistic electrocatalytic influence was revealed concerning the AuNPs and H-GNs. Enhanced electroanalytical parameters of the developed biosensor (high value of sensitivity, wide linear response range, good stability and reproducibility, fast response time, and good analyte specificity) were the result of the new sensor material properties (**Table 3**) [98].

Another interesting research was done by Zhou et al., who achieved a new H_2O_2 biosensor based on the modification of GCE using G, chitosan CS, Au, and HRP [23]. They prepared individual G sheets by the insertion of $-SO_3-$ radicals. By EDS (Energy Dispersive Spectroscopy) and TEM measurements, it was proved that the reduction and sulfonation techniques used for sensor material preparation did not produce the destruction of G morphology. As previously observed, the unaltered G structure is significant for the maintenance of the exceptional properties of G. By cyclic voltammetry techniques, the authors proved the existence of a direct electron transfer process among the fixed enzyme (HRP) and the surface of the electrode. Thus, a usual electrocatalytic reduction process of H_2O_2 occurs on the electrode surface. This work showed that in identical experimental conditions, the recorded current response of HRP/CS/GCE, Au/HRP/CS/GCE, sulfonated and reduced G/HRP/CS/GCE, and Au/sulfonated and reduced G/HRP/CS/GCE can significantly enhance

Electrodes	Analit	Linear range	Sensitivity	Detection limit	Ref.
Cu_2O/GNs	Glucose	0.3–3.3 mM	—	3.3 μM	[52]
	H_2O_2	0.3–7.8 mM	—	20.8 μM	
SPCE/rGO-HRP	H_2O_2	9–195 μM	0.09 A M^{-1} cm^{-2}	—	[97]
H-GNs/AuNPs/GCE	H_2O_2	0.3–1800 μM	2.77 A M^{-1} cm^{-2}	0.11 μM	[98]
H-GNs/GCE	H_2O_2	0.5–400 μM	—	0.2 μM	[99]
Au/G/HRP/CS/GCE	H_2O_2	5–5130 μM	—	1.7 μM	[23]
Pd/HRP/G	H_2O_2	25–3500 μM	0.092 A M^{-1} cm^{-2}	0.05 μM	[17]
Hb/AuNPs/ZnO/ rGO/GCE	H_2O_2	6–1130 μM	—	0.8 μM	[36]

GCE—glassy carbon electrodes, CS—chitosan, His—histidine, AChE—acetylcholinesterase, GOD—encapsulated glucose oxidase, AuNPs—gold nanoparticles, PPy—polypyrrole, GDH—glucose dehydrogenase, PTZ-O—phenothiazone, PAMAM—poly(amidoamine), GNs—G nanosheets, HRP—horseradish peroxidase, SPCE—screen-printed carbon electrode, H—hemin, Hb—hemoglobin.

Table 3.
Electroanalytical parameters of some G-based electrodes for H_2O_2 detection.

the sensitivity of the developed biosensor due to the G structures presence (**Table 3**). It was also stated that a wide linear range, low detection limit, and long-term stability represent other excellent characteristics of this type of sensor [23].

The easy, low-priced, and highly sensitive and selective amperometric assessments employed for H_2O_2 and glucose detection, based on hemin functionalized G nanosheets (H-GNs), were developed by Guo et al. [99]. The obtained H-GNs hybrid nanomaterial combines the G nanosheet properties of high electrical conductivity and high surface area value with hemin properties (i.e., exceptional electrocatalysis and synthetic enzyme simulation). The study showed that the H-GNs are able to compete with the natural enzymes. The advantages of the obtained H-GNs are their facile synthesis method, the strength of materials, and the stability of materials in irregular conditions. Moreover, due to their excellent recorded results (**Table 3**), the obtained biosensor presents favorable possible applications in different fields, including clinical diagnostics, biotechnology, and chemical or pharmaceutical industry [99].

Another electrochemical biosensor for H_2O_2 selective detection was established by Nandini et al. [17]. The preparation method of this new electrode was realized by a co-deposition process of palladium and HRP, on the functionalized G-modified graphite surface electrode. The fabricated biosensor revealed an increased electrocatalytic activity concerning the reduction process of H_2O_2 at 0.02 V *vs* SCE. It was found that Pd, by its presence, decreases the over-potential of H_2O_2 reduction and grows the active surface area of the modified electrode. Furthermore, the HRP and Pd co-deposition escapes the poisoning occurrence of the modified graphite-based electrode. The sensor performance exhibited a fast response within less than 2 s and exceptional linear concentration range under determinate optimal experimental conditions (**Table 3**). For the acquired properties of the biosensor (i.e., selectivity, repeatability, feasibility and stability), acceptable results were obtained. The presented technique could be applied to produce biosensor for a wide range of applications [17].

Based on the hemoglobin (Hb) immobilized on an rGO, flower-like ZnO, and AuNPs nanocomposite modified GCE (AuNPs/ZnO/rGO/GCE), a new amperometric H_2O_2 biosensor was proposed by Xie and co-workers [36]. Each biosensor component was found to have an important contribution to the H_2O_2 reduction process. Therefore, the ZnO flower-like nanoparticles exhibit good biocompatibility and conductivity. Then, the ZnO-aminopropyl triethylene silane (APS)-AuNPs composite was found to have good uniformity and be suitable for protein attachment. rGO possesses high specific surface area and can increase the ZnO-APS-AuNPs composite conductivity and its mechanical resistance. The authors showed that by combining the advantages of each component used for bioelectrical sensor preparation (i.e., nanosized ZnO, AuNPs, and rGO), the Hb/AuNPs/ZnO/rGO/GCE), the obtained modified electrode can lead to acceptable sensors performances of high sensitivity, satisfactory construction reproducibility, and good storage stability. Their study has shown that AuNPs/ZnO/rGO/GCE amperometric third-generation biosensor offers an advantageous application for nanoparticles-based electrode materials in order to be employed in the investigation of direct electron transfer of proteins and the improvement of biosensors (**Table 3**) [36].

It is also important to reveal that the study performed by Liu et al., which focused on glucose detection and H_2O_2 reduction, demonstrated that with the non-enzymatic electrochemical sensor fabricated from G nanosheet-wrapped Cu_2O nanocubes (Cu_2O/GNs), good sensor characteristics for glucose oxidation can be obtained [52]. Also, they proved that Cu_2O/GNs hybrids can be produced by an easy process at low temperature and that they exhibit excellent electrocatalytic activity toward

H_2O_2 reduction. Therefore, they tested the electrochemical detection capacity of glucose and H_2O_2 (**Table 3**). The developed sensor revealed good performances. An explanation for this behavior can be the enhanced electrocatalytic surface area of the electrode and the high conductivity of the G-hybrid nanostructure, reflected in the improved electron transfer ability through this matrix. Additionally, the authors stated that their synthesized Cu_2O/GNs structure revealed enhanced electrochemical stability in contrast with the Cu_2O nanocubes alone due to the G nanosheets that covered the oxide nanocubes. The advanced Cu_2O/GNs nanocomposite material opens up new opportunities for applications in several varieties of biosensors, bioelectronic devices, and catalysts [52].

Some advances in *in vivo* electrochemical analysis of H_2O_2 were reported by Deng and co-workers. Their interest was directed toward the electrochemical redox process of H_2O_2 at the electrode surface and on the diversity of catalysts (synthetic electrocatalysts or biomolecular electrocatalysts) used for improved electrochemical analysis of H_2O_2. A higher selectivity and sensitivity of the proposed biosensors were followed. The new technique, the photoelectrochemical (PEC) method for H_2O_2 detection was also discussed. The developments in the high selectivity analysis of H_2O_2 at the cellular and *in vivo* levels was highlighted [30]. Another review focused on the G as electrode materials for (bio)sensors applications was written by Shao et al. [32], who demonstrated the performances for hydrogen peroxide, NADH, dopamine, DNA, or heavy metals detection.

An important work dedicated to the significance of flexible sensors on H_2O_2 detection was realized by Giaretta et al. [87]. They compared the electroanalytical parameters for H_2O_2 detection obtained on different biosensors based on various material substrates. It was concluded that the carbonaceous materials are better in comparison with the polymeric substrates, which are the cheapest. The carbon materials-based biosensors results were based on the corroborated effect of the increased electrical conductivity, increased permeability, and increased porosity, respectively.

Different types of materials (noble metals, metal oxides, polymers, carbon materials, and other two-dimensional materials) employed in sensor development for H_2O_2 detection were presented in the review work of Yu and co-workers [100]. Additionally, their work presents the challenges and future prospects in the biological applications of electrochemical sensors for H_2O_2 detection.

3.1.3 Neurotransmitters and neurochemicals

Neurotransmitters are endogenous compounds that permit the transmission of nerve impulses between two neurons or between neuron and effector, named 'target' cell. Thus, the nervous system is based on the role of chemical couriers of the neurotransmitters, which transfer the information across the synapses by excitation or inhibition of the subsequent neural or effector cell. Neurotransmitters are ordered into monoamines (histamine, adrenaline, dopamine (DA), noradrenaline, serotonin, and melatonin), amino acids (aspartate, D-serine, glutamate, gamma-aminobutyric acid, and aminoacetic acid (glycine)), peptides (somatostatin, cocaine, and opioid), and other (including acetylcholine, adenosine, anandamide, and nitrogen monoxide) [101].

3.1.3.1 Dopamine, ascorbic acid, uric acid

It is well-known that monoamine neurotransmitters have in their structure one amino group connected by a chain of two atoms of carbon ($-CH_2-CH_2-$) with an aromatic ring. These types of neurotransmitters have a significant importance in secreting and producing neurotrophins via astrocytic glial cell. It is an essential local cellular source of trophic support, present both in healthy and unhealthy brain. Growth of neutrophins stimulates the survival of neurons and is known as neurotrophic factors. Based on the neuron behavior, the neurotransmitters act antagonistically, namely neurotransmitters that play an inhibitory role to relax the brain and those that play an excitatory role in stimulating the brain [102]. Dopamine (DA) owns both excitatory and inhibitory classification, being thus a unique neurotransmitter. DA vital roles lie in adjustable attention, motor control, cognition, executive functions, pleasure, motivation, arousal, reinforcement, reward, and hormonal processes. Also, dopamine is extensively dispersed in the main systems of the human body, such as central nervous, renal, hormonal, and cardiovascular.

Neurological problems in the human body can be due to the anomaly in the amount of dopamine. Thus, the illness such as Parkinson's disease (degenerative disorder of the central nervous system that mainly affects the motor system), restless leg syndrome (like Parkinson's disease, this is another long-term disease that manifests itself through the uncontrolled movement of the legs), attention deficit hyperactivity disorder (ADHD) (characterized by lack of ability to concentrate, hyperactivity and impulsivity), schizophrenia (a disease in the psychiatric spectrum in which episodes of psychosis occur), and infection with human immunodeficiency viruses (HIV) (a virus that over time can cause AIDS—the gradual collapse of the immune system leading to the onset of opportunistic infections and cancers) are strongly associated with a low level of dopamine. Also dopamine is also greatly correlated with the reward mechanism in the brain. On the other hand, the consumption of prohibited drugs or substance abuse leads to an increase in dopamine levels. Thus, the forbidden substances such as heroin or cocaine, and not compulsory substances such as nicotine or alcohol, block the DA carrying that inhibits the reuptake of dopamine. As was shown before, dopamine, a neurotransmitter, which is vital for message transfer functions, produces, in this case, an amplified risk of depression and drug dependence. Thus, the discovery of a reliable analytical technique is significant and necessary in order to estimate the disease evolution.

A simple and "green" technique to produce G flowers that were exploited to modify carbon fiber electrode (CFE) in order to detect AA, DA, and UC was employed by Du et al. [34]. The G flowers, well deposited on the surface of CFE, detect separately electroactive compounds as AA, DA and UC. Besides, this study proved that the obtained G flowers based modified electrodes, can detect simultaneously AA, DA and UC, with distinct signals from each other. Thus, great electrocatalytic activities of the electrode material toward the oxidation process of electroactive AA, DA and UC, good selectivity and sensitivity, were demonstrated (**Table 4**). Excellent performance was obtained with the modified electrode when the detection of real samples was proposed. The obtained results offer appreciable evidence for the use of G as a sensitive modifier material electrode in order to detect electroactive biomolecules [34].

Using a solution-gated G transistor (SGGT) with a G gate electrode, a very sensitive dopamine electrochemical sensor was achieved by Zhang et al. [103]. The detecting mechanism of the obtained system was ascribed to the electrooxidation process of DA at the gate electrode. This electrode changed the potential distribution at the boundaries between the G gate electrode and the G channel. A perfluorinated membrane with ionic properties and excellent selectivity for dopamine was obtained after the addition of a thin layer of Nafion to G gate electrode. The obtained sensor showed a limit of detection for dopamine down to 1 nM. This represents a worthy result for

Electrodes	Analyte	Linear range	Sensitivity	Detection limit	Ref.
GEF/CFE	AA	73.52–2305.53 μM	0.011 A M^{-1}	73.52 μM	[36]
	DA	1.36–125.69 μM	1.02 A M^{-1}	1.36 μM	
	UA	3.98–371.49 μM	0.074 A M^{-1}	3.98 μM	
SGGT	DA	10 nM–1 μM	—	1 nM	[103]
	AA	1 μM–100 μM	—	1 μM	
	UA	10 μM–100 μM	—	10 μM	
NG	DA	0.1–0.45·mM	0.044 A M^{-1}	0.93 μM	[53]
MIPs-G/CS	DA	1–100 nM 0.1–100 μM	188 A M^{-1} 0.94 A M^{-1}	10 pM	[44]
ZnO NWAs/3D-GF	UA	0.5–40 μM	4.25 A M^{-1}	1 nM	[51]
	DA	0.5–40 μM	4.81 A M^{-1}	1 nM	
	AA	0.5–80 μM	0.38 A M^{-1}	—	

CS—chitosan, AA—ascorbic acid, DA—dopamine, UA—uric acid, GEF/CFE—G flowers/carbon fiber electrode, SGGT—solution-gated G transistor, NG—nitrogen-doped G, MIP—molecularly imprinted polymer, ZnO NWAs/3D-GF—nanowire arrays fabricated on 3D G foam.

Table 4.
Electroanalytical parameters of some G-based electrodes for ascorbic acid, uric acid, and dopamine detection.

dopamine detection in medical uses. The dopamine interference measurements with AA and UA showed good selectivity of the device, proved by a recorded signal up to four orders of magnitude lower when compared with that obtained for dopamine. Since the channel and the gate of the device are both made of G, they have the advantage that they can be produced on different substrate materials (containing flexible, elastic, pliable, and stretchy ones) at low temperatures by suitable procedures. Centered on the equivalent mechanisms, various additional types of biosensors could be imagined and designed in the following years, and the whole-G SGGT represents a good solution for one-use, bendable, and very sensitive biosensors [103].

Li and co-workers demonstrated that nitrogen-doped rGO (N-rGO) with a very porous matrix and adjustable structure can be produced in three steps: first, the molecular functionalization; second, the fast thermal expansion–exfoliation, and third, the covalent binding [53]. The nitrogen-doped configurations, controlled by varying the temperature of expansion–exfoliation process, were used for the preparation of screen-printed electrodes. Judging by the potential value of the peak current recorded to the oxidation process of AA, DA, and UC among the nitrogen-doped sample, the pyrrolic-N revealed the maximum electrocatalytic activity. Nevertheless, the corresponding peak currents, of the oxidation of AA, DA, and UC, are related to the corroborated effect of the nitrogen-doped sample distribution, and structural properties (specific surface area and porosity) and the electroanalytical parameters presented in **Table 4**. The prepared SPEs exhibited high peak currents (for biomolecules' electrooxidation process), good selectivity (good peak separation), and sensibility for the detection of AA, DA, and UC from a blend [53]. A molecularly imprinted polymers (MIPs) based on Chi–G composite, as the functional matrix, was used by Liu et al. to develop a sensor for DA electrochemical detection [44]. Thus, the obtained MIPs-GR composite was used to modify GCE, in order to fabricate the sensor (MIPs-G/GCE). The improved sensitivity and low value of the detection limit of MIPs-G/GCE sensor for DA oxidation (**Table 4**) can be explained based on the special characteristics of G. The selectivity, stability, and reproducibility remained just as for the MIPs sensor. It was stated that the achieved information could offer a possible rapid and reliable method for DA determination in biological samples [44].

Another significant G-based sensor for dopamine was produced by Yue et al. [51]. Thus, the detection of three biomolecules (UA, DA, and AA) was realized by employing 3D G foam containing ZnO nanowires, vertically arranged at the electrode surface. Differential pulse voltammetry (DPV) technique was used for UA, DA, and AA electrochemical detection. The new structural design combined (i) the large mesoporous surface area 3D G structures that facilitated easy diffusion of ions through the electrode material, with (ii) the increased conductivity of 3D G foam, which led to a good electron transfer process, and (iii) the active sites of ZnO nanowires, which assure a high selectivity. The present study proved that the UA, DA, and AA selectivity was the result of thermal annealing of ZnO surface.

A high selectivity and a low value of detection limit for UA and DA were obtained with the optimized ZnO nanowire/3D G foam electrochemical sensor (**Table 4**). The obtained results were clarified by the gap variance among the LUMO (lowest unoccupied molecular orbitals) and HOMO (highest occupied molecular orbitals) of a biological molecule for a set of specified electrodes. For the Parkinson's disease test, the UA level was 25% lower than in healthy individuals. It was concluded that the reported work can open new perspectives for UA application as a biomarker for Parkinson's disease, which can offer better medical diagnostic control in addition to the possibility of tracking the disease [51].

3.1.3.2 Serotonin

Several chemically different types of G nanosheets were synthesized by Kim et al. with the aim of using them as electrocatalysts for serotonin (5-hydroxytryptamine, 5-HT) [24]. In order to estimate the G nanosheets surface morphologies, X-ray photoelectron spectroscopy (XPS) and field emission scanning electron microscopy (FE-SEM) techniques were used. By electrochemical impedance spectroscopy (EIS) technique the electrocatalytic activity was investigated. In this study, the three different arrangements of rGO obtained did not reveal some significant changes in the obtained XPS spectra and FE-SEM images but showed dissimilar electrochemical performance. Thus, EIS recorded spectra exhibited a dissimilarity in the electron transfer resistance. This result is in good agreement with the reducing agent. The obtained parameters for 5-HT determination and the acquired EIS results were in agreement as well. The rGO-based GCE sensor obtained for 5-HT detection exhibited high sensitivity, good selectivity, and smaller electron transfer resistance compared with previously obtained sensors. Between the evaluated G-modified GCEs, the best electrochemical sensor properties (i.e., lowermost detection limit, uppermost sensitivity and selectivity, broadest linear range, fastest response time, alongside the greatest defined peak of 5-HT) were obtained for rGO reduced using hydrazine and ammonia solution (**Table 5**). This study proved an irreversible diffusion-controlled electrode process for a 5-HT electrooxidation reaction mechanism [24].

A two-layered membrane sensing interface for serotonin detection was constructed by Xue et al. [37]. The production of the obtained device was based on the use of nanosized rGO/polyaniline (PANI) composites and molecularly imprinted polymers (MIPs) surrounded with AuNPs (AuNPs@MIPs). With the aim of obtaining a good sensitivity and selectivity of the elaborated device, the rGO/PANI nanocomposites were produced by the electrodeposition method. First, the protonated anilines were anchored by electrostatic adsorption on the rGO sheets. Afterward, the rGO/PANI nanocomposite film was made by cyclic voltammetry process on the surface of bare GCE. Over the nanosized rGO/PANI composites membrane, the AuNPs@MIPs were deposited. Interestingly, the obtained material interface showed improved properties for (i) selectivity to 5-HT, (ii) electric conductivity, and (iii) electrocatalytic activity (**Table 5**). The new AuNPs/PPy/RGO/GOD/chitosan-modified electrode was efficaciously applied toward 5-HT detection in human serum specimens. In the meantime, the interferences produced from AA, DA, UA, and epinephrine (EP) did not affect the 5-HT detection. Therefore, the approach was recommended by this research team for a sensitive and selective detection of targeted biomolecules in real sample [37].

3.1.3.3 Epinephrine

Adrenaline or epinephrine, manufactured by the suprarenal glands and certain neurons, is a hormone neurotransmitter. It is also a medication and fulfills a pivotal function in the fight-or-flight reaction. Thus, under its action, the blood circulation to muscle tissues, cardiac output, pupil dilation, and blood sugar increase. This happens because of the adrenaline binding to alpha and beta receptors. Due to its importance, many research works were focused on this subject [18, 25].

In order to prepare a sensor for epinephrine (EP), Cui et al. started from a method based on chemical reduction of both Au (III) and GO, to prepare rGO/Au nanocomposites [25]. Then, the prepared rGO/Au nanocomposites, used as modifier-sensitive material for GCE surface (rGO/Au/GCE) were found to confirm an improved electrochemical activity toward EP (**Table 5**). The role of gold nanoparticles (nano-Au) included in rGO matrix was that of a spacer in order to avoid the rGO sheets aggregation. The achieved rGO/Au/GCE proved high sensitivity for the detection of EP. The recorded CV in the presence of ascorbic acid (AA), which show total peak separation between EP and AA, demonstrated good electrochemical sensor selectivity. Furthermore, rGO/Au/GCE revealed exceptional electron transfer capacity

Electrodes	Analyte	Linear range	Detection limit	Ref.
rGO/PANI	Serotonin	0.2–10.0 M	11.7 nM	[37]
GO-S-$(CH_2)_4$-SH/GCE		1–100 μM	16 μM	[24]
GO-SH/GCE			0.38 nM	
rGO_3/GCE			0.52 nM	
GO/GCE			18 μM	
G/Au/GCE	Epinephrine	0.05–8.0 μM	7 nM	[25]
rGO/Thi/AuNPs	CEA	5.55×10^{-5}–1.66 nM	3.61 fM	[38]
rGO/PB/AuNPs	AFP	1.48×10^{-4}–4.44 nM	13.11 fM	
G/MBs–Ab1/CEA/Ab2–AuNPs–HRP	CEA	27.7–33.3 pM	27.7 pM	[39]

GCE—glassy carbon electrodes, GO—graphene oxide, rGO—reduced graphene oxide, AuNPs—gold nanoparticles, PANI—polyaniline, CEA—carcinoembryonic antigen, MBs—magnetic beads, Ab—antibody, AFP—α-fetoprotein, PB—Prussian Blue, Thi—thionine.

Table 5.
Electroanalytical parameters of some G-based electrodes for different types of analytes.

and exceptional electrocatalytic ability to additional biomolecules, including DA, AA, NADH, and pyrocatechol, opening up, thus, new perspective for applications improvement of rGO/Au nanocomposites in (bio)sensors [25].

3.1.4 Cancer and disease biomarkers

Cancer biomarker detection represents one of the most important achievements in the field of biosensor production. An interesting review work was done by Alsharabi et al., who highlighted the capacity of G and its various derivatives to conjugate their unique chemical structure and characteristic optical, electronic, mechanical, and thermal properties. These special materials may represent an excellent sensing platform for cancer biomarkers detection, representing a possible response to the important challenge in the diagnosis of cancer at the initial stages [29]. The G-based electrochemical biosensors for cancer biomarker detection were fabricated by using G surface that was improved with magnetic beads and enzyme-marked antibody-AuNPs [39]. One should emphasize that the development of an electrochemical biosensor for cancer biomarker determination remains essential for early discovery and diagnosis. In addition, the electrochemical sensor presents the advantages of improved characteristics (i.e., rapid, precise, and sensitive) in comparison with other investigation techniques.

In order to prepare a new biosensor for cancer-related biomarker, Jin and co-workers used a G platform that was prepared by CVD. Then, the MBs and enzyme-marked antibody-AuNP bioconjugate were added [39]. The attachment of MBs, covered with capture antibodies (Ab1), to the G sheet surface, was realized by applying an external magnetic field, with the aim to avoid the reduced G matrix conductivity. With the aim of increasing the sensitivity of the multi-nanomaterial-built biosensor, the AuNPs were improved with HRP and detection antibody (Ab2), forming the conjugate Ab2–AuNPs–HRP. In this electrode arrangement, the fast electron transfer between the multi-nanomaterial present on the electrode surface and analyte target was realized, and a low value for carcinoembryonic antigen detection limit was reached (**Table 5**). The acquired results evidenced a fast response and recovery time, which was more improved compared to that achieved when the old-style approaches were used. The good sensors achieved properties (sensitivity, specificity, simplicity of construction technique, ease of use, fast analysis, and reusability) confirm that the developed biosensor can be used for cancer's medical diagnosis [39].

A label-free electrochemical multiplexed immunosensor based on G nanocomposites was fabricated for the recognition of both carcinoembryonic antigen (CEA) and α-fetoprotein (AFP) by Jia et al. [38]. The indium tin oxide (ITO) sheets were used to be modified with anti-AFP fixed on G nanocomposite matrix. The electrode fabrication and voltametric sensing technique were centered on the electron transfer delay determined by the engineered antibody-antigen immunocomplex present on the ITO electrode surface. The obtained multiplexed immunosensor facilitated the concomitant detection of both CEA and AFP, and the found linear ranges are presented in **Table 5**. The limit of detection value for CEA and for AFP, are depicted in **Table 5**. A few aspects were pointed out as follows: (i) the fabricated immunosensor escaped the marking of both antigens or antibodies, making it easier and preventing the cross-talk among diverse analytes; (ii) G nanocomposites used as supporting scaffold were synthesized by a facile route, and the shapes and quantity of the immobilized AuNPs, by this method, could be easily controlled; (iii) the immunoassay having a worthy stability, large linear ranges received a good correlation with ELISA (enzyme-linked immunosorbent assay) and could be used in medical diagnosis. It was also stated that this simple approach could be adapted and combined for new biosensor applications [38].

4. Conclusions

Since its discovery, G has been tested in a large diversity of biosensing applications owing to its remarkable electrical, mechanical, and optical properties as well as its unique structure. If one compares G-based biosensors with conventional ones, clear benefits, such as high sensitivity and selectivity, low detection limit, reproducibility, stability, fast response, or easy miniaturization, make G-based biosensors a real candidate for a novel and efficient class of biosensors for medical applications.

Electrochemical biosensors offer an inexpensive, facile, fast, sensitive, and selective detection of biomolecules, and the use of G and its composites for biosensors development is due to its unique features combined with some peculiar properties. Moreover, the reusability of the biosensors is aimed. The present chapter summarized the G-based electrochemical sensors developed for sensing biomolecules and highlighted their significant advances. Thus, G and GO present a wide range of electrochemical potential and fast electron transfer rate. The drawback of using G in biosensors construction is its possible toxicity, as stated in the international nanotechnology guidelines. On the other hand, G can be easily functionalized/modified by electrodeposition, polymerization, electrochemical doping, or other methods, and even if G could be cytotoxic, biomedical applications can benefit from non-toxic, biocompatible and water-dispersible G layers obtained by chemical functionalization with different ligands. By biosensors incorporation into strong, transportable, and miniaturized devices, the detection of biomolecules and toxins for usability in clinical and diagnostic fields can be achieved.

Although there is a massive investment both in academia and industry, G-based biosensors are still at an incipient level, and commercial biosensors are yet to come. Nevertheless, there is a slow but promising translation into medical applications.

Acknowledgements

This work was supported by a grant from the Ministry of Research, Innovation, and Digitization, CNCS/CCCDI—UEFISCDI, project number PN-III-P2-2.1-PED-2021-3156, within PNCDI III.

List of abbreviations

AA	ascorbic acid
Ab	antibody
AChE	acetylcholinesterase
ADHD	attention deficit hyperactivity disorder
AFP	α-fetoprotein
APS	aminopropyl triethylene silane

CDH	cellobiose dehydrogenase
CEA	carcinoembryonic antigen
CS	chitosan
CVs	cyclic voltammograms
CVD	carbon vapor deposition
DA	dopamine
DC	direct current
DNA	deoxyribonucleic acid
DPV	differential pulse voltammetry
EDS	Energy Dispersive Spectroscopy
EIS	electrochemical impedance spectroscopy
ELISA	enzyme-linked immunosorbent assay
EP	epinephrine
FAD	flavin adenine dinucleotide
FE-SEM	field emission scanning electron microscopy
FGO	functionalized graphene oxide
G	graphene
GCE	glassy carbon electrodes
GDH	glucose dehydrogenase
GEF/CFE	G flowers/carbon fiber electrode
GO	graphene oxide
rGO	reduced graphene oxide
GOD	glucose oxidase
GNs	G nanosheets
H	hemin
Hb	hemoglobin
His	histidine
HIV	human immunodeficiency viruses
HOMO	highest occupied molecular orbitals
HRP	horseradish peroxidase
HR-TEM	High Resolution – Transmission Electron Microscopy
H-GNs	hemin-G nanosheets
ITO	indium tin oxide
LUMO	lowest unoccupied molecular orbitals
MBs	magnetic beads
MIPs	molecularly imprinted polymers
NADH	nicotinamide adenine dinucleotide reduced
NG	nitrogen-doped G
NPs	nanoparticles
NWAs/3D-GF	nanowire arrays fabricated on 3D G foam
PAMAM	poly(amidoamine)
PANI	polyaniline
PB	Prussian Blue
PEC	photoelectrochemical
PPy	polypyrrole
PTZ-O	phenothiazone
SEM	Scanning Electron Microscopy
SPE	screen-printed electrode
SPCE	screen-printed carbon electrode
SGGT	solution-gated G transistor

UC	uric acid
Thi	thionine
TEM	Transmission Electron Microscopy
XPS	X-ray photoelectron spectroscopy
5-HT	5-hydroxytryptamine-serotonin

Author details

Carmen Ioana Fort[1,2*], Liviu Cosmin Cotet[1,2], Lucian Cristian Pop[1,2,3], Monica Baia[2,3,4] and Lucian Baia[2,3,4*]

1 Faculty of Chemistry and Chemical Engineering, "Babes-Bolyai" University, Cluj-Napoca, Romania

2 Institute for Research-Development-Innovation in Applied Natural Sciences, "Babes-Bolyai" University, Cluj-Napoca, Romania

3 Interdisciplinary Research Institute on Bio-Nano-Sciences, "Babes-Bolyai" University, Cluj-Napoca, Romania

4 Faculty of Physics, "Babes-Bolyai" University, Cluj-Napoca, Romania

*Address all correspondence to: ioana.fort@ubbcluj.ro and lucian.baia@ubbcluj.ro

References

[1] Lawal AT. Progress in utilisation of graphene for electrochemical biosensors. Biosensors and Bioelectronics. 2018;**106**:149-178. DOI: 10.1016/j.bios.2018.01.030

[2] Karimi-Maleh H, Orooji Y, Karimi F, Alizadeh M, Baghayeri M, Rouhi J, et al. A critical review on the use of potentiometric based biosensors for biomarkers detection. Biosensors and Bioelectronics. 2021;**184**(15):113252. DOI: 10.1016/j.bios.2021.113252

[3] Mohankumar P, Ajayan J, Mohanraj T, Yasodharan R. Recent developments in biosensors for healthcare and biomedical applications: A review. Measurement. 2021;**167**(1):108293. DOI: 10.1016/j.measurement.2020.108293

[4] Goyal D, Mittal SK, Choudhary A, Dang RK. Graphene: A two dimensional super material for sensor applications. Materials Today Proceedings. 2021;**43**(1):203-208. DOI: 10.1016/j.matpr.2020.11.637

[5] Wang Z, Yu J, Gui R, Jin H, Xia Y. Carbon nanomaterials-based electrochemical aptasensors. Biosensors and Bioelectronics. 2016;**79**:136-149. DOI: 10.1016/j.bios.2015.11.093

[6] Fort CI, Pop LC. Heavy metal and metalloid electrochemical detection by composite nanostructures. In: Advanced Nanostructures for Environmental Health. Amsterdam, Netherlands: Elsevier; 2020. pp. 185-250. DOI: 10.1016/B978-0-12-815882-1.00005-7

[7] Ikebukuro K, Kiyohara C, Sode K. Novel electrochemical sensor system for protein using the aptamers in sandwich manner. Biosensors and Bioelectronics. 2005;**20**:2168-2172. DOI: 10.1016/j.bios.2004.09.002

[8] Wang M, Zhai S, Ye Z, He L, Peng D, Feng X, et al. An electrochemical aptasensor based on a TiO_2/three-dimensional reduced graphene oxide/PPy nanocomposite for the sensitive detection of lysozyme. Dalton Transactions. 2015;**44**:6473-6479. DOI: 10.1039/C5DT00168D

[9] Kumar R, Singh R, Hui D, Feo L, Fraternali F. Graphene as biomedical sensing element: State of art review and potential engineering applications. Composites Part B: Engineering. 2018;**134**:193-206. DOI: 10.1016/j.compositesb.2017.09.049

[10] Quach QH, Jung J, Kim H, Chung BH. A simple, fast and highly sensitive assay for the detection of telomerase activity. Chemical Communications. 2013;**49**:6596-6598. DOI: 10.1039/C3CC42571A

[11] Zhang Z, Sharon E, Freeman R, Liu X, Willner I. Fluorescence detection of DNA, Adenosine-5′-triphosphate (ATP), and telomerase activity by zinc(II)-protoporphyrin IX/G-quadruplex labels. Analytical Chemistry. 2012;**84**:4789-4797. DOI: 10.1021/ac300348v

[12] Farrokhnia M, Karimi S, Momeni S, Khalililaghab S. Colorimetric sensor assay for detection of hydrogen peroxide using green synthesis of silver chloride nanoparticles: Experimental and theoretical evidence. Sensors and Actuators B: Chemical. 2017;**246**:979-987. DOI: 10.1016/j.snb.2017.02.066

[13] Wolfbeis OS, Durkop A, Wu M, Lin Z. A europium-ion-based

luminescent sensing probe for hydrogen peroxide. Angewandte Chemie International Edition. 2002;**41**:4495-4498. DOI: 10.1002/1521-3773(20021202)41:23<4495::AID-ANIE4495>3.0.CO;2-I

[14] Wang L-J, Zhang Y, Zhang C-Y. Ultrasensitive detection of telomerase activity at the single-cell level. Analytical Chemistry. 2013;**85**:11509-11517. DOI: 10.1021/ac402747r

[15] Kurbanoglu S, Ozkan SA. Electrochemical carbon based nanosensors: A promising tool in pharmaceutical and biomedical analysis. Journal of Pharmaceutical and Biomedical Analysis. 2018;**147**:439-457. DOI: 10.1016/j.jpba.2017.06.062

[16] Khan A, DeVoe E, Andreescu S. Carbon-based electrochemical biosensors as diagnostic platforms for connected decentralized healthcare. Sensors and Diagnostics. 2023;**2**:529. DOI: 10.1039/d2sd00226d

[17] Nandini S, Nalini S, Manjunatha R, Shanmugam S, Melo JS, Suresh GS. Electrochemical biosensor for the selective determination of hydrogen peroxide based on the co-deposition of palladium, horseradish peroxidase on functionalized-graphene modified graphite electrode as composite. Journal of Electroanalytical Chemistry. 2013;**689**:233-242. DOI: 10.1016/j.jelechem.2012.11.004

[18] Silai IE, Fort IC, Casoni D, Turdean GL. Epinephrine detection at Pt-nanoparticles modified graphite electrode by square-wave voltammetry. Revue Roumaine de Chimie. 2015;**60**:689-696

[19] Fort CI, Cotet LC, Vasiliu F, Marginean P, Danciu V, Popescu IC. Methanol oxidation at carbon paste electrodes modified with (Pt–Ru)/carbon aerogels nanocomposites. Materials Chemistry and Physics. 2016;**172**:179-188. DOI: 10.1016/j.matchemphys.2016.01.061

[20] Fort CI, Ortiz R, Cotet LC, Danciu V, Popescu IC, Gorton L. Carbon aerogel as electrode material for improved direct electron transfer in biosensors incorporating cellobiose dehydrogenase. Electroanalysis. 2016;**28**:2311-2319. DOI: 10.1002/elan.201600219

[21] Fort CI, Cotet LC, Danciu V, Turdean GL, Popescu IC. Iron doped carbon aerogel – New electrode material for electrocatalytic reduction of H_2O_2. Materials Chemistry and Physics. 2013;**138**:893-898. DOI: 10.1016/j.matchemphys.2012.12.079

[22] Forţ CI, Coteţ LC, Turdean GL, Danciu V. Meldola blue immobilised on mesoporous carbon aerogel— New electrode material for NADH electrocatalytic oxidation. Studia Universitatis Babes-Bolyai Chemia. 2015;**60**(3):215-224

[23] Zhou K, Zhu Y, Yang X, Luo J, Li C, Luan S. A novel hydrogen peroxide biosensor based on Au–graphene–HRP–chitosan biocomposites. Electrochimica Acta. 2010;**55**(9):3055-3060. DOI: 10.1016/j.electacta.2010.01.035

[24] Kim SK, Kim D, Jeon S. Electrochemical determination of serotonin on glassy carbon electrode modified with various graphene nanomaterials. Sensors and Actuators B: Chemical. 2012;**174**:285-291. DOI: 10.1016/j.snb.2012.08.034

[25] Cui F, Zhang X. Electrochemical sensor for epinephrine based on a glassy carbon electrode modified with graphene/gold nanocomposites. Journal of Electroanalytical Chemistry.

2012;**669**:35-41. DOI: 10.1016/j.jelechem.2012.01.021

[26] Forț CI, Popescu IC. NADH oxidation at Meldola Blue modified glassy carbon electrodes. A comparative study. Studia Universitatis Babes-Bolyai Chemia. 2011;**4**:255-264

[27] Ladiu CI, Garcia R, Popescu IC, Gorton L. NADH electrocatalytic oxidation at glassy carbon paste electrodes modified with Meldola blue adsorbed on acidic alpha-titanium phosphate. Revista de Chimie. 2007;**58**:465-469

[28] Ladiu CI, Garcia R, Popescu IC, Gorton L. NADH electrocatalytic oxidation at glassy carbon paste electrodes modified with Meldola blue adsorbed on acidic alpha-zirconium phosphate. Revue Roumaine de Chimie. 2007;**52**:67-74

[29] Alsharabi RM, Rai S, Mohammed HY, Farea MA, Srinivasan S, Saxena PS, et al. A comprehensive review on graphene-based materials as biosensors for cancer detection. Oxford Open Materials Science. 2023;**3**(1):itac013. DOI: 10.1093/oxfmat/itac013

[30] Deng Z, Zhao L, Zhou H, Xu X, Zheng W. Recent advances in electrochemical analysis of hydrogen peroxide towards in vivo detection. Process Biochemistry. 2022;**115**:57-69. DOI: 10.1016/j.procbio.2022.01.025

[31] Goodrum R, Weldekidan H, Li H, Mohanty AK, Misra M. Graphene-based nanostructures from green processes and their applications in biomedical sensors. Advanced Industrial and Engineering Polymer Research. DOI: 10.1016/j.aiepr.2023.03.001

[32] Shao Y, Wang J, Wu H, Liu J, Aksay IA, Lina Y. Graphene based electrochemical sensors and biosensors: A review. Electroanalysis. 2010;**22**:1027-1036. DOI: 10.1002/elan.200900571

[33] Wang C, Yokota T, Someya T. Natural biopolymer-based biocompatible conductors for stretchable bioelectronics. Chemical Reviews. 2021;**121**:2109-2146. DOI: 10.1021/acs.chemrev.0c00897

[34] Du J, Yue R, Ren F, Yao Z, Jiang F, Yang P, et al. Novel graphene flowers modified carbon fibers for simultaneous determination of ascorbic acid, dopamine and uric acid. Biosensors and Bioelectronics. 2014;**53**:220-224. DOI: 10.1016/j.bios.2013.09.064

[35] Dhara K, Stanley J, Ramachandran T, Nair BG, Satheesh Babu TG. Pt-CuO nanoparticles decorated reduced graphene oxide for the fabrication of highly sensitive non-enzymatic disposable glucose sensor. Sensors and Actuators B: Chemical. 2014;**195**:197-205. DOI: 10.1016/j.snb.2014.01.044

[36] Xie L, Xu Y, Cao X. Hydrogen peroxide biosensor based on hemoglobin immobilized at graphene, flower-like zinc oxide, and gold nanoparticles nanocomposite modified glassy carbon electrode. Colloids and Surfaces: B Biointerfaces. 2013;**107**:245-250. DOI: 10.1016/j.colsurfb.2013.02.020

[37] Xue C, Wang X, Zhu W, Han Q, Zhu C, Hong J, et al. Electrochemical serotonin sensing interface based on double-layered membrane of reduced graphene oxide/polyaniline nanocomposites and molecularly imprinted polymers embedded with gold nanoparticles. Sensors and Actuators B: Chemical. 2014;**203**:412-416. DOI: 10.1016/j.snb.2014.01.100

[38] Jia X, Liu Z, Liu N, Ma Z. A label-free immunosensor based on graphene nanocomposites for simultaneous

multiplexed electrochemical determination of tumor markers. Biosensors and Bioelectronics. 2014;**53**:160-166. DOI: 10.1016/j.bios.2013.09.050

[39] Jin B, Wang P, Mao H, Hu B, Zhang H, Cheng Z, et al. Multi-nanomaterial electrochemical biosensor based on label-free graphene for detecting cancer biomarkers. Biosensors and Bioelectronics. 2014;**55**:464-469. DOI: 10.1016/j.bios.2013.12.025

[40] Luo Z, Yuwen L, Han Y, Tian J, Zhu X, Weng L, et al. Reduced graphene oxide/PAMAM-silver nanoparticles nanocomposite modified electrode for direct electrochemistry of glucose oxidase and glucose sensing. Biosensors and Bioelectronics. 2012;**36**(1):179-185. DOI: 10.1016/j.bios.2012.04.009

[41] Ladiu CI, Popescu IC, Gorton L. Electrocatalytic oxidation of NADH at carbon paste electrodes modified with meldola blue adsorbed on zirconium phosphate: Effect of Ca2+ and polyethyleneimine. Journal of Solid State Electrochemistry. 2005;**9**(5):296-303. DOI: 10.1007/s10008-004-0618-6

[42] Gorton L, Dominguez E. Electrocatalytic oxidation of NAD(P)H at mediator-modified electrodes. Reviews in Molecular Biotechnology. 2002;**82**(4):371-392. DOI: 10.1016/s1389-0352(01)00053-8

[43] Ruecha N, Rangkupan R, Rodthongkum N, Chailapakul O. Novel paper-based cholesterol biosensor using graphene/polyvinylpyrrolidone/polyaniline nanocomposite. Biosensors and Bioelectronics. 2014;**52**:13-19. DOI: 10.1016/j.bios.2013.08.018

[44] Liu B, Lian HT, Yin JF, Sun XY. Dopamine molecularly imprinted electrochemical sensor based on graphene–chitosan composite. Electrochimica Acta. 2012;**75**:108-114. DOI: 10.1016/j.electacta.2012.04.081

[45] Turdean GL, Fort IC, Simon V. In vitro short-time stability of a bioactive glass-chitosan composite coating evaluated by using electrochemical methods. Electrochimica Acta. 2015;**182**:707-714. DOI: 10.1016/j.electacta.2015.09.132

[46] Xue K, Zhou S, Shi H, Feng X, Xin H, Song W. A novel amperometric glucose biosensor based on ternary gold nanoparticles/polypyrrole/reduced graphene oxide nanocomposite. Sensors and Actuators B: Chemical. 2014;**203**:412-416. DOI: 10.1016/j.snb.2014.07.018

[47] Du D, Wang L, Shao Y, Wang J, Engelhard MH, Lin Y. Functionalized graphene oxide as a nanocarrier in a multienzyme labeling amplification strategy for ultrasensitive electrochemical immunoassay of phosphorylated p53 (S392). Analitical Chemistry. 2011;**83**(3):746-752. DOI: 10.1021/ac101715s

[48] Jain P, Das S, Chakma B, Goswami P. Aptamer-graphene oxide for highly sensitive dual electrochemical detection of Plasmodium lactate dehydrogenase. Analytical Biochemistry. 2016;**514**:32-37. DOI: 10.1016/j.ab.2016.09.013

[49] Liang S, Zhu J, Wang C, Yu S, Bi H, Liu X, et al. Fabrication of α-Fe2O3@graphene nanostructures for enhanced gas-sensing property to ethanol. Applied Surface Science. 2014;**292**:278-284. DOI: 10.1016/j.apsusc.2013.11.130

[50] Jiang Z, Wang J, Meng L, Huang Y, Liu L. A highly efficient chemical sensor material for ethanol: Al2O3/graphene nanocomposites fabricated from graphene oxide. Chemical

Communication (Cambridge). 2011;**47**(22):6350-6352. DOI: 10.1039/C1CC11711D

[51] Yue HY, Huang S, Chang J, Heo C, Yao F, Adhikari S, et al. ZnO nanowire arrays on 3D hierarchical graphene foam: Biomarker detection of Parkinson's disease. ACS Nano. 2014;**8**(2):1639-1646. DOI: 10.1021/nn405961p

[52] Liu M, Liu R, Chen W. Graphene wrapped Cu2O nanocubes: Non-enzymatic electrochemical sensors for the detection of glucose and hydrogen peroxide with enhanced stability. Biosensors and Bioelectronics. 2013;**45**:206-212. DOI: 10.1016/j.bios.2013.02.010

[53] Li S-M, Yang S-Y, Wang Y-S, Lien C-H, Tien H-W, Hsiao S-T, et al. Controllable synthesis of nitrogen-doped graphene and its effect on the simultaneous electrochemical determination of ascorbic acid, dopamine, and uric acid. Carbon. 2013;**59**:418-429. DOI: 10.1016/j.carbon.2013.03.035

[54] Szunerits S, Boukherroub R. Graphene-based nanomaterials in innovative electrochemistry. Current Opinion in Electrochemistry. 2018;**10**:24-30. DOI: 10.1016/j.coelec.2018.03.016ff

[55] Cotet LC, Fort CI, Pop LC, Baia M, Baia L. Insights into graphene-based materials as counter electrodes for dye-sensitized solar cells. In Book: Dye-Sensitized Solar Cells. London, United Kingdom: Academic Press, Elsevier; Jan 2019. pp. 341-396. DOI: 10.1016/B978-0-12-814541-8.00010-0

[56] Nag A, Mitra A, Mukhopadhyay SC. Graphene and its sensor-based applications: A review. Sensors and Actuators A: Physical. 2018;**270**:177-194. DOI: 10.1016/j.sna.2017.12.028

[57] Kuila T, Bose S, Khanra P, Mishra AK, Kim NH, Lee JH. Recent advances in graphene-based biosensors. Biosensors and Bioelectronics. 2011;**26**(12):4637-4648. DOI: 10.1016/j.bios.2011.05.039

[58] Foo ME, Gopinath SCB. Feasibility of graphene in biomedical applications. Biomedicine and Pharmacotheraphy. 2017;**94**:354-361. DOI: 10.1016/j.biopha.2017.07.122

[59] Bollella P, Fusco G, Tortolini C, Sanzo G, Favero G, Gorton L, et al. Beyond graphene: Electrochemical sensors and biosensors for biomarkers detection. Biosensors and Bioelectronics. 2017;**89**:152-166. DOI: 10.1016/j.bios.2016.03.068

[60] Cotet LC, Magyari K, Todea M, Dudescu MC, Danciu V, Baia L. Versatile self-assembled graphene oxide membranes obtained under ambient conditions by using a water-ethanol suspension. Journal of Materials Chemistry A. 2017;**5**(5):2132-2142. DOI: 10.1039/C6TA08898H

[61] Wick P, Louw-Gaume AE, Melanie K, Krug HF, Kostarelos K, Fadeel B, et al. Classification framework for graphene-based materials. Angewandte Chemie International Edition. 2014;**53**:7714-7718

[62] Ferrari AC, Bonaccorso F, Fal'ko V, Novoselov KS, Roche S, Boggild P, et al. Science and technology roadmap for graphene, related two-dimensional crystals, and hybrid systems. Nanoscale. 2015;7(11):4598-4810. DOI: 10.1039/C4NR01600A

[63] Paton KR, Varrla E, Backes C, Smith RJ, Khan U, O'Neill A, et al. Scalable production of large quantities of defect-free few-layer graphene by shear exfoliation in liquids. Nature Materials. 2014;**13**(6):624-630. DOI: 10.1038/nmat3944

[64] Cernat A, Tertiş M, Săndulescu R, Bedioui F, Cristea A, Cristea C. Electrochemical sensors based on carbon nanomaterials for acetaminophen detection: A review. Analytica Chimica Acta. 2015;**886**:16-28. DOI: 10.1016/j.aca.2015.05.044

[65] Hummers WS, Offeman RE. Preparation of graphitic oxide. Journal of the American Chemical Society. 1958;**80**(6):1339-1339. DOI: 10.1021/ja01539a017

[66] Dimiev AM, Eigler S, editors. Graphene Oxide: Fundamentals and Applications. John Wiley & Sons; 2016

[67] Marcano DC, Kosynkin DV, Berlin JM, Sinitskii A, Sun ZZ, Slesarev A, et al. Improved synthesis of graphene oxide. ACS Nano. 2010;**4**(8):4806-4814. DOI: 10.1021/nn1006368

[68] Perez del Pino A, Gyorgy E, Cotet C, Baia L, Logofatu C. Laser induced chemical transformation of free-standing graphene oxide membranes in liquid and gas ammonia environments. RSC Advances. 2016;**6**(55):50034-50042. DOI: 10.1039/C6RA07109K

[69] Peng L, Xu Z, Liu Z, Wei YY, Sun HY, Li Z, et al. An iron-based green approach to 1-h production of single-layer graphene oxide. Nature Communications. 2015;**6**:5716. DOI: 10.1038/ncomms6716

[70] Yu HT, Zhang BW, Bulin CK, Li RH, Xing RG. High-efficient synthesis of graphene oxide based on improved Hummers method. Scientific Reports. 2016;**6**:36143. DOI: 10.1038/srep36143

[71] Olszowska K, Pang J, Wrobel PS, Zhao L, Ta HQ, Liu Z, et al. Three-dimensional nanostructured graphene: Synthesis and energy, environmental and biomedical applications. Synthetic Metals. 2017;**234**:53-85. DOI: 10.1016/j.synthmet.2017.10.014

[72] Costinas C, Salagean CA, Cotet LC, Baia M, Todea M, Magyari K, et al. Insights into the stability of graphene oxide aqueous dispersions. Nanomaterials. 2022;**12**(24):4489-4505. DOI: 10.3390/nano12244489

[73] Worsley MA, Pauzauskie PJ, Olson TY, Biener J, Satcher JH, Baumann TF. Synthesis of graphene aerogel with high electrical conductivity. Journal of the American Chemical Society. 2010;**132**(40):14067-14069. DOI: 10.1021/ja1072299

[74] Shao Y, Zhang S, Engelhard MH, Li G, Shao G, Wang Y, et al. Nitrogen doped graphene and its electrochemical applications. Journal of Materials Chemistry. 2010;**20**(35):7491-7496. DOI: 10.1039/C0JM00782J

[75] Rusu MM, Vulpoi A, Maurin I, Cotet LC, Pop LC, Fort CI, et al. Thermal evolution of C–Fe–Bi nanocomposite system: From nanoparticle formation to heterogeneous graphitization stage. Microscopy and Microanalysis. 2022;**28**:317-329. DOI: 10.1017/S1431927622000241

[76] Chuquitarqui A, Cotet LC, Baia M, Gyorgy E, Magyari K, Barbu-Tudoran L, et al. New fabrication method for producing reduced graphene oxide flexible electrodes by using low-power visible laser diode engraving system. Nanotechnology. 2020;**31**:325402. DOI: 10.1088/1361-6528/ab8d67

[77] Mao S, Pu HH, Chen JH. Graphene oxide and its reduction: Modeling and experimental progress. RSC Advances. 2012;**2**(7):2643-2662. DOI: 10.1039/C2RA00663D

[78] Pei SF, Cheng HM. The reduction of graphene oxide. Carbon. 2012;**50**(9):3210-3228. DOI: 10.1016/j.carbon.2011.11.010

[79] Dreyer DR, Murali S, Zhu YW, Ruoff RS, Bielawski CW. Reduction of graphite oxide using alcohols. Journal of Materials Chemistry. 2011;**21**(10):3443-3447. DOI: 10.1039/C0JM02704A

[80] Moon IK, Lee J, Ruoff RS, Lee H. Reduced graphene oxide by chemical graphitization. Nature Communications. 2010;**1**:73. DOI: 10.1038/ncomms1067

[81] Zhang JL, Yang HJ, Shen GX, Cheng P, Zhang JY, Guo SW. Reduction of graphene oxide via L-ascorbic acid. Chemical Communications. 2010;**46**(7):1112-1114. DOI: 10.1039/B917705A

[82] De Silva KKH, Huang HH, Joshi RK, Yoshimura M. Chemical reduction of graphene oxide using green reductants. Carbon. 2017;**119**:190-199. DOI: 10.1016/j.carbon.2017.04.025

[83] Chen WF, Yan LF. Preparation of graphene by a low-temperature thermal reduction at atmosphere pressure. Nanoscale. 2010;**2**(4):559-563. DOI: 10.1039/B9NR00191C

[84] Stankovich S, Dikin DA, Dommett GHB, Kohlhaas KM, Zimney EJ, Stach EA, et al. Graphene-based composite materials. Nature. 2006;**442**:282. DOI: 10.1038/nature04969

[85] Dong X-C, Xu H, Wang X-W, Huang Y-X, Chan-Park MB, Zhang H, et al. 3D graphene–cobalt oxide electrode for high-performance supercapacitor and enzymeless glucose detection. ACS Nano. 2012;**6**(4):3206-3213. DOI: 10.1021/nn300097q

[86] Akhavan O, Ghaderi E, Rahighi R. Toward single-DNA electrochemical biosensing by graphene nanowalls. ACS Nano. 2012;**6**(4):2904-2916. DOI: 10.1021/nn300261t

[87] Giaretta JE, Duan H, Oveissi F, Farajikhah S, Dehghani F, Naficy S. Flexible sensors for hydrogen peroxide detection: A critical review. ACS Applied Materials & Interfaces. 2022;**14**:20491-20505. DOI: 10.1021/acsami.1c24727

[88] Guo W, Li H, Li M, Dai W, Shao Z, Wu X, et al. Synthesis of nickel nanosheet/graphene composites for biosensor applications. Carbon. 2014;**79**:636-645. DOI: 10.1016/j.carbon.2014.08.043

[89] Zhang H, Li ZF, Snyder A, Xie J, Stanciu LA. Functionalized graphene oxide for the fabrication of paraoxon biosensors. Analitica Chimica Acta. 2014;**827**:86-94. DOI: 10.1016/j.aca.2014.04.014

[90] Ravenna Y, Xia L, Gun J, Mikhaylov AA, Medvedev AG, Lev O, et al. Biocomposite based on reduced graphene oxide film modified with phenothiazone and flavin adenine dinucleotide-dependent glucose dehydrogenase for glucose sensing and biofuel cell applications. Analytical Chemistry. 2015;**87**(19):9567-9571. DOI: 10.1021/acs.analchem.5b02949

[91] Chu Z, Liu Y, Xu Y, Shi L, Peng J, Jin W. In-situ fabrication of well-distributed gold nanocubes on thiol graphene as a third-generation biosensor for ultrasensitive glucose detection. Electrochimica Acta. 2015;**176**:162-171. DOI: 10.1016/j.electacta.2015.06.123

[92] Fort CI, Cotet LC, Vulpoi A, Turdean GL, Danciu V, Baia L, et al. Bismuth doped carbon xerogel nanocomposite incorporated in chitosan matrix for ultrasensitive voltammetric detection of Pb(II) and Cd(II).

Sensors and Actuators B: Chemical. 2015;**220**:712-719. DOI: 10.1016/j.snb.2015.05.124

[93] Rusu MM, Fort CI, Cotet LC, Vulpoi A, Todea M, Turdean GL, et al. Insights into the morphological and structural particularities of highly sensitive porous bismuth-carbon nanocomposites based electrochemical sensors. Sensors and Actuators B: Chemical. 2018;**268**:398-410. DOI: 10.1016/j.snb.2018.04.103

[94] Neal CJ, Gupta A, Barkam S, Saraf S, Das S, Cho HJ, et al. Picomolar detection of hydrogen peroxide using enzyme-free inorganic nanoparticle-based sensor. Scientific Reports. 2017;**7**(1):1324. DOI: 10.1038/s41598-017-01356-5

[95] Fort CI, Rusu MM, Pop LC, Cotet LC, Vulpoi A, Baia M, et al. Preparation and characterization of carbon xerogel based composites for electrochemical sensing and photocatalytic degradation. Journal of Nanoscience and Nanotechnology. 2020;**20**:1-11. DOI: 10.1166/jnn.2021.18963

[96] Fort CI, Rusu MM, Cotet LC, Vulpoi A, Florea I, Tuseau-Nenez S, et al. Carbon xerogel nanostructures with integrated Bi and Fe components for hydrogen peroxide and heavy metal detection. Molecules. 2021;**26**:117. DOI: 10.3390/molecules26010117

[97] Selvakumar P, Binesh U, Chen SM. An amperometric biosensor based on direct immobilization of horseradish peroxidase on electrochemically reduced graphene oxide modified screen printed carbon electrode. International Journal Of Electrochemical Science. 2012;7:7935-7947. DOI: 10.1016/S1452-3981(23)17966-1

[98] Song H, Ni Y, Kokot S. A novel electrochemical biosensor based on the hemin-graphene nano-sheets and gold nano-particles hybrid film for the analysis of hydrogen peroxide. Analitica Chimica Acta. 2013;**788**:24-31. DOI: 10.1016/j.aca.2013.06.016

[99] Guo Y, Li J, Dong S. Hemin functionalized graphene nanosheets-based dual biosensor platforms for hydrogen peroxide and glucose. Sensors and Actuators B: Chemical. 2011;**160**(1):295-300. DOI: 10.1016/j.snb.2011.07.050

[100] Yu Y, Pan M, Peng J, Hu D, Hao Y, Qian Z. A review on recent advances in hydrogen peroxide electrochemical sensors for applications in cell detection. Chinese Chemical Letters. 2022;**33**:4133-4145. DOI: 10.1016/j.cclet.2022.02.045

[101] Lodish H, Berk A, Zipursky SL, Matsudaira P, Baltimore D, Darnell J. Neurotransmitters, synapses, and impulse transmission. In: Molecular Cell Biology. 4th ed. New York: W. H. Freeman; 2000

[102] Mele T, Čarman-Kržan M, Jurič DM. Regulatory role of monoamine neurotransmitters in astrocytic NT-3 synthesis. International Journal of Developmental Neuroscience. 2010;**28**(1):13-19. DOI: 10.1016/j.ijdevneu.2009.10.003

[103] Zhang M, Liao C, Yao Y, Liu Z, Gong F, Yan F. High-performance dopamine sensors based on whole-graphene solution-gated transistors. Advanced Functional Materials. 2014;**24**(7):978-985. DOI: 10.1002/adfm.201302359

Chapter 7

Application of Graphene in Lithium-Ion Batteries

Chuanlei Qi, Jiaran Wang, Shengping Li, Yuting Cao, Yindong Liu and Luhai Wang

Abstract

Graphene has excellent conductivity, large specific surface area, high thermal conductivity, and sp^2 hybridized carbon atomic plane. Because of these properties, graphene has shown great potential as a material for use in lithium-ion batteries (LIBs). One of its main advantages is its excellent electrical conductivity; graphene can be used as a conductive agent of electrode materials to improve the rate and cycle performance of batteries. It has a high surface area-to-volume ratio, which can increase the battery's energy storage capacities as anode material, and it is highly flexible and can be used as a coating material on the electrodes of the battery to prevent the growth of lithium dendrites, which can cause short circuits and potentially lead to the battery catching fire or exploding. Furthermore, graphene oxide can be used as a binder material in the electrode to improve the mechanical stability and adhesion of the electrodes so as to increase the durability and lifespan of the battery. Overall, graphene has a lot of potential to improve the performance and safety of LIBs, making them a more reliable and efficient energy storage solution; the addition of graphene can greatly improve the performance of LIBs and enhance chemical stability, conductivity, capacity, and safety performance, and greatly enrich the application backgrounds of LIBs.

Keywords: conductivity, electrochemical processes, electrode materials, graphene oxide, graphene materials, lithium storage, lithium-ion batteries

1. Introduction

During the Industrial Revolution, the world rapidly developed and economically prospered with clean, affordable and reliable energy [1]. The consumption of traditional fossil energy has attracted widespread attention from various industries, and environmental protection has driven research on alternative energy supplies, especially renewable energy, and efforts are being made to develop new renewable energy types while providing an effective environmental protection approach [2, 3].

In order to meet the growing energy demand and reduce greenhouse gas emissions, many countries have recently conducted extensive research on low-cost and environmentally friendly renewable energy sources, such as solar, tidal, wind, biomass, and geothermal [4]. At the same time, it is necessary not only to develop energy but also to maximize energy storage. In addition, there has been a renewed interest in electric vehicles as substitutes for internal combustion engine vehicles, which account for 25% of greenhouse gas emissions [5], making energy storage a priority for the new global energy management system.

Electrochemical energy storage technology has obvious advantages among the various energy storage technologies. The battery is not limited by geographical location, convenient and efficient charging and discharging processes, and higher efficiency [6]. As the main force of energy storage technology, electrochemical energy storage has received widespread attention in market development and scientific research fields. At present, mainstream electrochemical energy storage technologies include LIBs [7], lead batteries [8], and flow batteries [9]. Among them, the LIBs have the characteristics of long cycle characteristics, fast response speed, and high system comprehensive efficiency and are widely used in portable charging equipment, transportation systems, aerospace, and other fields [10, 11].

However, the energy storage processes of the LIB electrode systems are different from each other. With the demand for reliable and durable energy storage devices in portable electronic products and power grids, improving the power density and cycle life of LIBs has become an important goal [12, 13]. It is crucial to design and manufacture efficient electrode materials that can provide high specific capacity and energy density to meet the growing demand for high-performance electrical equipment [10, 14].

Since the energy density of LIBs depends largely on electrode materials, the research direction is aimed at high-specific capacity electrode materials. It has been recognized that nanostructured electrode materials with special electrochemical properties will be necessary to achieve the purpose. The dimensionality reduction of nanomaterials can shorten the diffusion time of Li^+ [15].

2. Overview of the graphene chemistry

Graphene and carbon nanotubes [16] have played important roles in nanomaterials, which can be applied to portable communication equipment, electric vehicles, and large-scale energy storage systems. Many research results have shown that energy storage technology could achieve a qualitative leap by breaking through the technical difficulty of electrode nanomaterials named graphene, a novel two-dimensional carbon material that was discovered by mechanical exfoliation of graphite in 2004 [17]. As illustrated in **Figure 1**, the typical structure of graphene [18] is composed of a carbon material 1 million times thicker than the diameter of a single hair comprising a hexagonal two-dimensional honeycomb lattice of sp^2 hybridized carbon atoms. The structure can be divided into single-layer, double-layer, and multilayer, with an ultra-high specific surface area of about 2630 $m^2\ g^{-1}$.

Graphene has many desirable properties, with very high electron mobility at room temperature and rapid heterogeneous electron transfer at the edges, with values exceeding 15,000 $cm^2\ V^{-1}S^{-1}$ [19]. Graphene also has a significantly high Young's modulus (1.0 TPa), high breaking strength, extraordinary mechanical strength, excellent thermodynamic, electrical conductivity ($5–6.4 \times 10^6$ S m^{-1}), optical transmittance activity (about 97.7%), material density (less than 1 g cm^{-3}), stability, catalytic properties, and other graphene tunable properties [20, 21].

The continuous two-dimensional conductive network formed by graphene can effectively improve the electron and ion transport kinetics of electrode materials, and

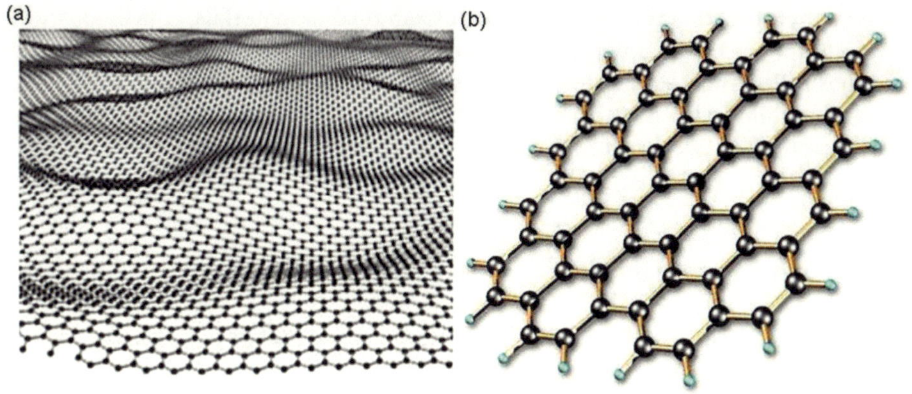

Figure 1.
(a) and (b) Typical structure of graphene [18].

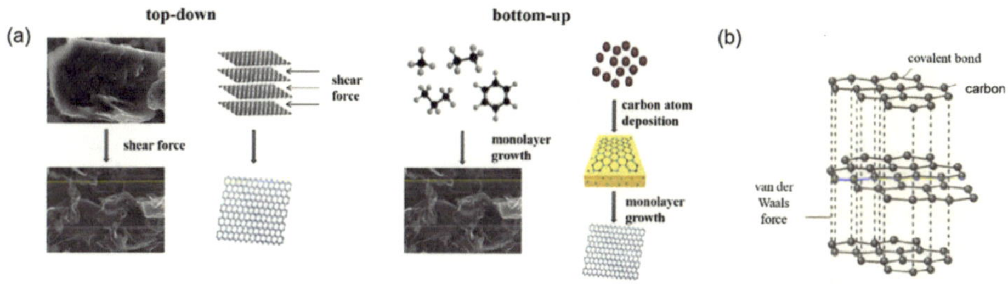

Figure 2.
(a) Schematic diagram of top-down and bottom-up approaches, (b) schematic diagram of graphitc structure [29].

graphene is used to improve the rate performance and cycle stability of LIBs because of its high specific surface area, stable chemical properties, excellent electrical and thermal conductivity [18]. With its excellent characteristics, graphene has lots of applications in electronic devices, photonic devices, photocatalysis, and advanced composite materials applicable in the military, aerospace, and other fields.

Although graphene shows excellent properties in chemical and mechanical aspects, its application in electronics and energy storage devices needs to be continuously explored [22]. Researchers have developed many methods in graphene applications, such as combining with other materials to develop graphene-based composites [23] including graphene/polymers [24], graphene/metals [25], and graphene/carbon nanotubes [26] composites for energy storage devices.

In addition, recent scientific advancements have allowed the development of various low-cost and environmentally friendly methods for preparing graphene. This is particularly important for large-scale production and application. The following analysis analyzes the application of graphene and graphene-based nanocomposites as electrode materials in LIBs, and provides possible development paths in the future.

The main production methods for graphene include bottom-up and top-down methods, and graphene properties have great differences in structural integrity, sheet size, and cost with different methods [27, 28]. As illustrated in **Figure 2**, the top-down approach refers to the method of obtaining a product by crushing or peeling off a large amount of material. The bottom-up approach, on the other hand, refers to the method of synthesizing the desired product from smaller materials, continuously growing graphene by breaking the chemical bonds of carbon-containing compounds and depositing carbon atoms on a suitable substrate [29–32]. Various forms of graphene nanomaterials have been prepared, including graphene spheres, graphene scrolls, graphene networks, graphene tubes, graphene cages, and other structures of graphene [21].

3. Lithium-ion batteries, LIBs

As shown in **Figure 3**, lithium-ion batteries (LIBs) consist of two electrodes and a separator impregnated with an electrolyte to provide the electrons and ions needed for electrochemically active nanomaterials [33].

The metal-foil anode is the negative electrode of the battery, and it is made of lithium-containing material. During discharge, lithium ions are released from the anode and travel through the electrolyte to the cathode, where they are intercalated (inserted) into the electrode material. The anode also serves as the current collector for the negative terminal of the battery. The cathode, or positive electrode, is made of an electron-rich material that can store the lithium ions after they are released from the anode. It is typically coated onto a metal foil current collector. During discharge, electrons flow through the external circuit to the cathode, where they combine with the lithium ions to form the discharge product. The separator is a thin film that separates the anode and cathode to prevent direct contact between them. It is made of a material that is permeable to lithium ions but impermeable to electrons, ensuring that the electrons flow through the external circuit during discharge and charge. The separator also helps to prevent short circuits within the battery. The electrolyte is a liquid or solid material that transports the lithium ions between the anode and cathode. It is typically a lithium salt dissolved in a non-aqueous solvent.

The LIBs are based on the movement of lithium ions between the anode and cathode during discharge and charge cycles. When the battery is discharged, electrons

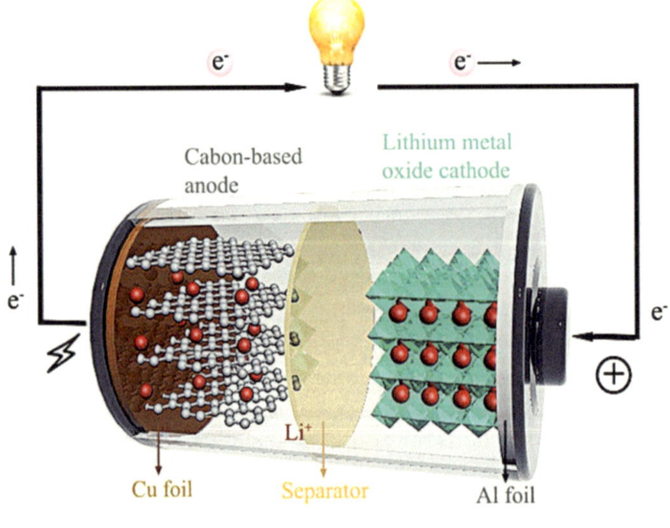

Figure 3.
Schematic diagram of the structure of commercial LIBs [33].

flow through the external circuit to the cathode, where they combine with lithium ions to form the discharge product. At the same time, lithium ions travel through the electrolyte.

At the moment, LIBs are regarded as one of the most promising energy storage technologies due to their high energy density, energy conversion efficiency, and output voltage [34]. With the development of information electronics, electric vehicles, and smart grids, there is a huge demand for LIBs with high energy density, long cycle life, and lower cost [35, 36]. However, the electrode materials of LIBs generally have poor conductivity, and the charge-discharge reaction cannot be completely carried out due to electrode polarization during the charging and discharging process. Moreover, the effective capacity of the electrode material cannot be fully exerted [37–39].

At present, the conductive agents on the market mainly include conductive carbon black, conductive graphite, and carbon nanotubes [40]. The excellent thermal conductivity of graphene can improve the thermal stability of LIBs. The "face" contact between graphene and electrode materials can improve the conductivity of electrodes [41]. Based on the special physical and chemical properties of graphene, and it has great potential as an electrode material for LIBs. LIBs are composed of four parts: cathode electrode material, anode electrode material, separator, and electrolyte, and the electrode material plays an important role in battery performance [42, 43]. According to application fields, the application of graphene mainly has three directions in LIBs: (1) graphene use as an active electrode material: graphene can be used as an anode material for LIBs to provide reversible storage space for Li^+, improving specific capacity and rapid charge and discharge efficiency [44]. (2) Graphene can be combined with cathode or anode materials to improve the performance of electrodes: the combination of graphene with a cathode active material enhances electrode conductivity [45]. In this respect, graphene can be compounded with anode-active materials to construct a three-dimensional structure and provide space for volume expansion [46]. (3) Graphene is used as a conductive additive to provide a fast channel for electron and Li^+ transport and improve the conductivity of the electrode [47, 48].

4. Graphene as LIBs electrode conductive agent

At present, the development of energy storage technology has made higher requirements for LIBs in terms of energy density, ion transport rate, and cycle performance [49]. However, the poor conductivity of electrode materials greatly limits the performance of LIBs. Adding a conductive agent can enhance the electron transport efficiency and reduce the polarization of the electrode. It is important to utilize the effective capacity of the active material. Compared with conductive agents such as commercial conductive agent carbon black (CB), graphene has higher conductivity and specific surface area, and studies have shown that advanced carbon materials, including carbon (one-dimensional) [50], graphene (two-dimensional) [51, 52], and 3D graphene backbones (three-dimensional) [53], have been used to build continuous conductive networks for LIBs [54].

Graphene is a powerful planar conductive additive, which is considered to be one of the most promising conductive additives due to its unique physicochemical properties including high aspect ratio, chemical resistance, excellent conductivity, and low dosage of effective characteristics [55–57]. Compared with the "point-to-point" contact mode constructed by the traditional graphite conductive agent, graphene can form a "point-to-surface" contact mode with the electrode material in the electrode, providing a long-range and fast conduction path for electrons and Li^+, reducing the amount of conductive agent is equivalent to increasing the content of the active material of the electrode and increasing the capacity of the electrode [54, 58]. Therefore, using a small amount of graphene as a conductive additive can greatly improve the electronic conductivity of the electrode.

It has been proposed as a simple and effective method to prepare graphene conductive slurry as a conductive agent for LIBs by combining mechanical stirring, ultrasonic dispersion, and dispersant modification [48]. **Figure 4a** shows the schematic diagram of graphene dispersion. Graphene slurry also exhibits excellent battery performance as a conductive agent for LIBs. At 100 mAg^{-1} current density, the first charge and discharge capacity are 1273.8 and 1723.7 $mAhg^{-1}$, respectively, and the coulombic efficiency is 73.9%. The capacity retention rate of the anode is 84% (1070.2 $mAhg^{-1}$) after 100 cycles at 200 mAg^{-1}. Another article reported that graphene nanosheets (GN) with different sizes as conductive additives can affect the electrochemical performance of $LiFePO_4$ (LFP) [61]. Compared with conventional conductive additives, GNs and Super P Conductive Carbon Black (SP) can construct an effective electronic conductive network and significantly improve the electrochemical performance of LFP as conductive additives. It also shows that with the increase of GN size, the specific capacity and rate performance of nanoscale LFP tended to deteriorate, which is due to the "barrier effect" of GN extending the length of the ion transport path and greatly reducing the ion conductivity. The small-size GN can effectively balance the rapid diffusion of Li^+ and the electron transport of nanoscale LFP. The effect of the amount of graphene used in different thicknesses electrode on the electrode performance was also studied [59]. The results showed that when the thickness of the electrode is thin, the cycle performance and rate performance of

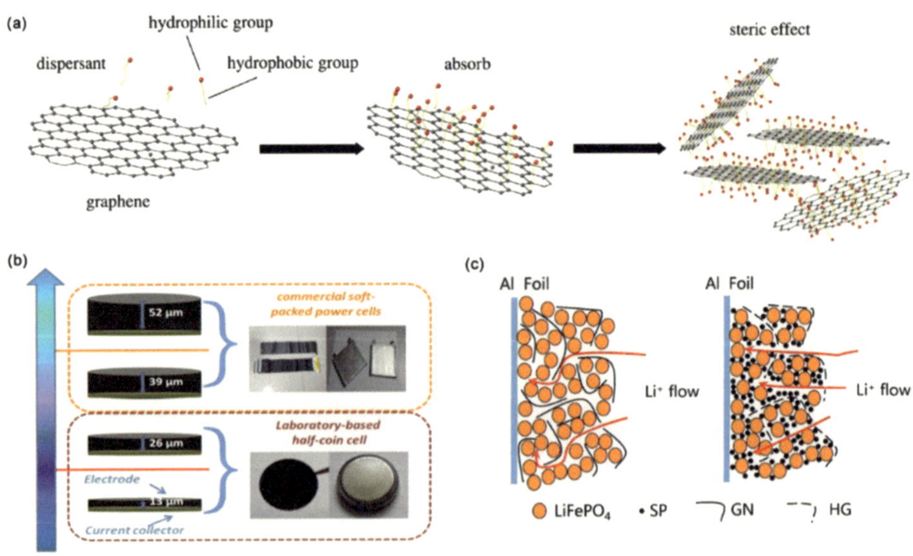

Figure 4.
(a) Schematic diagram of graphene dispersion, (b) schematic diagram of the evolution of electrode thickness from laboratory half-coin battery to commercial soft-packaged battery [59], and (c) schematic diagram of lithium-ion transport path in $LiFePO_4$ cathode with GN or HG + SP as conductive additive [60].

the electrode increase with the increase of graphene addition (**Figure 4b**). When the electrode thickness is higher, the rate performance of the electrode has a linear relationship with the amount of graphene. The Li^+ diffusion path is greatly extended, and the spatial effect of GN is amplified, causing slower ion transport kinetics. Different research results indicate that the conductivity of the electrode material is significantly improved when the amount of graphene is appropriate. However, when the size or the amount of the graphene nanosheet (GN) is too large in the higher-thickness electrode, the Li^+ diffusion path will be greatly extended. The steric hindrance effect of graphene on the diffusion of Li^+ is amplified, the polarizability is higher at high rates, and the rate of performance of the electrode is reduced. So, researchers should pay attention to maintaining the balance between ion diffusion kinetics and electron conduction when using graphene as a conductive agent.

In order to improve the conductivity and the rate performance of LIBs, researchers have studied a lot on graphene composite conductive agents and graphene modification. Graphene nanoribbons (GNBs) have been used as a conductive agent for LFP [62]. The cathode with olivine structures is used to promote rapid redox reactions and achieve high-rate cell performance. The results have shown that the cathodes with 5 wt% graphene nanoribbons and 10 wt% conductive carbon nanoparticles exhibited a capacity of 163.25 $mAhg^{-1}$ at 0.1 C and 130.60 $mAhg^{-1}$ at 2 C, the capacity retention rate is 98.21% after 100 cycles at 2 C. Graphene nanoribbons play the role of bridges creating connected networks to facilitate electron transport. Porous graphene (HG) has been prepared by KOH activation [60]. HG, with a large number of pores and a large specific surface area, greatly improved the electronic conductivity of the LFP electrode, but it did not affect the efficient transport of ions. LFP cells with traditional graphene additives exhibit lower rate performance because graphene with a planar structure hinders the transport of Li^+. Binary conductive additives containing only 1 wt% HG and 1 wt% carbon black (such as SP) can make LIBs obtain higher rate performance comparable to batteries containing 10 wt% SP, and the schematic diagram of the Li^+ transport path of the composite conductive agent is shown in **Figure 4c**. A small amount of SP complements the remote conductive network formed by HG, which can fully contact LFP so that the entire LFP electrode has excellent conductivity. The simultaneous use of HG and SP can achieve a balance of electron conductivity and ion diffusivity.

5. Application of graphene as LIBs cathode materials

As the cathode of LIBs, the electrode material should have the requirements of high reversible capacity, high stability potential, and lower manufacturing cost [63, 64]. At present, the cathode materials of LIBs are mostly $LiFePO_4$, $LiCoO_2$, $LiMn_2O_4$, $Li_3V_2(PO4)_3$, and $LiNi_xCo_yM_{1-x-y}O_2$, which have the characteristics of high specific capacity, non-toxicity, and low cost, but the conductivity is poor, and the mobility of lithium-ion is low [65–67].

When $LiFePO_4$ material is compounded with graphene, theoretically, the rate performance can be improved as conductivity improves [68, 69]. The LPF-graphene composite, LFP@C/G, was successfully synthesized, as illustrated in **Figure 5**, using the high energy ball milling-assisted rheological phase method [70]. The multilayer graphene film is not stacked on the carbon-coated $LiFePO_4$ nanospheres, thus forming rich mesopores forming unique 3D “ball-in-chip” and “ball-on-chip” conductive network structures. High conductivity and rich mesopores facilitate the transport of

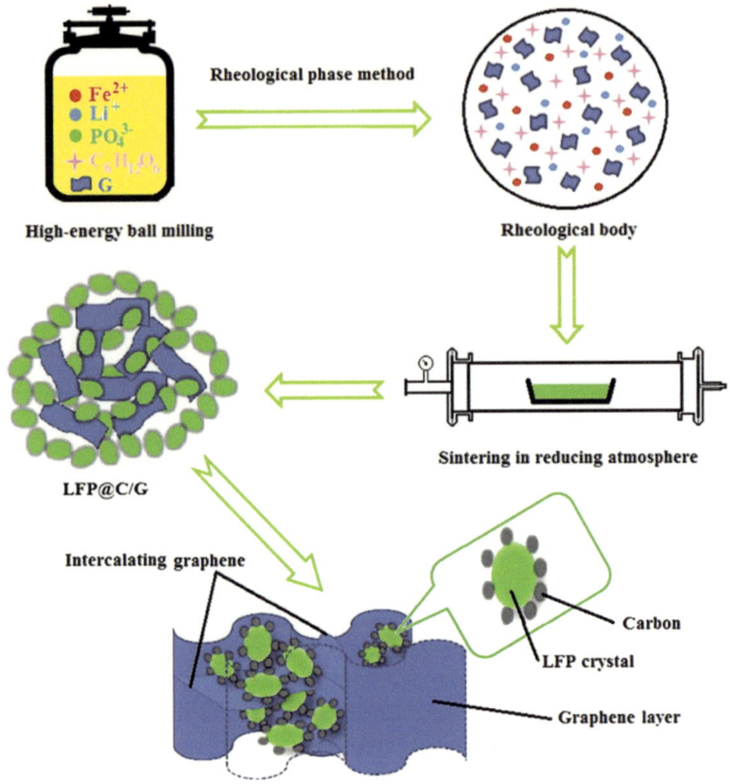

Figure 5.
Schematic diagram of the synthesis of LFP@C/G composites [70].

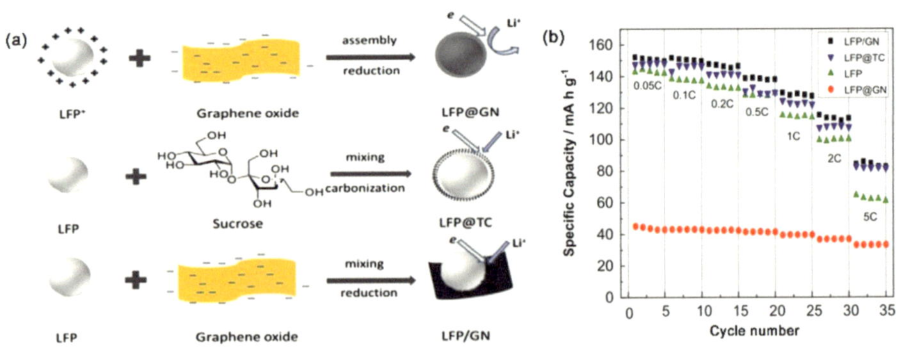

Figure 6.
(a) Preparation and electrochemical reaction protocols of LFP@TC and LFP/GN, (b) rates data of LFP@TC and FP/GN [66].

electrons and ions. The results show that the mixed materials with graphene content of about 3 wt% show excellent rate performance, and the initial discharge capacities were 163.8 and 147.1 $mAhg^{-1}$, respectively. In addition, the composites also showed excellent cycling stability, with a capacity decay of only 8% after 500 cycles at 10 C. However, there are also reports that graphene composite $LiFePO_4$ exhibits unfavorable electrochemical performance when used in cathode materials of LIBs. The effect of carbon coating on the electrochemical performance of $LiFePO_4$ has been discussed [71]. The schematic diagram of the synthesis of lFP@graphene nanosheets (LFP@GN), sucrose-derived amorphous carbon-coated LFP (LFP@TC), and LFP/GN (partial wrapping) is shown in **Figure 6a**. Graphene partially wrapped $LiFePO_4$ enhances the conductivity of LFP/GN material, but when the cathode material (LFP/GN) is fully wrapped in graphene, the ion transport efficiency decreases, leading to a decline in rate performance (**Figure 6b**). The results show that although graphene coating improved the conductivity of Li^+ and electrons, the complete and dense coating of high graphitic carbon is not conducive to the transport of electrons due to the influence of steric hindrance effect on Li^+ diffusion in the cathode material. Therefore, the ideal coating structure should maintain a balance between increased electron transport and rapid ion diffusion.

6. Application of graphene as LIBs anode materials

As the most widely used anode material for LIBs, graphite anode has the advantages of easy access to abundant raw materials and low costs, but its low specific capacity falls short of meeting the requirements of LIBs [72–74]. The lithium dendrite formed during the charging and discharging process of graphite anode makes it difficult for LIBs to achieve high rate and cycle life [75], in addition to the fact that they can cause short circuits and potentially lead to the battery catching fire or exploding. Therefore, many studies are devoted to the modification of graphite anodes and the development of new anode materials.

At present, the typical lithium-ion battery anode materials can be divided into three categories: intercalation reaction electrode materials, conversion reaction electrode materials, and alloy electrode materials. The intercalated electrode materials are mainly composed of carbon materials [76]. There are many studies on the application of carbon materials in LIBs, and the research on graphene application in anode materials is more extensive and in-depth than in cathode. Graphene can react with Li^+ on both sides of the graphene nanosheet to form LiC_3, and each C atom corresponds to about 0.33 Li^+, which is twice the amount of lithium intercalation in the traditional graphite electrode with a theoretical specific capacity of 744 $mAhg^{-1}$ [77, 78]. Porous graphene foam (GF) has been prepared and applied as the anode of lithium-ion batteries. It comprises a loosely porous three-dimensional network structure with excellent electrical conductivity and chemical stability. GF has high performance in specific capacity and cycle stability. However, graphene as the anode material for LIBs may cause stacking between graphene sheets, reducing the specific surface area of the material, resulting in a decrease in lithium storage and failure to achieve higher capacity as a single anode material.

The graphene-metal composite material produced by the graphene coating method can, however, improve the electrochemical performance of LIBs [79]. The flexible characteristics of graphene can effectively inhibit the metal electrode volume expansion during the charging and discharging process, and the morphology of graphene can change with changes in the preparation method [52, 80, 81]. With excellent electrical conductivity, graphene can establish a conductive network between particles, and the high specific surface area can also increase the storage capacity of lithium. Numerous studies have shown that the graphene-metal composite materials applied as anode materials can greatly improve the performance of LIBs [52, 80–82]. The bilayer graphene (BGra) was synthesized by the thermal evaporation-deposition

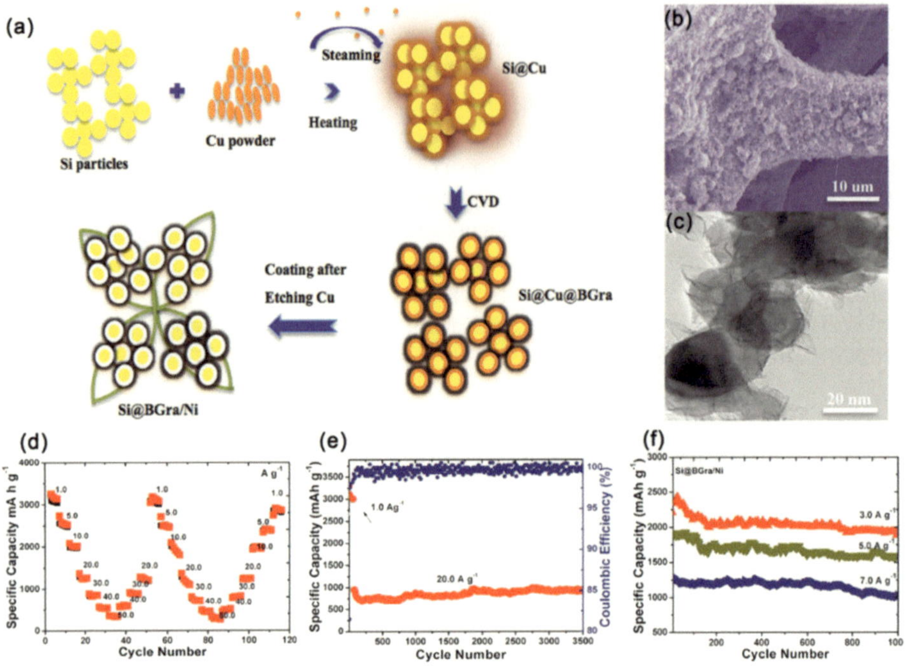

Figure 7.
(a) Schematic diagram of the preparation process of Si@BGra and Si@BGra/Ni (yellow represents silicon nanoparticles, orange is copper powder, gray is double-layer graphene, green is nickel foam), (b) SEM image of Si@BGra/Ni, (c) HRTEM image of Si@BGra/Ni, (d) Si@BGra/Ni magnification performance image at 1.0-50 Ag^{-1}, (e) Si@BGra/Ni cycle performance at 20 Ag^{-1}, and (f) cycling performance of Si@BGra/Ni at 3, 5, and 7 Ag^{-1} [83].

assisted chemical vapor deposition (CVD) method by coating on Si nanoparticles [83], as shown in the process diagram in **Figure 7a–c**. The results showed that the electrode composed of this material, Si@BGra, can provide a capacity of 2500 $mAhg^{-1}$ at the current density of 3 Ag^{-1}. At the same time, the capacity retention rate was 85% after 1000 cycles, exhibiting excellent cyclic stability (**Figure 7d–f**). The graphene-MnO_2 composite was also prepared as anode materials for LIBs [84]. The material also demonstrated excellent cycle stability at current rates of 2 C, 5 C, and 10 C after 500 charge-discharge tests.

The excellent cycle stability and higher rate performance of these composites can be attributed to the integration of graphene and mesoporous metallic material. Graphene, as an exceptional charge carrier, enhances the electronic conductivity of composite materials and enables complete reversible redox reactions in metallic materials. As in LIBs anode materials, graphene can also act as a buffer medium for large volume changes of the negative electrode material during the charging and discharging process and inhibit mechanical strain and the crushing of electrodes.

7. Application of other graphene derivatives in LIBs

Although graphene exhibits excellent electrochemical performance in electron and ion transport kinetics, its hydrophobic properties are challenging in electrode applications. As a strong and flexible carbon atom thin film, graphene offers a variety of possibilities for the modification and functionalization of its carbon backbone, such as chemical modification of H and O functional groups, so that its hydrophilic version of oxygen/hydrogen functionalization has recently gained popularity [85–87]. At present, the doping of heteroatoms into graphene to produce more Li^+ storage sites has been widely studied. Heteroatom doping produces a large number of defects on the graphene surface, which not only prevents the irreversible aggregation of the graphene layer but also provides a rich lithium reservoir [86]. At the same time, the surface of graphene derivatives such as graphene oxide (GO), nitrate graphene, and fluorographene, has a large number of functional groups, defects, and other active sites, which enhance the electronic and mechanical properties of graphene [88].

Graphene oxide is usually prepared by stripping of graphite oxide generated by chemical oxidation, and the use of strong oxidants to generate graphite oxide is the most common method for preparing graphene oxide [89]. At present, the production of GO and reduced graphene (r-GO) have been commercialized. They form the basic units for forming other 3D graphene complex material, which also has the problem of low relative conductivity [90, 91]. High-quality foamed graphene and vertical graphene prepared by chemical vapor deposition (CVD) and plasma-enhanced chemical vapor deposition (PECVD) have emerged in the latest report, but the preparation process involves high temperatures and complex techniques. It is inferred that high-end 3D graphene derivatives with few defects and high electronic conductivity are the development direction in the future [92, 93].

Many researchers have devoted themselves to the study of graphene derivatives as lithium-ion cathode materials. A series of reduced porous graphene oxide (rhG-x) has been synthesized by the chemical oxidation and annealing reduction process of porous waste graphite (hSG) [94], and the schematic diagram of the process is shown in **Figure 8**. Because of the unique oxygen-containing groups and electronic conductivity, the rhG-x series with "worm-like" segments and porous structure diagrams shows excellent lithium storage performance as the cathode for LIBs.

Few-layer graphene (FLG) has been prepared by a simple method to modify FLG using nitrogen doping. With the doping of heteroatoms, the rate performance and cycle stability of graphene were significantly improved [95].

In addition, graphene-based compounds are also widely used in LIBs. Uniformly dispersed Cu-hexahydroxytriphenyl (HHTP)/graphene (G) composites were synthesized using an in-situ growth strategy, and electrochemical performance was studied

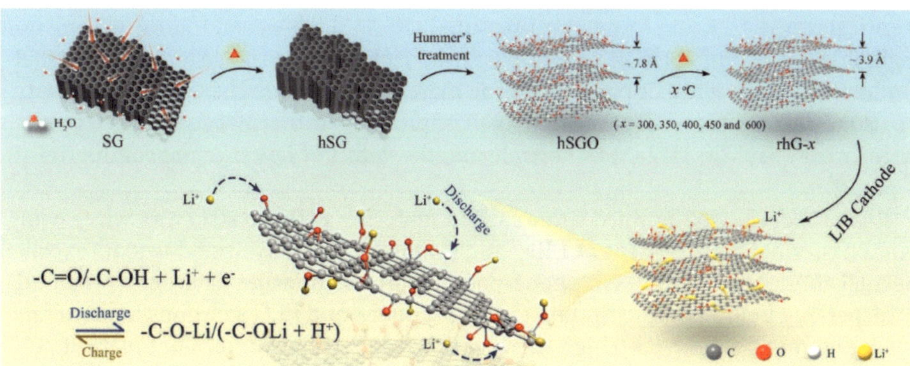

Figure 8.
Schematic diagram of rhG-x synthesis process [94].

for the first time as anode for LIBs [96]. It was observed that graphene formed a two-dimensional network of conductivity in the composites and effectively improved the energy storage of Cu-HHTP.

8. Application of graphene in thermal management of LIBs

Lithium-ion batteries have a wide range of applications in mobile communications, automobiles, and aerospace. With the rise of electric and hybrid electric vehicles (HEVs), there is another push for battery technology [97]. The battery and its management system are two of the three main technologies of electric vehicles, and the thermal management technology of the battery is an important part of the battery system. The operation life and efficiency of batteries are affected by high temperatures during working, thereby affecting the maintenance, life, and cost of electric vehicles [98–100]. Extending battery life and safe use of batteries requires controlling the operating temperature of LIBs within a safe range, and high or low temperatures will create adverse effects on LIBs. In addition, the electrolyte could solidify and fail to transport electrons at a lower temperature. Thus, the LIBs could experience thermal runaway or battery rupture or even explosion at high temperatures.

Many studies have shown that high, low, or uneven temperatures affect the charge-discharge efficiency and cycle stability of the power battery [101–103]. Therefore, it is necessary to design a reasonable battery thermal management system to effectively control and maintain a stable and uniform temperature of the battery in the battery pack. At present, the cooling methods of lithium-ion battery thermal management systems are mainly divided into three cooling methods: air cooling, liquid cooling, and phase change material (PCM) [104, 105].

A common method of thermal management of lithium-ion battery packs is based on the utilization of phase change materials (PCM) [106]. Phase change materials are a special class of functional materials that, in the phase change process, keep a small temperature change range or constant temperature and can absorb or release a large amount of latent heat [107, 108]. During the technological development of PCMs, researchers have studied many different kinds of materials, including inorganic systems (salts and salt hydrates), organic compounds (such as paraffins or fatty acids), and polymeric materials (such as polyethylene glycol) [109]. Paraffin wax is widely used as a phase change material in LIBs due to its excellent characteristics such as safety and non-toxicity, low price, small volume change during phase change, stable chemical properties, and low vapor pressure [110, 111]. However, there are some non-negligible shortcomings in the use of paraffin, such as easy leakage and low thermal conductivity during use of paraffin. Some materials with high thermal conductivity are usually combined with paraffin wax to improve their thermal conductivity and performance as PCM [112–114], considering the defect of low thermal conductivity of paraffin.

Researchers have combined some carbon materials with paraffin wax while using it in the thermal management of LIBs, and graphene is widely used due to its excellent thermal conductivity [115]. Graphene-epcm hybrid composites have been prepared by dispersing liquid phase stripping (LPE) graphene and FLG solutions in paraffin [116], and mixing them with high shear with a magnetic stirrer on a hot plate at 70°C. It is proved that graphene and FLG as fillers in organic phase change materials can improve their thermal conductivity by more than two orders of magnitude while maintaining their latent heat storage capacity. Graphene-coated nickel foam was prepared using chemical vapor deposition technology, and paraffin wax was used as a phase change material to penetrate into the voids of graphene-coated nickel foam. The thermal characteristics of saturated paraffin graphene-coated nickel foam and its application in the thermal management of LIBs were studied [117]. The results showed that: (1) the thermal conductivity of graphene @Ni/saturated paraffin wax was increased by 23 times compared with pure paraffin; (2) the melting temperature and freezing temperature of graphene-coated nickel foam composites of nickel foam/saturated paraffin were higher than paraffin wax and lower than paraffin, respectively.

9. Conclusions and perspectives

In conclusion, the application of graphene in lithium-ion batteries has shown significant potential in improving battery performance. Graphene's exceptional electrical conductivity, high specific surface area, and excellent mechanical properties make it an ideal candidate for enhancing the capabilities of these batteries. The various approaches, graphene derivatives, and graphene-based electrodes have been successfully utilized to improve the capacity, rate capability, cycling stability, and thermal stability of lithium-ion batteries.

However, there are still challenges that need to be addressed for the widespread application of graphene in lithium-ion batteries. One of the main challenges is the production of high-quality graphene in a scalable manner. The development of efficient and cost-effective methods for the synthesis of graphene is crucial for its commercialization in battery applications. Additionally, understanding the mechanisms behind the electrochemical performance of graphene-based electrodes is crucial for optimizing their properties.

Future perspectives in this field include exploring new applications of graphene beyond electrodes, such as in separators, electrolytes, and other battery components. The combination of graphene with other materials, such as metal oxides, carbon nanotubes, or polymers, may lead to the development of novel electrode architectures with improved performance. Furthermore, the development of 3D-printed graphene composites for battery applications is a promising direction that could lead to the production of customized battery components with improved mechanical properties and conductivity.

Overall, the application of graphene in lithium-ion batteries holds great promise for the development of next-generation energy storage devices with higher energy density, longer cycle life, and better rate capability. Continuing research efforts in this field are expected to lead to further advancements in the field of energy storage and pave the way for a sustainable future.

Author details

Chuanlei Qi[1], Jiaran Wang[2], Shengping Li[1], Yuting Cao[1], Yindong Liu[1] and Luhai Wang[1*]

1 PetroChina Company Limited, Petrochemical Research Institute, Changping, Beijing, P.R. China

2 State Key Laboratory of Heavy Oil Processing, China University of Petroleum, Beijing, Changping, P.R. China

*Address all correspondence to: wlh459@petrochina.com.cn

References

[1] Chu S, Majumdar A. Opportunities and challenges for a sustainable energy future. Nature. 2012;**488**:294-303. DOI: 10.1038/nature11475

[2] He W, Li Z, Liu T, Liu Z, Guo X, Du J, et al. Research progress and application of deep learning in remaining useful life, state of health and battery thermal management of lithium batteries. Journal of Energy Storage. 2023;**70**:107868. DOI: 10.1016/j.est.2023.107868

[3] Mand Khan B, Chun Oh W, Nuengmatch P, Ullah K. Role of graphene-based nanocomposites as anode material for Lithium-ion batteries. Materials Science and Engineering: B. 2023;**287**:116141. DOI: 10.1016/j.mseb.2022.116141

[4] Larcher D, Tarascon JM. Towards greener and more sustainable batteries for electrical energy storage. Nature Chemistry. 2015;**7**:19-29. DOI: 10.1038/nchem.2085

[5] AbdelHamid AA, Mendoza-Garcia A, Ying JY. Advances in and prospects of nanomaterials' morphological control for lithium rechargeable batteries. Nano Energy. 2022;**93**:106860. DOI: 10.1016/j.nanoen.2021.106860

[6] Wang H, Dai J, Wei H, Lu Q. Understanding technological innovation and evolution of energy storage in China: Spatial differentiation of innovations in lithium-ion battery industry. Journal of Energy Storage. 2023;**66**:107307. DOI: 10.1016/j.est.2023.107307

[7] Li G, Huang B, Pan Z, Su X, Shao Z, An L. Advances in three-dimensional graphene-based materials: Configurations, preparation and application in secondary metal (Li, Na, K, Mg, Al)-ion batteries. Energy & Environmental Science. 2019;**12**:2030-2053. DOI: 10.1039/c8ee03014f

[8] Yin J, Lin H, Shi J, Lin Z, Bao J, Wang Y, et al. Lead-carbon batteries toward future energy storage: From mechanism and materials to applications. Electrochemical Energy Reviews. 2022;**5**:2. DOI: 10.1007/s41918-022-00134-w

[9] He H, Tian S, Tarroja B, Ogunseitan OA, Samuelsen S, Schoenung JM. Flow battery production: Materials selection and environmental impact. Journal of Cleaner Production. 2020;**269**:121740. DOI: 10.1016/j.jclepro.2020.121740

[10] Dehghani-Sanij AR, Tharumalingam E, Dusseault MB, Fraser R. Study of energy storage systems and environmental challenges of batteries. Renewable and Sustainable Energy Reviews. 2019;**104**:192-208. DOI: 10.1016/j.rser.2019.01.023

[11] Ma Q, Xie H, Qu J, Zhao Z, Zhang B, Song Q, et al. Tailoring the polymer-derived carbon encapsulated silicon nanoparticles for high-performance lithium-ion battery anodes. ACS Applied Energy Materials. 2019;**3**:268-278. DOI: 10.1021/acsaem.9b01463

[12] Bruce PG, Scrosati B, Tarascon JM. Nanomaterials for rechargeable lithium batteries. Angewandte Chemie (International Edition in English). 2008;**47**:2930-2946. DOI: 10.1002/anie.200702505

[13] Lu Y, Rong X, Hu Y-S, Chen L, Li H. Research and development of advanced battery materials in China. Energy Storage Materials. 2019;**23**:144-153. DOI: 10.1016/j.ensm.2019.05.019

[14] Xu J, Liu Y, Xu C, Li J, Yang Z, Yan H, et al. Aqueous non-metallic ion batteries: Materials, mechanisms and design strategies. Coordination Chemistry Reviews. 2023;**474**:214867. DOI: 10.1016/j.ccr.2022.214867

[15] Chen S, Xing W, Duan J, Hu X, Qiao SZ. Nanostructured morphology control for efficient supercapacitor electrodes. Journal of Materials Chemistry A. 2013;**1**:2941-2954. DOI: 10.1039/c2ta00627h

[16] Chen X, Ma Y. Wearable lithium ion batteries based on carbon nanotubes and graphene. Advanced Materials Technologies. 2018;**3**:1800041. DOI: 10.1002/admt.201800041

[17] Novoselov KS, Geim AK, Morozov SV, Jiang D, Zhang Y, Dubonos SV, et al. Electric field effect in atomically thin carbon films. Science. 2004;**306**:666-669. DOI: 10.1126/science.1102896

[18] Chen X, Tian Y. Review of graphene in cathode materials for lithium-ion batteries. Energy & Fuels. 2021;**35**:3572-3580. DOI: 10.1021/acs.energyfuels.0c04191

[19] Chen F, Tao NJ. Electron transport in single molecules: From benzene to graphene. Accounts of Chemical Research. 2009;**42**:429-438. DOI: 10.1021/ar800199a

[20] Ali A, Shen PK. Nonprecious metal's graphene-supported electrocatalysts for hydrogen evolution reaction: Fundamentals to applications. Carbon Energy. 2019;**2**:99-121. DOI: 10.1002/cey2.26

[21] Ji L, Meduri P, Agubra V, Xiao X, Alcoutlabi M. Graphene-based nanocomposites for energy storage. Advanced Energy Materials. 2016;**6**:1502159. DOI: 10.1002/aenm.201502159

[22] Hossain S, Abdalla AM, Suhaili SBH, Kamal I, Shaikh SPS, Dawood MK, et al. Nanostructured graphene materials utilization in fuel cells and batteries: A review. Journal of Energy Storage. 2020;**29**:101386. DOI: 10.1016/j.est.2020.101386

[23] Huang X, Qi X, Boey F, Zhang H. Graphene-based composites. Chemical Society Reviews. 2012;**41**:666-686. DOI: 10.1039/c1cs15078b

[24] Mao H-n, Wang X-g. Use of in-situ polymerization in the preparation of graphene/polymer nanocomposites. New Carbon Materials. 2020;**35**:336-343. DOI: 10.1016/s1872-5805(20)60493-0

[25] Ali A, Shen PK. Recent Progress in graphene-based nanostructured electrocatalysts for overall water splitting. Electrochemical Energy Reviews. 2020;**3**:370-394. DOI: 10.1007/s41918-020-00066-3

[26] Kinloch IA, Suhr J, Lou J, Young RJ, Ajayan PM. Composites with carbon nanotubes and graphene: An outlook. Science. 2018;**362**:547-553. DOI: 10.1126/science.aat7439

[27] Madurani KA, Suprapto S, Machrita NI, Bahar SL, Illiya W, Kurniawan F. Progress in graphene synthesis and its application: History, challenge and the future outlook for research and industry. ECS Journal of Solid State Science and Technology. 2020;**9**:093013. DOI: 10.1149/2162-8777/abbb6f

[28] Wang X-Y, Narita A, Müllen K. Precision synthesis versus bulk-scale fabrication of graphenes. Nature Reviews Chemistry. 2017;**2**:0100. DOI: 10.1038/s41570-017-0100

[29] Wen Y, Liu H, Jiang X. Preparation of graphene by exfoliation and its application in lithium-ion batteries. Journal of Alloys and Compounds. 2023;**961**:170885. DOI: 10.1016/j.jallcom.2023.170885

[30] Sattar T. Current review on synthesis, composites and multifunctional properties of graphene. Topics in Current Chemistry (Cham). 2019;**377**:10. DOI: 10.1007/s41061-019-0235-6

[31] Urade AR, Lahiri I, Suresh KS. Graphene properties, synthesis and applications: A review. JOM. 1989;**2023**(75):614-630. DOI: 10.1007/s11837-022-05505-8

[32] Choi SH, Yun SJ, Won YS, Oh CS, Kim SM, Kim KK, et al. Large-scale synthesis of graphene and other 2D materials towards industrialization. Nature Communications. 2022;**13**:1484. DOI: 10.1038/s41467-022-29182-y

[33] Zhang L, Qin X, Zhao S, Wang A, Luo J, Wang ZL, et al. Advanced matrixes for binder-free nanostructured electrodes in lithium-ion batteries. Advanced Materials. 2020;**32**:1908445. DOI: 10.1002/adma.201908445

[34] Van Noorden R. The rechargeable revolution: A better battery. Nature. 2014;**507**:26-28. DOI: 10.1038/507026a

[35] Guo H, Ping H, Hu J, Song X, Zheng J, Pan F. Controllable synthesis of $LiFePO_4$ in different polymorphs and study of the reaction mechanism. Journal of Materials Chemistry A. 2017;**5**:14294-14300. DOI: 10.1039/c7ta03369a

[36] Amaraweera THNG, Wijayasinghe A, Mellander BE, Dissanayake MAKL. Development of $Li(Ni_{1/3}Mn_{1/3}Co_{1/3-x}Na_x)O_2$ cathode materials by synthesizing with glycine nitrate combustion technique for Li-ion rechargeable batteries. Ionics. 2017;**23**:3001-3011. DOI: 10.1007/s11581-017-2098-6

[37] Shi Y, Wen L, Pei S, Wu M, Li F. Choice for graphene as conductive additive for cathode of lithium-ion batteries. Journal of Energy Chemistry. 2019;**30**:19-26. DOI: 10.1016/j.jechem.2018.03.009

[38] Song G-M, Wu Y, Xu Q, Liu G. Enhanced electrochemical properties of $LiFePO_4$ cathode for Li-ion batteries with amorphous NiP coating. Journal of Power Sources. 2010;**195**:3913-3917. DOI: 10.1016/j.jpowsour.2009.12.089

[39] Zou Y, Chen S, Yang X, Ma N, Xia Y, Yang D, et al. Suppressing Fe-Li Antisite defects in $LiFePO_4$/carbon hybrid microtube to enhance the lithium ion storage. Advanced Energy Materials. 2016;**6**:1601549. DOI: 10.1002/aenm.201601549

[40] Kang MS, Heo I, Kim S, Yang J, Kim J, Min S-J, et al. High-areal-capacity of micron-sized silicon anodes in lithium-ion batteries by using wrinkled-multilayered-graphenes. Energy Storage Materials. 2022;**50**:234-242. DOI: 10.1016/j.ensm.2022.05.025

[41] Jiao X, Kirianova AV, Xu X, Kapitanova OO, Krivchenko VA, Napolskiy FS, et al. Conductive additives for improving the rate capability of cathode materials in secondary lithium batteries. ACS Applied Energy Materials. 2023;**6**:2855-2862. DOI: 10.1021/acsaem.2c03772

[42] Liu H, Cheng X, Chong Y, Yuan H, Huang J-Q, Zhang Q. Advanced electrode processing of lithium ion batteries: A review of powder technology in battery fabrication. Particuology. 2021;**57**:56-71. DOI: 10.1016/j.partic.2020.12.003

[43] Liu Y, Zhang R, Wang J, Wang Y. Current and future lithium-ion battery manufacturing. iScience. 2021;**24**:102332. DOI: 10.1016/j.isci.2021.102332

[44] Al Hassan MR, Sen A, Zaman T, Mostari MS. Emergence of graphene as a promising anode material for rechargeable batteries: A review. Materials Today Chemistry. 2019;**11**:225-243. DOI: 10.1016/j.mtchem.2018.11.006

[45] Hong J, Zhang J, Li X, Guo Y, Zhou X, Liu Z. Graphene-wrapped composites of Si nanoparticles, carbon nanofibers, and pyrolytic carbon as anode materials for lithium-ion batteries. ACS Applied Nano Materials. 2023;**6**:10138-10147. DOI: 10.1021/acsanm.3c00911

[46] Zhang Q, Yang Y, Wang D, Zhang R, Fan H, Feng L, et al. A silicon/carbon/reduced-graphene composite of honeycomb structure for high-performance lithium-ion batteries. Journal of Alloys and Compounds. 2023;**944**:169185. DOI: 10.1016/j.jallcom.2023.169185

[47] Wei X, Guan Y, Zheng X, Zhu Q, Shen J, Qiao N, et al. Improvement on high rate performance of $LiFePO_4$ cathodes using graphene as a conductive agent. Applied Surface Science. 2018;**440**:748-754. DOI: 10.1016/j.apsusc.2018.01.201

[48] YongJian WU, RenHeng T, WenChao LI, Ying W, Ling H, LiuZhang O. A high-quality aqueous graphene conductive slurry applied in anode of lithium-ion batteries. Journal of Alloys and Compounds. 2020;**830**:154575. DOI: 10.1016/j.jallcom.2020.154575

[49] Zhao X, Jiaqiang E, Wu G, Deng Y, Han D, Zhang B, et al. A review of studies using graphenes in energy conversion, energy storage and heat transfer development. Energy Conversion and Management. 2019;**184**:581-599. DOI: 10.1016/j.enconman.2019.01.092

[50] Shi C, Owusu KA, Xu X, Zhu T, Zhang G, Yang W, et al. 1D carbon-based nanocomposites for electrochemical energy storage. Small. 2019;**15**:1902348. DOI: 10.1002/smll.201902348

[51] Ampong DN, Agyekum E, Agyemang FO, Mensah-Darkwa K, Andrews A, Kumar A, et al. MXene: Fundamentals to applications in electrochemical energy storage. Discover Nano. 2023;**18**:3. DOI: 10.1186/s11671-023-03786-9

[52] Gao S, Abduryim E, Chen C, Dong C, Guan X, Guo S, et al. Twin-graphene: A promising anode material for lithium-ion batteries with ultrahigh specific capacity. The Journal of Physical Chemistry C. 2023;**127**:14065-14074. DOI: 10.1021/acs.jpcc.3c01872

[53] Ji X, Mu Y, Liang J, Jiang T, Zeng J, Lin Z, et al. High yield production of 3D graphene powders by thermal chemical vapor deposition and application as highly efficient conductive additive of lithium ion battery electrodes. Carbon. 2021;**176**:21-30. DOI: 10.1016/j.carbon.2021.01.128

[54] Fang R, Chen K, Yin L, Sun Z, Li F, Cheng HM. The regulating role of carbon nanotubes and graphene in lithium-ion and lithium-Sulfur batteries. Advanced Materials. 2019;**31**:1800863. DOI: 10.1002/adma.201800863

[55] Hsu TH, Liu WR. Effects of graphene nanosheets with different lateral sizes as conductive additives on the electrochemical performance of $LiNi_{0.5}Co_{0.2}Mn_{0.3}O_2$ cathode materials for Li ion batteries. Polymers

(Basel). 2020;**12**:1162. DOI: 10.3390/polym12051162

[56] Ye Z, Zou Z, Jiang C. Enhancing the rate and cycle performance of graphite anode for Li-ion batteries by constructing a multidimensional conducting network. Diamond and Related Materials. 2023;**136**:109911. DOI: 10.1016/j.diamond.2023.109911

[57] Chi T, Wang X, Zeng L, Qin Z, Zhou X, Liu Z. Unraveling the effect of conductive additives on Li-ion diffusion using electrochemical impedance spectroscopy: A case study of graphene vs carbon black. Journal of the Electrochemical Society. 2023;**170**:040515. DOI: 10.1149/1945-7111/accb71

[58] Lv W, Li Z, Deng Y, Yang Q-H, Kang F. Graphene-based materials for electrochemical energy storage devices: Opportunities and challenges. Energy Storage Materials. 2016;**2**:107-138. DOI: 10.1016/j.ensm.2015.10.002

[59] Ke L, Lv W, Su F-Y, He Y-B, You C-H, Li B, et al. Electrode thickness control: Precondition for quite different functions of graphene conductive additives in $LiFePO_4$ electrode. Carbon. 2015;**92**:311-317. DOI: 10.1016/j.carbon.2015.04.064

[60] Xu L, Lv W, Shi K, Xiao S, You C, He Y-B, et al. Holey graphenes as the conductive additives for $LiFePO_4$ batteries with an excellent rate performance. Carbon. 2019;**149**:257-262. DOI: 10.1016/j.carbon.2019.04.025

[61] Liu T, Sun S, Zang Z, Li X, Sun X, Cao F, et al. Effects of graphene with different sizes as conductive additives on the electrochemical performance of a $LiFePO_4$ cathode. RSC Advances. 2017;**7**:20882-20887. DOI: 10.1039/c7ra02155k

[62] Nguyen TT, Nguyen NQ, Thai D, Tieu TD, Tran VM, Le MLP. Enabling stable and high-rate of an olivine-type cathode $LiFePO_4$ for Li-ion batteries by using graphene nanoribbons as conductive agent. Advances in Natural Sciences: Nanoscience and Nanotechnology. 2023;**14**:015009. DOI: 10.1088/2043-6262/acc734

[63] Murdock BE, Toghill KE, Tapia-Ruiz N. A perspective on the sustainability of cathode materials used in lithium-ion batteries. Advanced Energy Materials. 2021;**11**:2102028. DOI: 10.1002/aenm.202102028

[64] Zhou L, Yang H, Han T, Song Y, Yang G, Li L. Carbon-based modification materials for lithium-ion battery cathodes: Advances and perspectives. Frontiers in Chemistry. 2022;**10**:914930. DOI: 10.3389/fchem.2022.914930

[65] Hu J, Huang W, Yang L, Pan F. Structure and performance of the $LiFePO_4$ cathode material: From the bulk to the surface. Nanoscale. 2020;**12**:15036-15044. DOI: 10.1039/d0nr03776a

[66] Liu Z, Zhang R, Xu F, Gao Y, Zhao J. Structure and electrochemical performance of $LiFePO_4$ cathode materials modified with carbon coating and metal doping. Journal of Solid State Electrochemistry. 2022;**26**:1655-1665. DOI: 10.1007/s10008-022-05198-8

[67] Chen S-P, Lv D, Chen J, Zhang Y-H, Shi F-N. Review on defects and modification methods of $LiFePO_4$ cathode material for lithium-ion batteries. Energy & Fuels. 2022;**36**:1232-1251. DOI: 10.1021/acs.energyfuels.1c03757

[68] Wang C, Yuan X, Tan H, Jian S, Ma Z, Zhao J, et al. Three-dimensional carbon-coated $LiFePO_4$ cathode with improved Li-ion battery performance.

Coatings. 2021;**11**:1137. DOI: 10.3390/coatings11091137

[69] Liu R, Chen J, Li Z, Ding Q, An X, Pan Y, et al. Preparation of $LiFePO_4$/C cathode materials via a green synthesis route for lithium-ion battery applications. Materials (Basel). 2018;**11**:2251. DOI: 10.3390/ma11112251

[70] Wang X, Feng Z, Huang J, Deng W, Li X, Zhang H, et al. Graphene-decorated carbon-coated $LiFePO_4$ nanospheres as a high-performance cathode material for lithium-ion batteries. Carbon. 2018;**127**:149-157. DOI: 10.1016/j.carbon.2017.10.101

[71] Wei W, Lv W, Wu M-B, Su F-Y, He Y-B, Li B, et al. The effect of graphene wrapping on the performance of $LiFePO_4$ for a lithium ion battery. Carbon. 2013;**57**:530-533. DOI: 10.1016/j.carbon.2013.01.070

[72] Zhang Y, Wang P, Zheng T, Li D, Li G, Yue Y. Enhancing Li-ion battery anode performances via disorder/order engineering. Nano Energy. 2018;**49**:596-602. DOI: 10.1016/j.nanoen.2018.05.018

[73] Roy K, Banerjee A, Ogale S. Search for new anode materials for high performance Li-ion batteries. ACS Applied Materials & Interfaces. 2022;**14**:20326-20348. DOI: 10.1021/acsami.1c25262

[74] Li Z, Fu N, Yang Z. Particulate modification of lithium-ion battery anode materials and electrolytes. Particuology. 2023;**83**:129-141. DOI: 10.1016/j.partic.2023.02.010

[75] Selis LA, Seminario JM. Dendrite formation in silicon anodes of lithium-ion batteries. RSC Advances. 2018;**8**:5255-5267. DOI: 10.1039/c7ra12690e

[76] Kumar Prajapati A, Bhatnagar A. A review on anode materials for lithium/sodium-ion batteries. Journal of Energy Chemistry. 2023;**83**:509-540. DOI: 10.1016/j.jechem.2023.04.043

[77] Smyth CJ, Rennie AJ, Inkson BJ. Modification of graphene anode morphologies via wet and dry milling. Energy Procedia. 2018;**151**:168-173. DOI: 10.1016/j.egypro.2018.09.043

[78] Song H, Na R, Hong C, Zhang G, Li X, Kang Y, et al. In situ measurement and mechanism analysis of the lithium storage behavior of graphene electrodes. Carbon. 2022;**188**:146-154. DOI: 10.1016/j.carbon.2021.11.066

[79] Zhang Z, Liu X, Lu Y, Li H. Exploration of the lithium ions storage performance and mechanism of ZrN@reduced graphene oxide composite. Journal of Alloys and Compounds. 2023;**959**:170448. DOI: 10.1016/j.jallcom.2023.170448

[80] Bu F, Shakir I, Xu Y. 3D graphene composites for efficient electrochemical energy storage. Advanced Materials Interfaces. 2018;**5**:1800468. DOI: 10.1002/admi.201800468

[81] Xin W, De Rosa IM, Ye P, Severino J, Li C, Yin X, et al. Graphene template-induced growth of single-crystalline gold nanobelts with high structural tunability. Nanoscale. 2018;**10**:2764-2773. DOI: 10.1039/c7nr07514f

[82] Yang J, Zeng D, Zheng H, Xie Q, Huang J, Xiao L, et al. 3D graphene encapsulated ZnO-NiO-CuO double-shelled hollow microspheres with enhanced lithium storage properties. Journal of Alloys and Compounds. 2018;**765**:1158-1166. DOI: 10.1016/j.jallcom.2018.06.253

[83] Ding X, Wang Y. Bilayer-graphene-coated Si nanoparticles as advanced anodes for high-rate lithium-ion

batteries. Electrochimica Acta. 2020;**329**:134975. DOI: 10.1016/j.electacta.2019.134975

[84] Ette PM, Bosubabu D, Ramesha K. Graphene anchored mesoporous MnO_2 nanostructures as stable and high-performance anode materials for Li-ion batteries. Electrochimica Acta. 2022;**414**:140164. DOI: 10.1016/j.electacta.2022.140164

[85] Albers PW, Leich V, Ramirez-Cuesta AJ, Cheng Y, Hönig J, Parker SF. The characterisation of commercial 2D carbons: Graphene, graphene oxide and reduced graphene oxide. Materials Advances. 2022;**3**:2810-2826. DOI: 10.1039/d1ma01023a

[86] Bi J, Du Z, Sun J, Liu Y, Wang K, Du H, et al. On the road to the Frontiers of lithium-ion batteries: A review and outlook of graphene anodes. Advanced Materials. 2023;**35**:2210734. DOI: 10.1002/adma.202210734

[87] Sidoli M, Magnani G, Fornasini L, Scaravonati S, Morenghi A, Vezzoni V, et al. Defective graphene decorated with TiO_2 nanoparticles as negative electrode in Li-ion batteries. Journal of Alloys and Compounds. 2023;**958**:170420. DOI: 10.1016/j.jallcom.2023.170420

[88] Naraprawatphong R, Chokradjaroen C, Thiangtham S, Yang L, Saito N. Nanoscale advanced carbons as an anode for lithium-ion battery. Materials Today Advances. 2022;**16**:100290. DOI: 10.1016/j.mtadv.2022.100290

[89] Araújo MP, Soares OSGP, Fernandes AJS, Pereira MFR, Freire C. Tuning the surface chemistry of graphene flakes: New strategies for selective oxidation. RSC Advances. 2017;**7**:14290-14301. DOI: 10.1039/c6ra28868e

[90] Latorrata S, Balzarotti R. Advances in graphene and graphene-related materials. Applied Sciences. 2023;**13**:8929. DOI: 10.3390/app13158929

[91] Cao Y, Li X, Aksay IA, Lemmon J, Nie Z, Yang Z, et al. Sandwich-type functionalized graphene sheet-sulfur nanocomposite for rechargeable lithium batteries. Physical Chemistry Chemical Physics. 2011;**13**:7660. DOI: 10.1039/c0cp02477e

[92] Strativnov E, Khovavko A, Nie G, Ji PG. Graphite-graphene composite as an anode for lithium-ion batteries. Applied Nanoscience. 2023;**13**:7531-7536. DOI: 10.1007/s13204-023-02904-4

[93] He P, Chen S. Vertically oriented graphene Nanosheets for electrochemical energy storage. ChemElectroChem. 2021;**8**:758. DOI: 10.1002/celc.202100102

[94] Min W, Chen X, Huang S, Liao Y, Liang Z, Lei Y, et al. High performance lithium ion battery cathode based reduced holey graphene oxides from spent lithium ion batteries. Carbon. 2023;**210**:118038. DOI: 10.1016/j.carbon.2023.118038

[95] Ting P-M, Huang J-Y, Muruganantham R, Liu W-R. Nitrogen-doping effects on few-layer graphene as an anode material for lithium-ion batteries. Materials Today Communications. 2022;**31**:103498. DOI: 10.1016/j.mtcomm.2022.103498

[96] Yin X, Li Y, Cai W, Fan C, Liu W, Wang N, et al. In-situ synthesis of Cu-based conductive metal organic frameworks on graphene layers for high-performance lithium and potassium ion batteries. Applied Surface Science. 2023;**624**:157124. DOI: 10.1016/j.apsusc.2023.157124

[97] Ye J, Chen H, Wang Q, Huang P, Sun J, Lo S. Thermal behavior and failure

mechanism of lithium ion cells during overcharge under adiabatic conditions. Applied Energy. 2016;**182**:464-474. DOI: 10.1016/j.apenergy.2016.08.124

[98] Alami AH, Maghrabie HM, Abdelkareem MA, Sayed ET, Yasser Z, Salameh T, et al. Potential applications of phase change materials for batteries' thermal management systems in electric vehicles. Journal of Energy Storage. 2022;**54**:105204. DOI: 10.1016/j.est.2022.105204

[99] Al-Zareer M, Dincer I, Rosen MA. A review of novel thermal management systems for batteries. International Journal of Energy Research. 2018;**42**:3182-3205. DOI: 10.1002/er.4095

[100] Abdelkareem MA, Maghrabie HM, Abo-Khalil AG, Adhari OHK, Sayed ET, Radwan A, et al. Thermal management systems based on heat pipes for batteries in EVs/HEVs. Journal of Energy Storage. 2022;**51**:104384. DOI: 10.1016/j.est.2022.104384

[101] Ma S, Jiang M, Tao P, Song C, Wu J, Wang J, et al. Temperature effect and thermal impact in lithium-ion batteries: A review. Progress in Natural Science: Materials International. 2018;**28**:653-666. DOI: 10.1016/j.pnsc.2018.11.002

[102] Ruffa F, De Capua C, Morello R, Liu Z. Temperature sensing and evaluation of thermal effects on battery packs for automotive applications. IEEE Sensors Journal. 2019;**19**:11634-11645. DOI: 10.1109/jsen.2019.2933901

[103] Paarmann S, Schuld K, Wetzel T. Inhomogeneous aging in lithium-ion batteries caused by temperature effects. Energy Technology. 2022;**10**:2200384. DOI: 10.1002/ente.202200384

[104] Kim J, Oh J, Lee H. Review on battery thermal management system for electric vehicles. Applied Thermal Engineering. 2019;**149**:192-212. DOI: 10.1016/j.applthermaleng.2018.12.020

[105] Ghaeminezhad N, Wang Z, Ouyang Q. A review on lithium-ion battery thermal management system techniques: A control-oriented analysis. Applied Thermal Engineering. 2023;**219**:119497. DOI: 10.1016/j.applthermaleng.2022.119497

[106] Agyenim F, Hewitt N, Eames P, Smyth M. A review of materials, heat transfer and phase change problem formulation for latent heat thermal energy storage systems (LHTESS). Renewable and Sustainable Energy Reviews. 2010;**14**:615-628. DOI: 10.1016/j.rser.2009.10.015

[107] Sharma A, Tyagi VV, Chen CR, Buddhi D. Review on thermal energy storage with phase change materials and applications. Renewable and Sustainable Energy Reviews. 2009;**13**:318-345. DOI: 10.1016/j.rser.2007.10.005

[108] Lawag RA, Ali HM. Phase change materials for thermal management and energy storage: A review. Journal of Energy Storage. 2022;**55**:105602. DOI: 10.1016/j.est.2022.105602

[109] Safari A, Saidur R, Sulaiman FA, Xu Y, Dong J. A review on supercooling of phase change materials in thermal energy storage systems. Renewable and Sustainable Energy Reviews. 2017;**70**:905-919. DOI: 10.1016/j.rser.2016.11.272

[110] Kenar JA. Latent heat characteristics of biobased oleochemical carbonates as potential phase change materials. Solar Energy Materials and Solar Cells. 2010;**94**:1697-1703. DOI: 10.1016/j.solmat.2010.05.031

[111] Pielichowska K, Pielichowski K. Phase change materials for thermal energy storage. Progress in Materials Science. 2014;**65**:67-123. DOI: 10.1016/j.pmatsci.2014.03.005

[112] Jiang T, Zhang Y, Olayiwola S, Lau C, Fan M, Ng K, et al. Biomass-derived porous carbons support in phase change materials for building energy efficiency: A review. Materials Today Energy. 2022;**23**:100905. DOI: 10.1016/j.mtener.2021.100905

[113] Wu S, Yan T, Kuai Z, Pan W. Thermal conductivity enhancement on phase change materials for thermal energy storage: A review. Energy Storage Materials. 2020;**25**:251-295. DOI: 10.1016/j.ensm.2019.10.010

[114] Singh P, Sharma RK, Khalid M, Goyal R, Sarı A, Tyagi VV. Evaluation of carbon based-supporting materials for developing form-stable organic phase change materials for thermal energy storage: A review. Solar Energy Materials and Solar Cells. 2022;**246**:111896. DOI: 10.1016/j.solmat.2022.111896

[115] Shi J, Qin M, Aftab W, Zou R. Flexible phase change materials for thermal energy storage. Energy Storage Materials. 2021;**41**:321-342. DOI: 10.1016/j.ensm.2021.05.048

[116] Goli P, Legedza S, Dhar A, Salgado R, Renteria J, Balandin AA. Graphene-enhanced hybrid phase change materials for thermal management of Li-ion batteries. Journal of Power Sources. 2014;**248**:37-43. DOI: 10.1016/j.jpowsour.2013.08.135

[117] Hussain A, Abidi IH, Tso CY, Chan KC, Luo Z, Chao CYH. Thermal management of lithium ion batteries using graphene coated nickel foam saturated with phase change materials. International Journal of Thermal Sciences. 2018;**124**:23-35. DOI: 10.1016/j.ijthermalsci.2017.09.019

Section 4

Graphene Toxicities

Chapter 8

Graphene Oxide Nanotoxicity: A Comprehensive Analysis

Mohammad Mahdi Sepahi and Marzieh Azizi

Abstract

Graphene oxide (GO) has emerged as a promising nanomaterial with physicochemical properties that make it a candidate for a wide range of applications. However, the potential toxicity of GO is a significant threat that must be addressed before GO's safe use in biomedical and environmental applications can fully be realized. Numerous studies have demonstrated that GO has the ability to induce oxidative stress, inflammation, genotoxicity, and cytotoxicity in cell types and animal models. Importantly, the toxicity of GO is influenced by its size, morphology, charge, and surface functional groups. The current review summarizes recent research findings on the toxicity of GO by focusing on its cellular uptake, biodistribution, and biological effects. It provides an overview of the current understanding of GO nanotoxicity and highlights the need for additional research to assess its safety in various applications. By addressing these concerns and developing appropriate safety measures, we can fully exploit the potential of GO while ensuring its safe utilization in biomedical and environmental settings. Thus, the comprehensive evaluation of GO's potential risks is crucial for its full exploitation and utilization.

Keywords: biodistribution, cellular uptake, genotoxicity, graphene oxide, inflammation, nanotoxicity, oxidative stress

1. Introduction

Graphene oxide (GO) has garnered significant attention as a highly promising nanomaterial due to its exceptional physicochemical properties, including its large surface area, high mechanical strength, excellent electrical conductivity, and remarkable optical properties [1–4]. These characteristics make GO an attractive candidate for a wide range of applications in the fields of biomedicine, electronics, energy storage, and environmental remediation among others [5, 6]. However, numerous in vitro and in vivo studies using different cell types and animal models have demonstrated that GO can induce various adverse effects, including oxidative stress, inflammation, genotoxicity, and cytotoxicity, among other toxicities [2, 7, 8]. These toxic effects have raised concerns about the safe use of GO in biomedical and environmental applications highlighting the need to investigate its potential toxicity (**Figure 1**).

The mechanisms underlying the toxic effects of GO involve its interactions with cellular components such as proteins, lipids, and nucleic acids. These interactions can disrupt cellular functions and cell signaling pathways, leading to cellular damage and other adverse biological effects [9–13]. The toxicity of GO is influenced by

IntechOpen

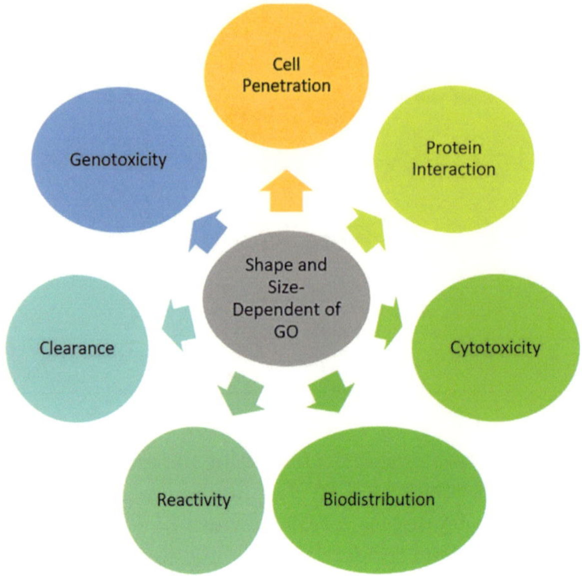

Figure 1.
Shape and size-dependent nanotoxicity of GO.

its physicochemical properties, including size, morphology, charge, and surface functional groups [14–16]. Ou et al., for example, reviewed the toxicity of graphene family nanomaterials (GFNs) and found that the nanotoxicity of the material was influenced by, among other factors, the size, surface structure, functionalization, charge, impurities, aggregations, and corona effect. Different mechanisms have been identified to explain GFN toxicity, including physical destruction, oxidative stress, DNA damage, inflammatory response, apoptosis, autophagy, and necrosis. Toll-like receptors (TLRs), transforming growth factor β (TGF-β), and tumor necrosis factor-alpha (TNF-α) dependent pathways are involved in the signaling pathway network, with oxidative stress playing a crucial role in these pathways [17]. Understanding the relationship between these properties and the toxicity of GO is crucial for designing safer nanomaterials and minimizing the risks of their use.

Thus, in order to assess the toxicity of GO comprehensively, it is essential to evaluate its cellular uptake, its biodistribution through the various biological matrices, and the accompanying biological effects. The understanding of how GO is internalized by cells, and how it gets distributed within the body leading to the specific biological responses that it elicits can provide valuable insights into its potential hazards and, therefore, guide the development of appropriate mitigation measures [18]. Moreover, the full analysis of the current knowledge of GO nanotoxicity can help identify knowledge gaps and highlight areas where further research is required in assessing its safety in various applications.

By addressing these concerns and developing effective strategies to minimize the adverse effects associated with GO, we can exploit its immense potential while ensuring its safe applications. The current analysis, therefore, aims to summarize recent research findings on the toxicity of GO and underscore the need for additional research to assess its safe use in different fields. The physicochemical properties of GO are important in understanding its nanotoxicity because they determine the interactions between the nanomaterials and living organisms. The combined effect of size and shape in nanotoxicity studies is paramount, influencing cellular uptake, biodistribution, clearance, and overall interactions with living organisms [19]. The size and shape of GO nanoparticles are linked to their reactivity and cellular uptake in biological systems. The potential toxicity of GO nanoparticles can affect the nervous tissues and other crucial body systems [20] and understanding these factors, therefore, is essential for designing safer and clinically relevant GO-based nanosystems. Furthermore, the different shapes of nanomaterials, such as sheets [10], flakes, and nanoribbons, can influence the speed and extent of their internalization and interactions with biological structures. GO nanosheets, for example, have a higher potential for membrane penetration than the other GO morphologies, leading to increased cytotoxicity. In contrast, GO nanoribbons have been implicated in inducing greater genotoxic effects due to their ability to cause DNA damage [9, 12]. Additionally, GO sheets with irregular shapes are cleared more slowly from the body, resulting in prolonged exposure and increased toxicity over time. Conversely, GO nanoparticles with a more spherical shape are cleared more quickly from the body, reducing their potential toxicity over time [21]. These findings emphasize the critical role of the shape of the GO nanoparticles in controlling the toxicity and safety of GO-based materials. Consequently, the size of GO materials is another critical determinant of its cytotoxicity [18, 22].

Surface functionalization of GO has been widely studied in its performance in biomedical applications, such as drug delivery and diagnostics. This technique allows tailoring the properties of nanomaterials to improve their interactions with biological systems for specific applications [23, 24]. Surface functionalization involves modifying the surface properties of nanomaterials by attaching various molecules to alter their stability, solubility, and interactions with biological systems [25, 26]. Different functional groups can influence the toxicity of nanomaterials by affecting cellular uptake, biodistribution, and biocompatibility [27]. However, surface functionalization can also affect particle agglomeration and enhance stability [28], which can impact their behavior and toxicity in biological systems. Also, functionalization of GO nanoparticles can alter the physicochemical properties of GO, influencing its toxicity [16]. For instance, functionalization with hydrophilic groups can reduce aggregation and enhance biocompatibility, while functionalization with targeting ligands can improve the specificity of GO-based drug delivery systems [29]. By carefully selecting appropriate surface functionalization strategies, researchers can mitigate potential nanotoxicity risks associated with GO.

In the realm of biomedical applications, surface functionalization emerges as a promising strategy in enabling targeted delivery of therapeutic agents using GO nanoparticles. However, the selection of functional groups and their stability is a crucial consideration in ensuring both the safety and effectiveness of targeted applications. This multifaceted approach to utilizing GO, in conjunction with a comprehensive understanding of its concentration-dependent effects on cell viability, paves the way for innovative and tailored applications with potentially far-reaching benefits.

The resulting surface GO nanoparticles control their interactions with biological systems and it plays a critical role in determining the toxicity of GO in cells and tissues by impacting their stability, aggregation, and interactions with biological systems [12, 30–32] as well as their cellular uptake pathways (**Figure 2**). Positively charged sheets are taken up through phagocytosis and clathrin-mediated endocytosis, while negatively charged sheets cross the membrane through phagocytosis and sulfate-receptor-mediated endocytosis [33]. Also, the charge of nanoparticles controls their affinity for cell membranes, leading to variations in cellular uptake and toxicity [34]. Furthermore, the surface charge of nanoparticles can lead to an imbalance in the

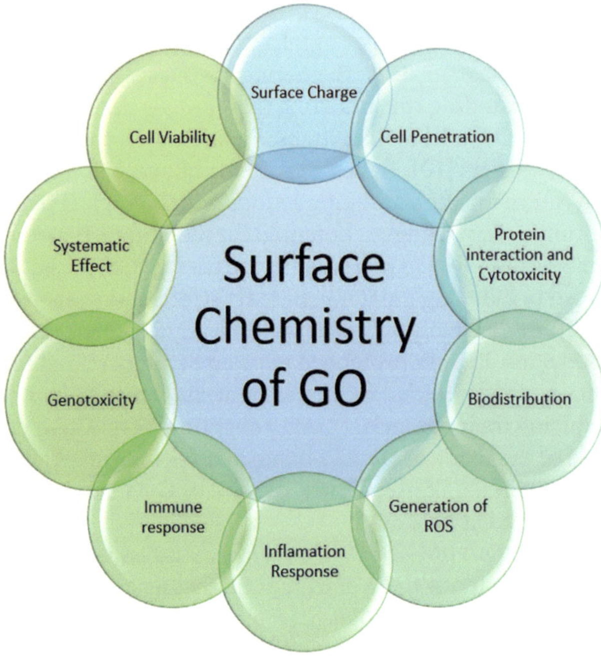

Figure 2.
Key points regarding the effect of GO surface functional group on nanotoxicity.

redox state of cells, which, in turn, can cause malfunctions in the internal metabolism of cells and in pathologies at the organismal level [8].

Whichever the case, the surface chemistry of GO has a significant influence on its nanotoxicity, affecting various aspects of their behavior and interactions with biological systems. Understanding the role of surface chemistry in GO nanotoxicity is therefore essential for assessing potential risks and developing strategies to mitigate toxicity.

2. Biodistribution

2.1 Effects of GO size and shape on biodistribution

The effects of particle size on biodistribution have different aspects, impacting circulation time, organ accumulation, and the kinetics of clearance. Smaller nanoparticles tend to have a longer circulation time within the body and can accumulate in specific organs or tissues, magnifying their toxic potential. On the other hand, larger nanoparticles are cleared more rapidly from the body, reducing their overall toxicity. For example, nanomaterials smaller than 20 nm tend to accumulate in the kidney, while those in the range of 20–100 nm preferentially deposit in the liver [35].

Nonetheless, different shapes, such as spheres, rods, or sheets, can exhibit varying degrees of cellular uptake, distribution, and toxicity. GO sheets with irregular shapes are cleared more slowly from the body compared to those with more regular shapes. Prolonging the exposure period through slower rates of clearance intensifies the toxicities of the particles over time [36]. Conversely, GO nanoparticles with a more spherical shape are cleared more rapidly from the body, reducing their potential toxicity over time. **Figure 3** shows the clearance and biodistribution of GO in the different study.

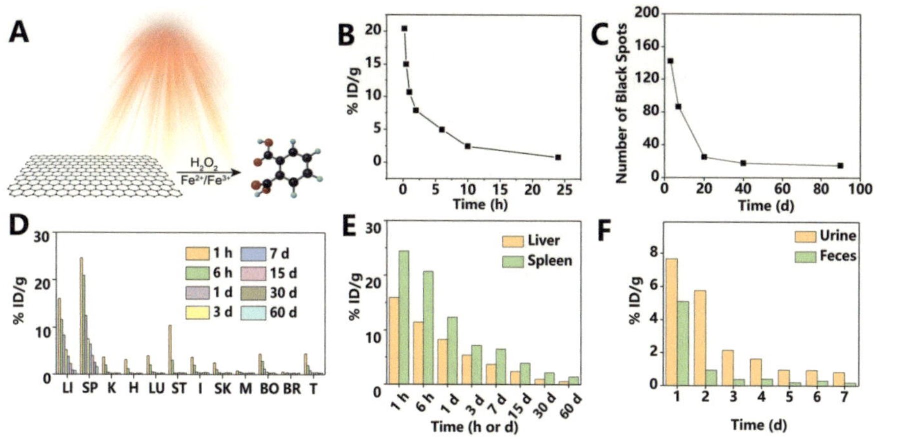

Figure 3.
Clearance and biodistribution of GO. (A) Proposed pathway for the photo-Fenton reaction, which contributes to the degradation of GO, potentially leading to the formation of oxidized polycyclic aromatic hydrocarbons [18]. (B) The blood circulation curve of GO is depicted, highlighting its behavior in the bloodstream. (C) The number of black spots observed in liver sections at different time points post-injection (PI) indicates the accumulation of GO in the liver. (D) The time-dependent distribution of GO in different organs, including the LI: liver; SP: spleen; K: kidney; H: heart; LU: lungs; ST: stomach; I: intestine; SK: skin; M: muscle; BO: bone; BR: brain; and T: thyroid, is presented. (E) The signal of GO in the liver and spleen is shown at different time points PI. (F) The presence of GO in urine and feces is demonstrated at different time points PI [18].

2.2 Functional group effects on biodistribution

Functionalization of GO nanoparticles strongly influences their biodistribution. For example, intravenous administration of dextran-functionalized GO led to extensive fecal excretion [37, 38], whereas PEGylated GO showed lesser feces excretion suggesting that the choice of functional groups can impact the biodistribution and clearance pathways of GO nanoparticles [39]. This also indicated that dextran-functionalized GO might be excreted through the gastrointestinal system, while PEGylated GO may be cleared differently [40]. **Table 1** provides the clearance and biodistribution of 10–30 nm GO with 1 nm thickness and the corresponding tracking methods and the influencing factors.

Functional group	Organ of biodistribution	Clearance speed	Tracking method
PEG [39]	Spleen [39]	Slow[1]	Radiolabeling [39]
Cyclic R10 peptide [41]	Liver [39]		
DNA [42]	Stomach [39]		TEM, confocal Raman mapping [43]
MicroRNA [44]	Lungs [39]		
Boron doping [45]	Kidneys [39]		Mass spectrometry imaging [46]
Liposomes [47]			

[1]*Slow biodegradation means that over half of the material degrades more than 1 month.*

Table 1.
Clearance and biodistribution of 10–30 nm GO with 1 nm thickness and the corresponding tracking methods and the influencing factors [18].

In a study by Yang and colleagues, PEGylated GO biodistribution was tracked over time. The pharmacokinetics analysis revealed a two-compartment model: the central compartment with fast material exchange (e.g., blood, heart, and liver) and the peripheral compartment with slow material exchange (e.g., muscle, bone, and fat). The blood circulation of GO was found to consist of a rapidly distributed component and a slower elimination component. The half-lives of the blood circulation in the first and second phases were 0.39 ± 0.097 and 6.97 ± 0.62 hour, respectively [18]. At 1-hour post-injection, GO was distributed across different organs of the mice. Strong signals were observed in the bone marrow, which may be due to macrophage uptake of GO into the bone marrow. The radiation signal was mainly concentrated in the mononuclear phagocyte system (MPS), such as the liver and spleen. This experience evidenced the influence of circulation time of GO nanoparticles in various body compartments. They were found to distribute across different organs, including the bone marrow, liver, and spleen. It showed that the surface chemistry of GO nanoparticles influences their interactions with macrophages in the MPS, affecting their accumulation in these organs [18].

Long-term biodistribution studies found that GO was significantly present in the liver, indicating accumulation, which was attributed to enzymatic and photo-Fenton degradation. However, a small amount of GO remained in the liver even after 20 days, raising concerns about potential long-term toxicity [48]. GO also accumulated in the kidneys and intestines, suggesting its excretion through urine and feces [18].

2.3 Organ toxicity

The biodistribution of GO in various organs, such as the liver, lungs, and spleen, depends on the route of administration and dosage. Higher doses of GO can lead to organ toxicity, characterized by inflammation, oxidative stress, and histopathological changes. These findings spotlight the importance of dose and route of administration in organ toxicity GO [39, 49].

Animal studies have been conducted to assess the in vivo toxicity of GO. Intravenous injection of GO of various sizes in mice resulted in acute and chronic damage to the lung and kidney. This highlights the potential adverse effects of GO on vital organs and the importance of considering toxicity in animal models. Similar in vivo studies have shown that blood exposure to GO can cause accidental death in mammals, including non-human primates and mice, due to acute anaphylactic reactions. The severity of GO's toxic effects on living cells and organs varies based on factors such as administration route, dose, synthesis method, and physicochemical properties. Smaller GO nanoplatelets have shown lower toxicity and potential as drug carriers while enhancing the anticancer effect of chemotherapeutic drugs. This emphasizes the need for further studies to fully understand the potential toxicity of GO and its impact on various aspects, including behavior, reproduction, development, and genotoxicity [49].

2.4 Clearance

The size and shape of GO nanoparticles also play a crucial role in their clearance from the body, impacting the resident contact time with tissues and long-term potential toxicity. Smaller and thinner 2D GO sheets exhibit distinct clearance patterns in the body. Well-dispersed and smaller GO sheets have the capability to pass through the lung capillaries and get excreted through the glomerular filtration barrier (GFB) in the kidney, a process facilitated by their structural features, such as thinness, flexibility, and orientation in flow [50]. The urinary excretion of GO sheets following systemic circulation administration has been reported in several studies [37, 39, 51, 52]. In an animal study using a direct bladder puncture method, about 50–60% of GO was excreted in the urine 2 hours after injection [53, 54] and no significant changes in kidney function or structural damage to the glomerular and tubular regions were found after injection of well-dispersed GO nanosheets. GO's ability to pass through the kidney's filtration barrier is remarkable, considering that most GO nanosheets are larger than the pore size in the kidneys. Nevertheless, they can traverse these pores through rolling, crumpling, or folding, facilitated by their mechanical flexibility. This process allows thin graphene with high mechanical flexibility to be easily discharged from the body (**Figure 4**) [18].

It is noteworthy that the feces excretion of GO has also been observed, although to a lesser extent. The extent of feces elimination has been linked to specific functionalization moieties and routes of administration. For instance, intravenous administration of dextran-functionalized GO [37, 38] resulted in extensive feces excretion, while PEGylated GO showed it to a lesser extent [39]. Oral administration of GO similarly led to extensive feces elimination [50].

Although most GO nanosheets are excreted in the urine, a small but significant part remains isolated in the spleen at approximately 1.6% ID/g at 60 days after intravenous injection [51]. GO accumulates in the marginal zone macrophages of the mouse spleen [43]. Over time, the quantity of GO in the spleen decreases, but a portion remains inaccessible for removal due to slow degradation, a process taking over 270 days [50]. The spleen-residing GO gradually accumulates defects, resulting in structural changes and a transition from the initial sp2-hybridized carbon structure to mainly tetrahedral sp3 amorphous carbon [18, 43]. The physiological environment of the spleen is essential for the degradation process of GO, as GO incubated in vitro without exposure to light at 37°C did not exhibit similar structural changes [50].

This extensive exploration of GO's distribution and elimination in the body emphasizes the complex relationship between its size, shape, and to a lesser extent, mode of administration and surface functionalization, and the body's clearance processes. Smaller and thinner as well as more regular-shaped GO sheets exhibit unique excretion pathways and faster clearance, while larger, thicker sheets with more irregular morphology tend to accumulate in specific body organs, especially the spleen, leading to a prolonged clearance process and simultaneous toxic damage.

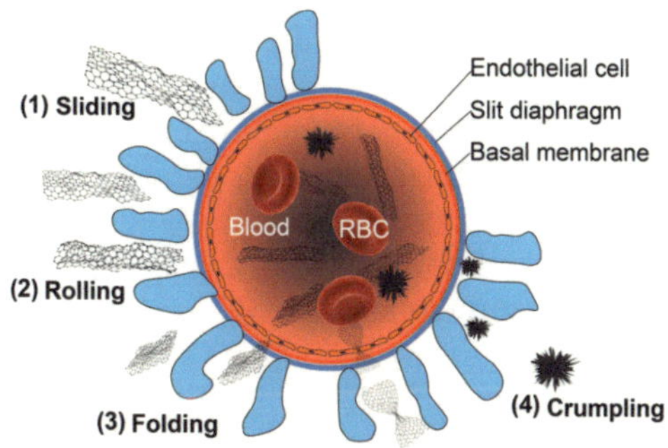

Figure 4.
Morphological reconfiguration of GO nanosheets during entrance to the glomerular filtration barrier that enables the thin and flexible GO nanosheets to cross the GFB and be excreted in the urine [18].

3. Penetration

The size of GO also plays a crucial role in its interaction with living organisms. Smaller nanoparticles exhibit a larger surface area-to-volume ratio, enhancing their reactivity and cellular uptake over those of the larger particles.

3.1 Influence of size and shapes

This is because the smaller nanoparticles experience less steric hindrance and have an increased propensity to penetrate biological barriers, which can result in greater interaction and higher toxicity levels. Also, smaller particles possess a larger surface area-to-volume ratio, enhancing their potential for interactions with biomolecules [31]. It has been reported that both sizes of GO (a ~500 nm protein-coated commercial GO and GO sheets of 1 μm particle range) crossed over into the lysosomes for excretion and that the smaller nanoparticles induce more pronounced cellular stress responses and oxidative damage exacerbating their cytotoxic effects compared to their larger ones [8]. In a study on mouse mesenchymal progenitor C_2C1_2 cells, a ~500 nm protein-coated commercial GO was internalized through clathrin-mediated endocytosis, while larger GO sheets of 1 μm particle range were taken up through phagocytosis. Enhanced cellular uptake can result in higher toxicity, as internalized nanoparticles interact with cellular components and induce adverse biological responses. On the other hand, different shapes of GO materials possess varying degrees of surface area, surface charge, and functional groups, all of which influence their binding affinity and interactions with biomolecules [7, 19, 31]. Rod-shaped nanoparticles, for instance, have been observed to exhibit greater cellular internalization relative to spherical nanoparticles, although some studies have yielded contradictory conclusions [55]. This effect, which is influenced by shape, underscores the importance of different shapes of GO in determining the safety and nanotoxicity of the material. Moreover, studies involving GO nanoflakes have shown their uptake by both cancerous and non-cancerous cell lines through cellular uptake pathways tracked by label-free Raman imaging [56].

Mendes et al., illustrated during a 12-hour incubation period, human macrophages (differentiated THP-1 cells) were exposed to nano-graphene oxide (NGO) at different concentrations. The data from the study did not show a clear difference in internalization between the concentrations or sizes of NGO flakes. Regardless of the conditions, the NGO was observed as large collections of flakes rather than individual flakes inside the cells. Notably, at a concentration of 100 μg/mL, the cells were able to internalize agglomerations of NGO with diameters of up to 4 μm. The study also conducted a detailed transmission electron microscopy (TEM) analysis to differentiate the NGO flakes from other intracellular structures, such as cellular membranes and endosome membranes. The TEM analysis confirmed the ability to distinguish NGO flakes from other internal cellular structures (**Figure 5**).

It illustrates a key concept that the surface area-to-volume ratio of small nanoparticles enhances their reactivity and propensity for cellular uptake. Furthermore, the shape of nanomaterials significantly influences their interactions with cells and tissues.

3.2 Influence of surface chemistry

The charge of GO nanoparticles also has a strong impact on their cellular uptake mechanisms and, in turn, their potential toxic effects [19, 33, 43, 55]. The electrostatic interactions between the charge of GO and cell membranes are essential determinants

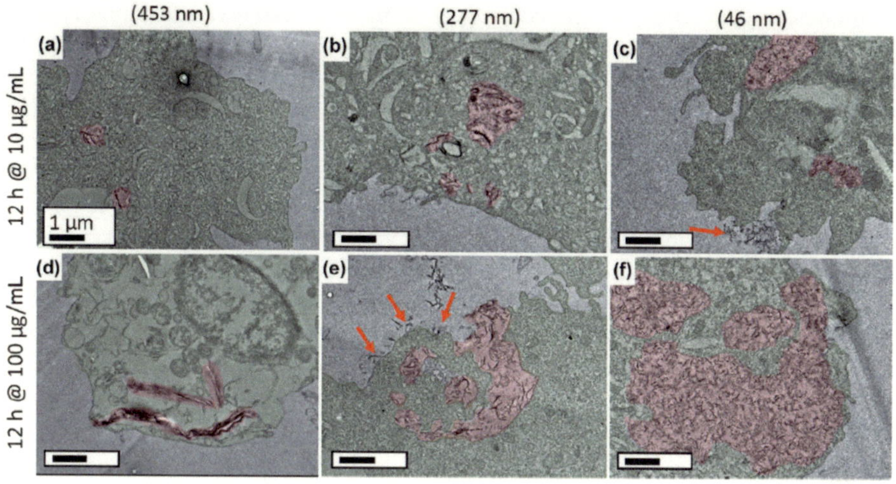

Figure 5.
Differentiated THP-1 cells that treated by NGO flakes for 12 hours. The NGO flakes had an average diameter of 453, 277, and 46 nm and were used at concentrations of 10 μg/mL (a–c) and 100 μg/mL (d–f). The red-shaded areas indicate the internalized NGO flakes or clusters. Additionally, the red arrows in (c) and (e) point to the NGO flakes and clusters that are in contact with the cellular membrane.

of the cellular uptake of GO nanoparticles. This means that the surface charge of GO nanoparticles directly affects their ability to internalize into cells and the specific cellular endocytosis pathways they utilize. In particular, positively charged GO nanoparticles exhibit a strong electrostatic attraction to negatively charged cell membranes, making them more readily absorbed by cells compared to negatively charged and neutral GO nanoparticles. This difference in cellular uptake mechanisms can have profound implications for the potential nanotoxicity of GO [31]. Studies have shown that negatively charged GO nanoparticles have a lower rate of endocytosis and do not typically utilize the clathrin-mediated endocytosis pathway, which is a common mechanism for cellular uptake. In contrast, positively charged GO nanoparticles internalize more rapidly via the clathrin-mediated pathway. This difference in uptake mechanisms can have implications for the intracellular fate of GO nanoparticles and the potential biological responses they trigger [57]. For example, research by Yang et al. in 2017 confirmed that positively charged GO nanoparticles were internalized more efficiently compared to negatively charged or neutral GO nanoparticles in human lung epithelial cells (A549). Once internalized, GO nanoparticles interface with intracellular components and trigger complex inflammatory responses. This observation highlights the importance of the surface charge of GO in influencing its cellular uptake, which can subsequently impact its nanotoxicity [58]. The extent of cellular uptake and internalization is, however, tied to the physicochemical characteristics of GO, underlining the significance of these properties in shaping the toxic outcomes [31].

4. Protein interaction

4.1 Effect of size and shape on GO protein interactions

Recent research has illuminated the profound influence of the GO shape on its cellular interactions. For instance, GO micropatterns with different shapes, including triangles and squares, impact differently the cell migration parameters, such as distance, speed, and directionality, underlining the critical role of shape-dependent effects of GO in living organisms [33]. However, the impact of shape and size on GO nanotoxicity extends beyond cellular interactions and encompasses its influence on protein interactions. The size and shape of GO nanomaterials play a critical role in governing their interactions with biomolecules, particularly proteins and DNA. GO size and shape play a pivotal role in dictating its interactions with subcellular components in living organisms, leading to the induction of functional perturbations in cells and tissues. The shape and size-dependent GO nanotoxicity is, therefore, a critical aspect to be considered when examining the potential hazards associated with GO nanomaterials. In this context, the size of GO nanoparticles assumes significance, as smaller nanoparticles exhibit an enlarged surface area-to-volume ratio, intensifying their reactivity in biological systems. Consequently, the size of GO nanoparticles emerges as an important factor in shaping their interactions with biomolecules. The shape of GO materials also affects their surface area, which, in turn, has a direct bearing on their proteins and other bio-molecular pharmacokinetics [59]. GO nanoflakes, characterized by a higher surface area, have been found to adsorb a greater amount of proteins compared to GO sheets with a lower surface area, significantly enhancing their biological activities and toxicity over those of the nanosheets [33]. Moreover, the shape of nanomaterials significantly determines their interactions with cells and tissues. Nanomaterials characterized by sharp edges or high aspect ratios have been associated with more pronounced damage to cells and tissues [22, 34]. Thus, comprehending the influence of nanomaterial shape on cellular interactions is vital for designing safer and more clinically relevant nanosystems. This highlights the significance of shape-dependent interactions with cells, irrespective of their physiological state.

4.2 Effect of surface chemistry on GO protein interactions

The functional groups and the resulting surface charge of GO nanoparticles strongly influence GO's interactions with proteins and, in turn, the extent of its cytotoxicity. The charge of GO nanoparticles can lead to damage of cells through its effects on cellular proteins and nucleic acids. Positively charged nanoparticles have been shown to disrupt cell membranes, induce oxidative stress, and trigger inflammatory responses. In contrast, negatively charged nanoparticles interfere with cellular processes and induce genotoxic effects. Overall, positively charged GO nanoparticles are often found to be more cytotoxic compared to negatively charged or neutral GO nanoparticles [60]. This was illustrated in a study by Peng et al. in 2018; the effect of charge on GO's protein adsorption was investigated using human serum albumin (HSA). The study revealed that positively charged GO nanoparticles adsorbed more HSA compared to negatively charged or neutral GO nanoparticles. This observation was attributed to the electrostatic attraction between positively charged GO and negatively charged HSA, highlighting the role of charge in influencing protein interactions with GO [61].

The charge of GO nanoparticles not only affects their interactions with proteins but also their compatibility with blood components. Again, positively charged nanoparticles tend to interact more with blood proteins, leading to the formation of protein coronas [62]. These protein coronas can significantly impact the behavior of nanoparticles, including their circulation time and potential for clearance from the bloodstream. These interactions play a crucial role in influencing the overall toxicity of GO nanoparticles. A study by Zhang et al. in 2018 investigated the effect of charge on GO's hemolytic activity using red blood cells (RBCs) [63]. The study found that positively charged GO nanoparticles were more hemolytic compared to negatively charged or neutral GO nanoparticles. This observation suggested a greater electrostatic interaction between positively charged GO and negatively charged RBC membranes [63].

In addition to charge, the surface functional groups of GO nanoparticles also play a vital role in modulating the cellular proteins' interactions with GO. These functional groups can affect protein adsorption, leading to the formation of a protein corona. The composition and structure of the protein corona have a significant impact on the biological effect, fate, and toxicity of GO nanoparticles [62]. This aspect provides opportunities to control protein corona formation and mitigate potential adverse effects, making it an important consideration in the design and evaluation of GO-based nanomaterials. Furthermore, the surface functional groups alter the surface charge, hydrophobicity, and other properties of GO nanoparticles. These modifications affect interactions with various blood components and different organs inducing such changes as in circulation time, rate and extent of accumulation in specific tissues, and clearance from the body, which bear direct implications for the overall toxicity profile of GO nanoparticles [18, 62].

The intricate reciprocity between the surface chemistry of GO and its potential to induce oxidative stress, specifically reactive oxygen species (ROS) generation, is a multifaceted phenomenon with significant implications for nanotoxicology and cellular responses. GO nanoparticles have been widely recognized for their propensity to cause oxidative stress, which is linked to genotoxicity and cytotoxicity [60, 64]. While several studies have elucidated the role of surface chemistry in modulating these outcomes, the impact of GO's surface area and functionalization on ROS generation is an intriguing facet that merits further exploration. Notably, a study conducted with SH-SY5Y cells revealed that GO with a surface area of 750 m^2/g exhibited an unexpectedly lower potential to induce oxidative stress through lipid peroxidation, as evidenced by significantly decreased malondialdehyde (MDA) levels [65]. For the 150 m^2/g graphene, no significant increase in MDA was observed at any concentration or time compared to the control group ($p > 0.05$). However, a significant decrease in MDA levels was observed at a concentration of 400 μg/mL after 24 and 48 hours compared to the cell control group ($p \leq 0.01$, $p \leq 0.001$) [65]. **Figure 6** shows the impact of two different types of graphene on MDA levels in SH-SY5Y cells [25, 65].

This unexpected finding challenges traditional notions about the pro-oxidative potential of GO and highlights the importance of GO surface area considerations. Furthermore, the observed increase in cellular glutathione (GSH) levels following exposure to GO, a known antioxidant marker, hints at a potential antioxidant role for certain GO formulations (**Figure 7**) [26, 65].

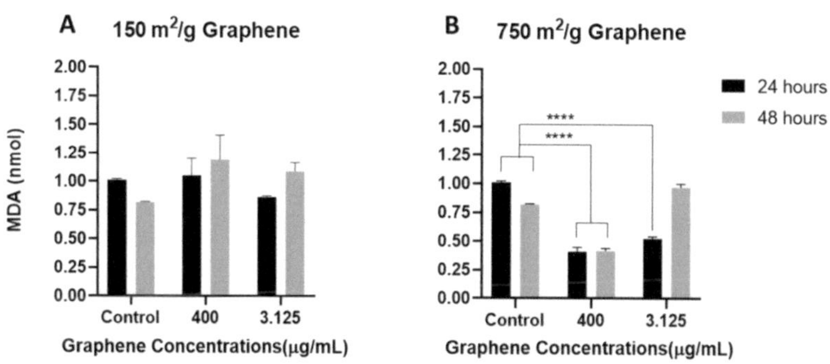

Figure 6.
The impact of two different types of graphene, (A) 150 m^2/g and (B) 750 m^2/g, on MDA levels in SH-SY5Y cells [65].

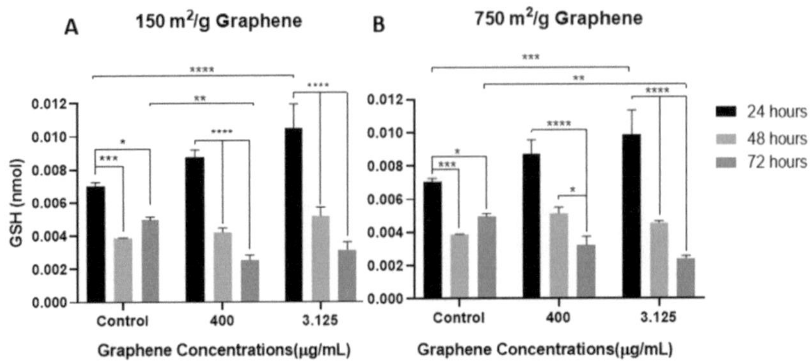

Figure 7.
The impact of two types of graphene, (A) 150 m^2/g and (B) 750 m^2/g, on total GSH production was evaluated. Statistical analysis showed that $^*p < 0.5$, $^{**}p \leq 0.01$, $^{***}p$ 0.001, and $^{****}p \leq 0.0001$ for the observed effects [65].

However, it is essential to note that the pro- or anti-oxidant behavior of GO is not solely dictated by surface area, but also by surface charge and functionalization. A study by Liu et al. demonstrated that single-layered GO (SLGO) induced lower levels of ROS production compared to multi-layered GO (MLGO), highlighting the consequence of the GO layer structure and oxidation levels [66]. Additionally, pristine graphene, characterized by lower surface oxygen content, was found to exhibit reduced cytotoxicity, and magnitude of the role of surface chemistry in modulating toxicological outcomes [11].

Furthermore, the generation of ROS by GO coatings has been recognized as an effective mechanism for preventing bacterial colonization on various surfaces, demonstrating the potential utility of ROS generation in GO applications. Researches demonstrated that the toxicity of graphene nanomaterials (GNMs) to bacteria is linked to surface functionalization, but the exact mechanisms involved are not fully understood. One study investigated the toxic mechanisms of differently functionalized GNMs on bacteria, focusing on physical interaction, oxidative damage, and cell autolysis. Three basic functionalizations of GNMs, including carboxylation (G-COOH), hydroxylation (G-OH), and amination (G-NH2), were examined. G-COOH (66% viability vs. CT group) and G-OH (54%) graphene exhibited higher toxicity to *Escherichia coli* (*E. coli*) than G-NH2 (96%) within 3 hours at a concentration of 50 mg/L. The three materials displayed distinct physical interaction modes with bacterial cells. G-COOH and G-OH made contact with the cell membrane via their sharp edges, causing more damage than G-NH2, which covered the bacteria by attaching along the basal plane. The three GNMs demonstrated similar radical generation capacities, indicating that direct radical generation is not the mechanism causing the toxicity. Instead, the GNMs can oxidize the cellular antioxidant glutathione (GSH), leading to oxidative damage. The oxidative capacity followed the order: G-COOH > G-OH > G-NH2, which correlated with the antibacterial activity. Cell autolysis, the degradation of the cell wall component peptidoglycan, was found to be a new mechanism inducing bacterial death. G-COOH and G-OH caused more cell autolysis than G-NH2, which partially accounts for the different toxicity of the three GNMs [26].

5. Cytotoxicity and genotoxicity

The cytotoxicity of GO nanoparticles is intricately influenced by their size and shape, both of which play effective roles in determining the extent of cellular damage and the specific mechanisms underlying it. This complex relationship is crucial for understanding the potential hazards associated with GO and its applications. Graphene family nanomaterials (GFNs) can enter cells through various means, which can lead to the generation of ROS, increase in lactate dehydrogenase (LDH) and MDA, and release of Ca^{2+} (**Figure 8**). As a result, GFNs can cause different types of cell damage, including harm to the cell membrane, inflammation, DNA damage, mitochondrial disorders, and cell death through apoptosis or necrosis [17].

5.1 Effect of GO size and shape on its cytotoxicity

The higher surface area graphene samples, linked to smaller-size graphene nanoparticles, induces greater cell damage, illuminating the correlation between surface area and cytotoxicity [65]. Smaller GO particles can readily penetrate cellular membranes and interact with various cellular components to trigger a cascade of events, including oxidative stress, inflammation, and ultimately cell death. These particles induce these responses, in part, through the generation of intracellular reactive oxygen species (ROS) and the activation of the mitochondrial pathway, ultimately leading to apoptosis [67]. The larger GO particles, on the other hand, tend to exert their cytotoxic effects through physical damage to cells and the induction of mechanical stress. They may not

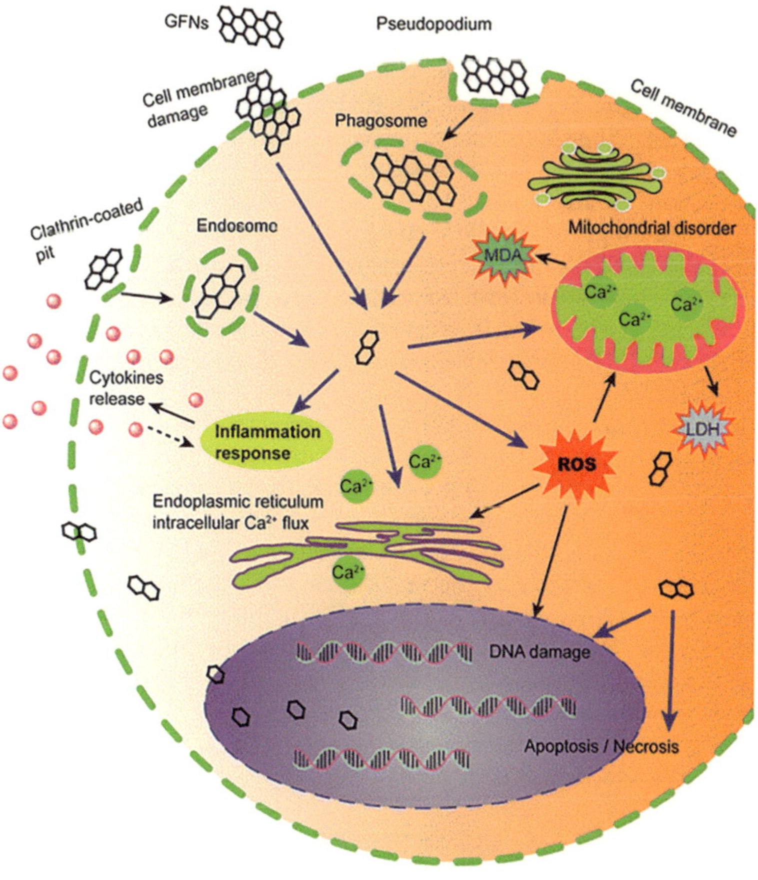

Figure 8.
GFNs cytotoxicity mechanism [17].

penetrate cells as easily as their smaller counterparts, but their sheer size and presence can disrupt cellular structures and induce mechanical stress, leading to cell damage and death [67]. The size of NGO flakes, for example, was found to impact cell viability, with larger flakes showing more pronounced reductions in cell viability compared to smaller flakes. Moreover, the size of NGO flakes was found to impact cell viability, with larger flakes showing more pronounced reductions in cell viability compared to smaller flakes [68]. Elsewhere, GO's lateral size has been identified as a significant parameter affecting its cytotoxic effects. Larger lateral sizes have been associated with stronger cytotoxic impacts on liver cells, emphasizing the importance of size in nanotoxicity [32]. Several studies have shown that GO nanosheets possess a higher potential for membrane penetration, leading to increased cytotoxicity and nanomaterials with sharp edges or high aspect ratios have been associated with greater damage to cells and tissues [22, 34].

5.2 Effect of GO size and shape on its genotoxicity

Investigating the genotoxicity and cytotoxicity of GO in the context of its physicochemical properties is of paramount importance to comprehensively assess its safety and potential impacts on cellular health and genetic material. GO's physicochemical attributes can exert a significant influence on its genotoxic and cytotoxic effects.

Recent investigations have delved into the impact of GO nanoparticle size and surface area on genetic materials. Smaller nanoparticles have displayed heightened interactions with cellular components and genetic material, culminating in escalated genotoxic effects [65]. This phenomenon is further corroborated by research that examined the neurotoxicity of graphene with varying surface areas in a dopaminergic neuron model [65]. While both sizes of graphene demonstrated an intriguing trend of increased cell viability as concentration decreased, an unexpected twist unfolded concerning the surface area aspect, which is a function of the GO particle size as well as shape. These findings collectively shed light on the knotty relationship between nanoparticle characteristics and their biological consequences.

Furthermore, studies have revealed that the shape of nanomaterials plays a crucial role in influencing proliferation, differentiation, and metabolism in both mammalian and microbial cells [31]. Domenech et al., discussed that graphene-based materials (GBM), like other nanomaterials, can induce genotoxicity through primary or secondary mechanisms (**Figure 9**). The primary genotoxicity arises from the direct interaction of GBMs with DNA or its associated histones or by GBMs interfering with DNA replication and/or repair processes, such as by triggering lipid peroxidation or the production of ROS. On the other hand, secondary genotoxicity occurs when GBMs interact with inflammatory cells, leading to downstream effects in the target cells. It is important to note that standard in vitro assays typically focus on non-inflammatory cell monocultures, limiting their ability to detect genotoxic effects caused by secondary mechanisms. In contrast, in vivo experiments conducted on whole organisms provide a broader perspective and can capture genotoxic effects induced by both primary and secondary mechanisms [69].

5.3 Effect of GO surface chemistry on its cytotoxicity and genotoxicity

The surface chemistry of GO, characterized by the presence of functional groups, emerges as a key player in its cytotoxic and genotoxic potential. The oxygen-containing functional groups, such as hydroxyl, carboxyl, and epoxy groups, interact with cellular components and genetic material, leading to genotoxic effects [70].

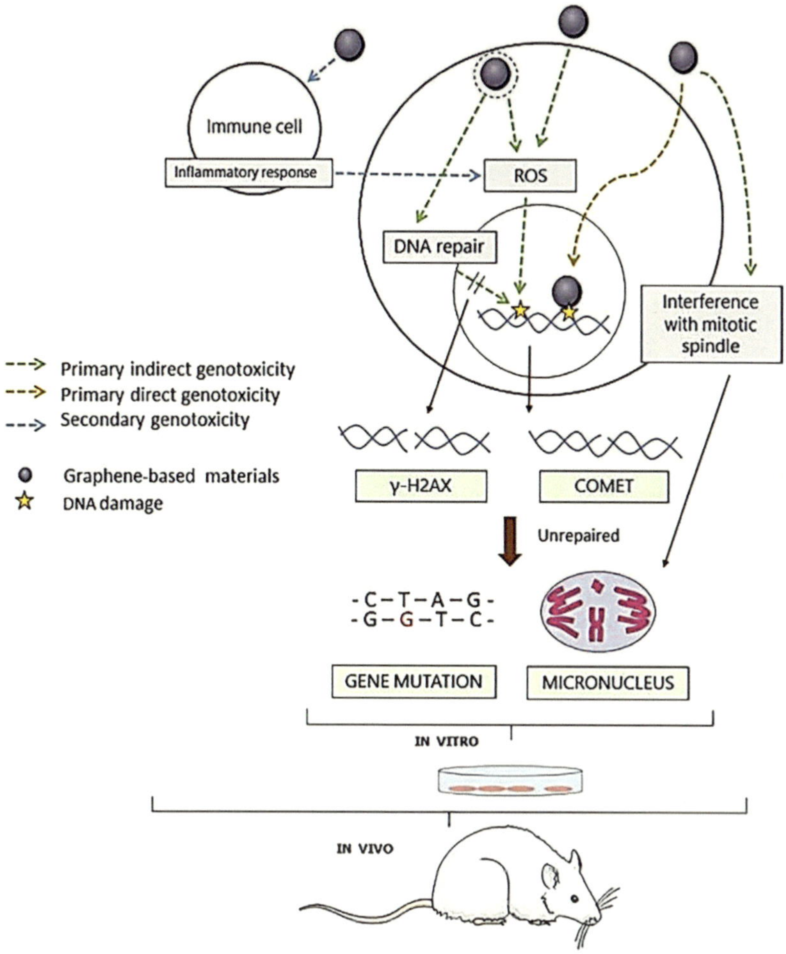

Figure 9.
Genotoxicity mechanism of graphene-based materials, and the methods used to identify these effects both within cells and living organisms [69].

Furthermore, the presence of these functional groups significantly influences the cellular uptake and internalization of GO, thereby modulating its cytotoxicity. Studies reveal that non-covalent conjugates based on GO and an alkylating agent demonstrated significant cytotoxicity toward specific cell lines [34].

The electronic charge of GO nanoparticles plays a prominent role in determining their genotoxic and cytotoxic effects. Positively charged nanoparticles are associated with higher cellular uptake and internalization, potentially leading to amplified genotoxicity and cytotoxicity. In contrast, negatively charged or neutral nanoparticles exhibit different interactions with cellular components and genetic material, resulting in varying levels of genotoxic and cytotoxic effects [12]. The zeta potential distribution of various GO derivatives, such as LA-PEG-GO, PEG-GO, and PEI-GO, influences their genotoxicity and cytotoxicity profiles (**Table 2**). Notably, the surface charge of GO derivatives has a profound effect on cell damage, and GO derivatives with lower positive electronic charges appear to mitigate their toxic effect on cells [12].

Furthermore, pristine graphene, a closely related material to GO, has been shown to induce cytotoxicity in various cell types. It can deplete mitochondrial

Type of GO	Functional group	Size (nm)	Lateral width (nm)	Zeta potential (mV)
LA-PEG-GO	Lactobionic acid-polyethylene glycol (LA-PEG)	100–200	2	18.4 ± 4.85
PEG-GO	Polyethylene glycol (PEG)	50–150	1.9	−8.86 ± 6.93
PEI-GO	Polyethylenimine (PEI)	200–500	2.5	+60.5 ± 5.27
GO	—	200–500	1	−65.1 ± 8.66

Table 2.
Zeta potential distribution of GO, lactobionic acid-polyethylene glycol (LA-PEG)-GO, PEG-GO, and polyethylenimine (PEI)-GO [12].

membrane potential, increase intracellular ROS generation, and trigger apoptosis through the activation of the mitochondrial pathway [71]. Additionally, GO has been found to generate ROS, leading to inflammation and apoptosis in alveolar macrophages and alveolar epithelial cells, further accentuating its influence on cytotoxicity [72].

The aggregation state of GO nanoparticles holds significance in determining their genotoxic and cytotoxic potential. Aggregated nanoparticles may experience reduced cellular uptake and internalization, which can lead to lower genotoxicity and cytotoxicity compared to dispersed nanoparticles. The formation of stable colloid dispersions, especially in hydrophilic graphene nanomaterials, is critical for avoiding aggregation and enhancing cellular internalization and removal [67].

In addition, exposure to higher concentrations of GO has been shown to result in genotoxic effects, encompassing DNA damage, chromosomal aberrations, and alterations in gene expression. The genotoxic potential of GO has been evaluated using techniques such as the comet assay and micronucleus assay, which have become standard methods for assessing the genotoxic effects of GO. Intriguingly, GO nanosheets have been demonstrated to have the capacity to intercalate into DNA molecules, hinting at their ability to interact with genomic DNA. These findings reinforce the importance of studying genotoxicity not just in isolation but in concert with cytotoxicity, as GO nanoparticles have demonstrated dose-dependent DNA damage, alterations in gene replication, and changes in gene expression patterns [58].

This complex relationship between GO's physicochemical properties and its genotoxic and cytotoxic effects highlights the multifaceted nature of evaluating the safety and potential applications of GO in various biological contexts. It emphasizes the need for thorough characterization, the modulation of surface properties, and an acute awareness of dose-dependent effects of the genotoxicity and cytotoxicity of GO in diverse cell types and biological systems. It underlines the need for further research and vigilance concerning the safety aspects of graphene nanoparticles.

In vitro studies have demonstrated that the concentration of GO plays a pivotal role in determining its cytotoxicity and impact on cell viability [33, 57, 60, 73]. One such study by Zhang et al. exemplified the concentration-dependent cytotoxicity of GO. Utilizing purified natural graphite as a starting material, the authors synthesized GO with irregular morphology and small-sized sheet structures. The cytotoxicity of GO was assessed using the Cell Counting Kit-8 (CCK8) assay on MDA-MB-231 human breast cancer cells, with GO concentrations ranging from 100 to 500 μg/mL. The results unequivocally revealed dose-dependent cytotoxicity, emphasizing that higher concentrations of GO correlate with reduced cell viability [74]. This outcome serves as a fundamental illustration of the direct relationship between GO concentration and its impact on cell survival.

Ribeiro et al., illustrated the nuanced interaction between GO surface chemistry and cell viability by investigating the interaction of GO with human breast cancer cells. The study explored the effects of 24, 48, and 72-hour exposure to GO, GO modified with DAB-AM-16 (GOD), and PAMAM dendrimers (GOP) at a concentration of 2.4 μg/cm^2. Intriguingly, the results demonstrated that at this particular concentration, there was no detectable toxicity in the cells during the specified exposure times. However, when higher concentrations of GO were employed, low cytotoxic effects became evident, revealing the sensitivity of cell viability to GO concentration [74]. Moreover, the research revealed that the affinity of GO, GOD, and GOP particles to cellular membranes induced morphological alterations and modifications in the organization of microfilaments and microtubules. However, when cells were treated with 24 μg/cm^2 of GO, GOD, or GOP for 24 hours and then allowed to recover for an additional 24-hour period in normal medium, the nanoparticles persisted in the cytoplasm of some cells, without causing significant alterations in cellular morphology. This observation aligned with the results of the cell proliferation experiment, which indicated that the cells remained viable for up to 72 hours (**Figure 10**). These findings accentuate the dynamic nature of GO-cell interactions and the significance of concentration-dependent effects.

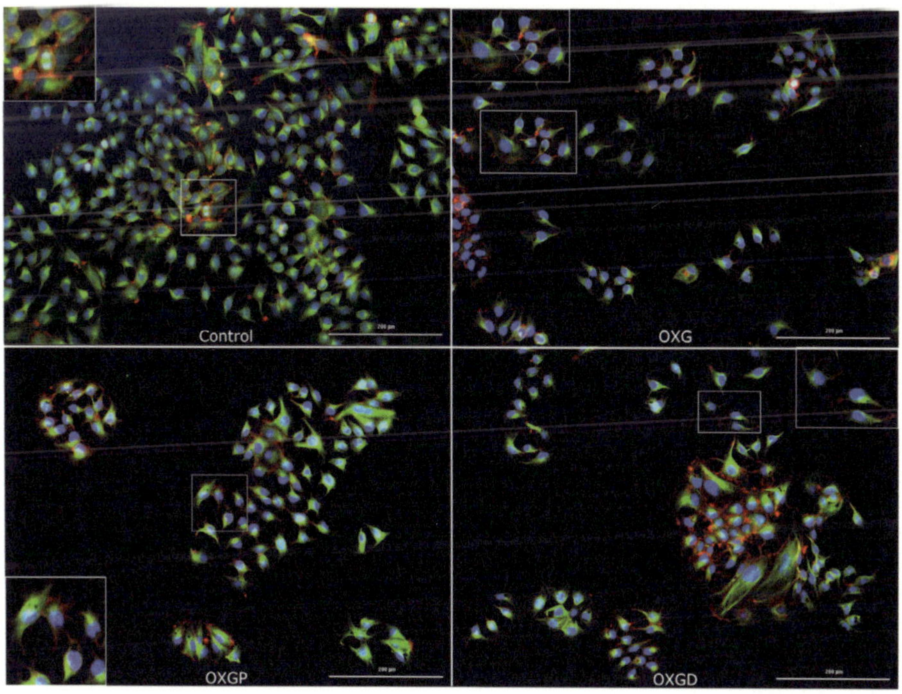

Figure 10.
The immunofluorescence reaction of GOX, GOXP, and GOXD treated and untreated MCF-7 cells at a concentration of 2.4 μg/cm^2 over a 48-hour period. The scale bar is 200 μm. The nuclei were stained with DAPI (blue), actin was stained with Alexa 633-phalloidin (red), and microtubules were evidenced by a monoclonal antibody against α and β tubulin and anti-mouse FITC-antibody (green). The figure also includes zoomed-in inserts of selected areas (white rectangles) [74].

6. Inflammation and other immune response

6.1 Inflammation response

In the realm of size and surface area, the dimensions of GO nanoparticles emerge as decisive determinants in modulating their interaction with immune cells, which, in turn, has a fundamental bearing on the ensuing inflammatory response. Smaller GO nanoparticles, owing to their larger surface area per unit mass, exhibit an enhanced proclivity for interaction with immune cells, which translates into a heightened activation of inflammatory pathways and the subsequent release of pro-inflammatory cytokines. The increased contact area with immune cells substantiates their pro-inflammatory potential, marking a notable contribution to the overall inflammation induced by GO [75].

6.2 Immune responses

The immune response elicited by GO exposure has become a crucial element of its biocompatibility and its possible biomedical applications. GO elicits dose-dependent immune responses characterized by the activation of immune cells and the subsequent release of pro-inflammatory cytokines. Higher doses of GO have been correlated with more pronounced immune reactions and notable alterations in immune components. These alterations extend to changes in the populations of different hepatic macrophage types and platelet infiltration in vital organs such as the liver and spleen. The uptake of GO nanomaterials by macrophages within these organs plays a central role in promoting the secretion of pro-inflammatory cytokines and has been robustly associated with inflammation [8, 75, 76]. Studies have pointed to the secretion of pro-inflammatory cytokines, including interleukin-6 (IL-6), IL-12, tumor necrosis factor-β (TNF-β), and interferon-gamma (IFN-γ), as a significant outcome of GO nanomaterial uptake by macrophages, thus magnitude the intricate link between GO exposure and the immune response. Moreover, the dose-dependent cytotoxicity of GO nanomaterials has been widely observed, underscoring the importance of balancing the immune response and cytotoxicity in determining the overall impact of GO [75]. Strikingly, the interplay between GO exposure, immune response, and inflammation has illuminated a novel avenue for mitigating inflammation through the modulation of autophagy. GO nanomaterials have been shown to disrupt lysosomal function, thus impeding autophagy flux. Paradoxically, this disruption of autophagy has been linked to a decrease in the levels of pro-inflammatory cytokines such as IL-6, IL-8, TLR4, and CXCL2, while concomitantly enhancing the levels of the anti-inflammatory cytokine IL-10. This results in the amelioration of conditions such as colitis [77]. In addition to these intriguing insights, GO nanostructures have exhibited multifaceted properties, including the ability to suppress the migration and metastasis of cancer cells and to inhibit prostaglandin-mediated inflammatory responses. These findings not only highlight the extensive potential of GO in biomedical applications but also emphasize the significant impact of dose-dependent immune responses on shaping its safety and effectiveness profiles [75].

Overall, the intricate balance between GO exposure and the immune response holds significant implications for the safe and efficacious utilization of GO in diverse biomedical applications. The collaboration between GO, the immune system, and inflammation unveils a dynamic and multifaceted relationship that warrants further exploration and offers new avenues for the development of strategies aimed at optimizing the potential of GO in various medical and therapeutic contexts. These findings collectively spotlight the importance of fine-tuning GO dosages and its effects on the immune system to harness its full potential while mitigating potential immune-mediated side effects.

6.3 Role of GO physicochemical properties on inflammatory and immune responses

Inflammation is a key indicator of the nanotoxicity exhibited by GO. It serves as a multifaceted orchestrator of cellular and tissue responses, setting in motion the intricate cascade of events leading to the production of pro-inflammatory cytokines and recruitment of immune cells. Several studies have elucidated the role of GO's physicochemical properties, including its charge, in mediating the magnitude and duration of the ensuing inflammatory responses [34]. It is imperative to explore this connection to harness a comprehensive understanding that would empower the development of strategies aimed at mitigating the inflammatory consequences of GO and improving its biocompatibility. The presence of oxygen-containing functional groups on the GO surface, such as hydroxyl, carboxyl, and epoxy groups, governs its interaction with immune cells and the subsequent induction of pro-inflammatory cascade responses. These GO functional groups engage immune cell receptors, initiating intricate signaling pathways that culminate in the production of pro-inflammatory cytokines. The density and distribution of these functional groups on the GO surface critically influence its pro-inflammatory potential [75]. ROS generation, incited by GO exposure, adds another layer of complexity to the inflammation response. GO's surface chemistry and size wield profound influence over its capacity to induce ROS generation, and ROS, which in turn, activate inflammatory pathways and amplify the release of pro-inflammatory cytokines. This intricate relationship between GO's physicochemical attributes and ROS generation reinforces the dynamic and multifaceted nature of GO-induced inflammation [76]. Moreover, the interplay between GO nanoparticles and proteins in biological systems serves as an additional regulator of inflammatory reactions. The binding of proteins to the GO surface can profoundly change their structure and function, leading to immune responses and inflammation. Furthermore, the interaction between GO and immune-related proteins or receptors can stimulate inflammatory signaling pathways, introducing another level of intricacy to the reciprocal impact between GO's physical and chemical properties and inflammation [61].

7. Dose-dependent nanotoxicity of GO

7.1 Effect of concentration

The dose-dependent nanotoxicity of GO represent a key aspect of its safety profile and efficacy in various biomedical, electronics, energy storage, and environmental remediation applications. It has been consistently demonstrated that GO elicits a dose-dependent response, where higher concentrations correlate with increased cytotoxicity and adverse effects, both in in vitro and in vivo settings. In in vitro experiments, the impact of GO on cells has been investigated, revealing the dose-dependent nature of its nanotoxicity. High concentrations of GO have been associated with the inducement of oxidative stress, DNA damage, and cell death [64].

Doses of GO exceeding 50 μg/mL have been shown to induce cytotoxicity in human fibroblast cells, leading to decreased cell adhesion and promoting cell apoptosis [78]. Furthermore, GO's impact on immune cells has been demonstrated, with dendritic cells experiencing decreased functional activity and immune proteasome unit down-regulation in response to GO exposure [11]. Commercial pristine graphene dispersed in 1% Pluronic F108 at concentrations up to 20 mg/mL interacted with murine macrophage-like RAW 264.7 cells. The thickness of graphene dispersed in Pluronic ranged from 2–3 to 500–1000 nm. The cells underwent apoptosis in a dose-dependent manner, which was associated with a decrease in mitochondrial potential and an increase in ROS levels. Autophagy was induced in murine macrophage-like RAW 264.7 cells when incubated with approximately 100 μg/mL of GO, and Toll-like receptors associated with inflammatory responses were expressed. Dendritic cells treated with up to 25 μg/mL of GO showed a decrease in functional activity. Specifically, antigen inhibition and down-regulation of intracellular levels of an immune proteasome unit responsible for antigen processing in these cells have been reported [11]. Similarly, the cytotoxicity of GO-PEG nanoparticles was concentration-dependent in the regulation of human myeloid suppressor cells (MDSC) [73]. Nonetheless, doses of up to 100 μg/mL of both GO sizes GO (a ~500 nm protein-coated commercial GO and GO sheets of 1 μm particle range) did not inhibit cell proliferation [19, 33].

Similarly, in vivo studies have reinforced the significance of dose-dependent nanotoxicity of GO. Exposure to elevated concentrations of GO has been shown to result in dose-dependent lung inflammation and organ damage [49, 79]. Elsewhere, the effects of NGO sheets on the viability and motility of spermatozoa were investigated in vivo. **Figure 11**. shows the dose-dependent effects, indicating that the viability and motility of spermatozoa decreased from approximately 75% to 40% as the concentration of NGO sheets increased from 0 to 2000 μg/mL. At concentrations of ≤20 μg/mL, there was only a slight reduction (~10%) in viability and motility, suggesting minimal

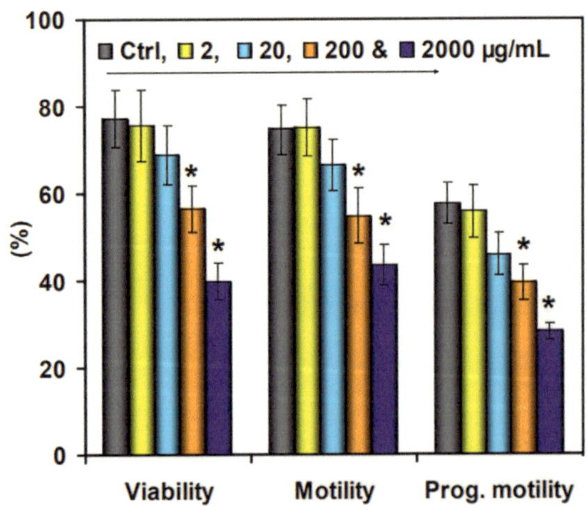

Figure 11.
The viability, motility, and progressive motility of spermatozoa in male Balb/C mice after approximately ~8 weeks of weekly injections of NGO sheets at different concentrations ranging from ~2 to 2000 μg/mL. The control samples were untreated mice spermatozoa. The study used at least three mice per group (N = 3), and significant results are indicated with asterisks (*) for p-values less than 0.05 [80].

toxicity. However, significant in vivo toxicity was observed at injected concentrations ≥200 μg/mL, with a reduction of approximately 45% in sperm viability and motility at 2000 μg/mL or ~4 mg/kg of body weight [80].

However, it is essential to note that not all cell types and concentrations exhibit high cytotoxicity when exposed to GO. For example, GO nanoparticles of different sizes demonstrated low cytotoxicity on mouse neural stem cells (mNSCs) at a concentration of 20 μg/mL [81].

Nevertheless, these findings underline the importance of understanding the dose-response relationship to establish safe exposure limits and to design appropriate dosing strategies when considering GO for biomedical applications [57, 60, 73]. These observations are crucial in understanding the potential systemic effects of GO and are instrumental in determining safe exposure limits in clinical applications [49, 79]. The establishment of appropriate dosing strategies and safe exposure limits will ultimately facilitate the development of GO-based biomedical interventions with a high degree of precision and effectiveness.

7.2 Time-dependent nanotoxicity

While GO exhibits dose-dependent toxicity, time-dependent factors play a substantial role in understanding the persistence of GO within biological systems and its potential for long-term biocompatibility. Time-dependent nanotoxicity of GO provide crucial insights into its fate and potential long-term effects within biological systems. One noteworthy observation is the gradual degradation of GO within living organisms. In vivo studies have revealed dynamic changes in GO behavior over time, shedding light on its biodegradability, biodistribution, and clearance mechanisms. Research has shown that GO can be broken down and metabolized over time, as evidenced by the disappearance of the Raman signal associated with GO in the spleen after 9 months [43]. The temporal aspects of GO's toxicity are, therefore, linked to its clearance mechanisms. Studies have indicated that GO particles can be rapidly excreted through the urinary system, with evidence of GO presence in urine within just 2-hours post-injection. This signifies that GO has efficient excretion pathways, contributing to its time-dependent behavior within the body [53]. The interaction between time-dependent nanotoxicity and GO's physiological fate is of great significance in evaluating its safety profile and potential biomedical applications. The dynamic degradation and clearance mechanisms offer insights into how GO may be processed by the body, which is critical in determining the long-term effects of GO exposure.

Siqueira et al., discuss the potential environmental impact of residual traces of GO and rGO in aquatic environments and their effects on zebrafish hepatocytes. The study found that GO and rGO nanosheets were internalized by the fish cells, and exposure for 72 hours caused harmful effects such as higher ROS production and stopping cell replication. Both GO and rGO were found to be toxic after 72 hours, but with different mechanisms of toxicity [82]. Akhavan et al., investigated the time-dependent biodistribution of NGO sheets in Balb/C mice after intravenous injection. The results showed high uptake of NGO sheets by residual organs like the liver and spleen, indicating effective filtering of the blood circulation and the kidneys also exhibited high uptake, suggesting regular renal clearance. Additionally, the thyroid and testis showed significant uptake, indicating potential effects on hormone secretion and spermatozoa performance (**Figure 12**) [80].

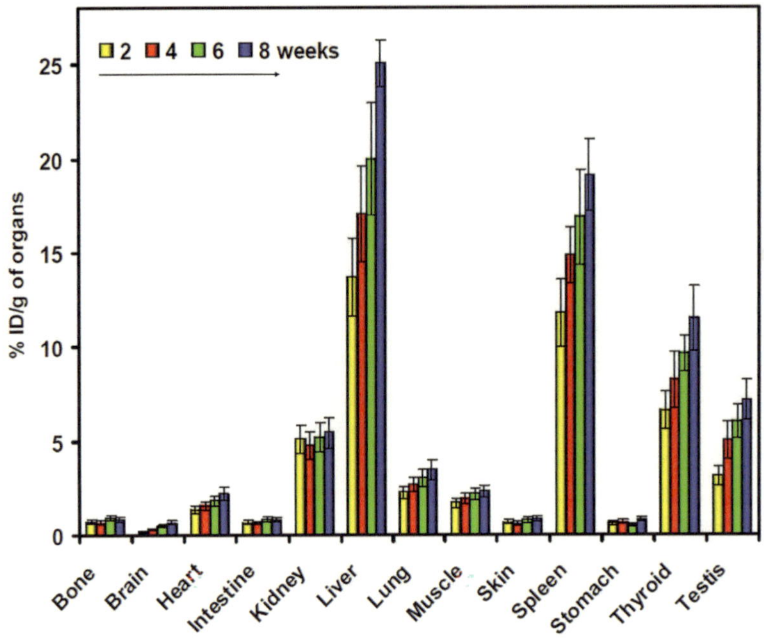

Figure 12.
Time-dependent biodistribution of fluorescent-labeled NGO sheets in male mice treated with NGO at an injected concentration of 2000 μg/mL. The measurements were taken at various post-injection times, starting from the first weekly injection. Error bars were included in the figure, representing the standard deviations calculated from three mice per group (N = 3) [80].

Overall, the time-dependent nanotoxicity of GO serve as a valuable aspect in the comprehensive evaluation of its safety, efficacy, and translational potential. These findings highlight the dynamic nature of GO's interaction with biological tissues and its potential to undergo transformation within the body [43]. However, these findings point to the need for further exploration of the kinetics and dynamics of GO within living organisms, facilitating the development of safer and more effective applications in biomedicine.

7.3 Systemic effects

The widespread impact of GO exposure on the body has attracted significant interest because of its implications for the safe use of GO in medical and related applications. Significant systemic effects have been linked to high doses of GO, making it a crucial factor in assessing its safety. Research has identified various systemic factors affected by GO, emphasizing the importance of careful dose selection and careful monitoring in different applications [19, 42, 74, 75, 83]. Hematological parameters have surfaced as a key area of impact, with findings suggesting that GO nanoparticles contribute to a decrease in red blood cell count, hemoglobin concentration, and mean corpuscular volume in a dose-dependent manner [83]. Such alterations in hematological parameters underline the systemic effects of GO and their potential consequences for biological systems. Furthermore, GO's influence extends to biochemical parameters, encompassing cholesterol, triglyceride, glucose, protein, albumin, creatinine, and bilirubin levels. These changes, when elucidated, could provide further insights into the systemic ramifications of GO exposure although specific changes in these parameters following GO treatment still remain inconclusive [83].

Liver enzymes, including aspartate aminotransferase (AST), alanine aminotransferase (ALT), alkaline phosphatase (ALP), and lactate dehydrogenase (LDH), have also been scrutinized in the context of GO exposure, albeit without specified alterations [83]. The impact on liver enzymes hints at potential liver damage induced by GO, emphasizing the importance of assessing hepatic function. Mendonca et al., demonstrated that the intravenous administration of rGO at a single dose of 7 mg/kg resulted in minor indications of toxicity, not only in the blood, but also in the liver, and the kidney after 7 days, with no evidence of an ongoing inflammatory process. These effects were temporary and did not cause permanent harm. Despite a previous study showing disruption of the blood-brain barrier and the presence of rGO in the hippocampus, no noticeable changes were observed in the response of hippocampal neurons or astrocytes. Additionally, the behavior of both control and rGO-treated animals in their cages was similar, with no observable clinical signs of toxicity. The data suggested that systemic GO administration posed no significant health risk for rats under these experimental conditions. However, it was evident that the interactions between graphene-based materials and biological systems depend on the experimental design and the physicochemical properties of the nanomaterial. Therefore, comprehensive toxicological investigations are therefore necessary for understanding the biological effects and addressing safety concerns before the practical application of graphene-based materials in clinical settings. This study provided a foundation for further toxicological examinations of rGO following long-term in vivo exposure and it encouraged research into the mechanisms underlying the interaction between rGO and biological systems [84, 85].

Renal dysfunction has emerged as another facet of GO-induced systemic effects, with elevated levels of blood urea nitrogen (BUN) and creatinine observed in animal models post-GO exposure [86]. These markers of renal function underscore the potential nephrotoxicity associated with GO exposure.

Nonetheless, it is essential to note that the systemic effects of GO are complexly dependent on factors such as dose, exposure duration, and the specific biological system under investigation [19, 42, 74, 75, 83]. Thus, while certain studies have reported substantial systemic alterations, others have found GO nanoparticles to be relatively benign, with no significant impact on lung, liver, kidney, and blood parameters in rats. Notably, in these studies, GO nanoparticles were retained in the body, primarily as agglomerates, without adverse effects on systemic parameters. Presence of nanoparticles was detected in the body tissues of all groups treated with GO. The largest solid aggregates, which could be up to 10 mm in diameter, were found near the injection site in the stomach serous membrane, between the connective tissues of the abdominal skin, muscles, and peritoneum. There were also many smaller, spherical-shaped aggregates with diameters around 2 mm that were lodged among the mesentery. The mesentery in the GO was almost black due to the high concentration of nanoparticles. Smaller aggregates were also observed in abdominal lipid tissue near the injection site and mesentery (**Figure 13**). This suggested the possibility of GO application in drug delivery systems without causing systemic harm [86].

These contrasting findings emphasize the necessity for in-depth research into the dose-dependent systemic effects of GO, clarifying the conditions under which it may

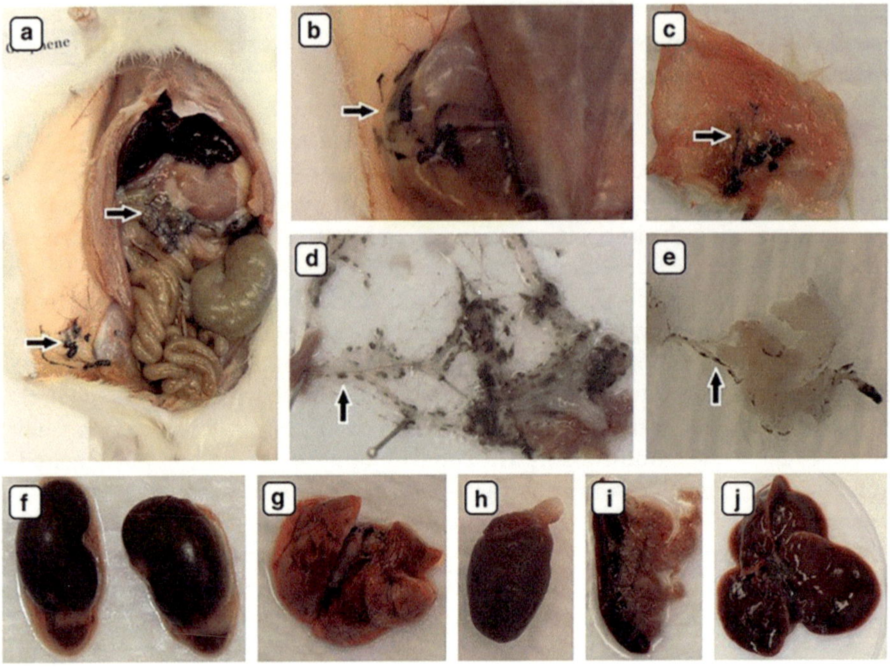

Figure 13.
Systemic effects of GO nanoparticles after multiple intraperitoneal injections into rats taken by digital camera. Solid aggregates were found in injected body regions (a, b, c) and mesentery (d). Average-sized dots of GO nanoparticle aggregates were also localized in abdominal lipid tissue (e). Black arrows were used to indicate the presence of the GO nanoparticles. The macroscopic structure of the kidney (f), lungs (g), heart (h), spleen (i), and liver (j) was also examined. No pathological features were found [86].

be safely utilized in various biomedical applications. Moreover, these investigations highlight the paramount importance of establishing precise dose thresholds and safety protocols to ensure the effective and secure application of GO-based materials in medicine and other fields.

8. Conclusion

Over the past few years, graphene-based nanomaterials have become popular for a variety of technological and biomedical applications, including biosensing, drug delivery, tissue engineering, and diagnosis. However, it is essential to establish the potential risks and safety levels of using graphene materials in biomedical applications. Currently, there is limited information available about the in vitro and in vivo toxicity of graphene and GO, and more studies are needed. The toxicity of graphene and GO depends on various factors, including the surface, size, number of layers, cell type, administration route, dose, time of exposure, and synthesis methods. Reactive oxygen species generation in target cells is the most important cytotoxicity mechanism of graphene. Small and hydrophilic graphene nanomaterials tend to form a stable colloid dispersion and are more likely to be internalized and removed/excreted from the application site.

Understanding the complex relationship between these properties and graphene and GO's potential toxicity is essential for the safe and effective use of graphene and

GO in biomedicine. By considering these properties and their effects on oxidative stress, inflammation, genotoxicity, and cytotoxicity, researchers can develop strategies to minimize the risks associated with graphene-based materials.

Graphene-based materials are promising candidates for biomedical applications due to their unique properties and lack of metallic impurities. However, toxicological studies must consider the purity of the sample and the non-molecular behavior of graphene-based nanomaterials. With further research and advancements, graphene and GO hold great promise in revolutionizing biomedicine while ensuring patient safety.

Author details

Mohammad Mahdi Sepahi[1,2*] and Marzieh Azizi[3,4]

1 Department of Biotechnology, Faculty of Advanced Sciences and Technologies, University of Isfahan, Isfahan, Iran

2 Department of Chemistry "Sapienza" University of Rome, Rome, Italy

3 Laboratory of Bioanalysis, Institute of Biochemistry and Biophysics, University of Tehran, Tehran, Iran

4 Department of Chemistry, University of Oslo, Oslo, Norway

*Address all correspondence to: m.m.sepahi@ut.ac.ir

References

[1] Wang Y, Li Z, Wang J, Li J, Lin Y. Graphene and graphene oxide: Biofunctionalization and applications in biotechnology. Trends in Biotechnology. 2011;**29**(5):205-212

[2] Ge C, Du J, Zhao L, Wang L, Liu Y, Li D, et al. Binding of blood proteins to carbon nanotubes reduces cytotoxicity. Proceedings of the National Academy of Sciences of the United States of America. 2011;**108**(41):16968-16973

[3] Lu C-H, Yang H-H, Zhu C-L, Chen X, Chen G-N. A graphene platform for sensing biomolecules. Angewandte Chemie International Edition. 2009;**48**(26):4785-4787

[4] Wang H, Zhang Q, Chu X, Chen T, Ge J, Yu R. Graphene oxide–peptide conjugate as an intracellular protease sensor for caspase-3 activation imaging in live cells. Angewandte Chemie International Edition. 2011;**50**(31):7065-7069

[5] Peng C, Hu W, Zhou Y, Fan C, Huang Q. Intracellular imaging with a graphene-based fluorescent probe. Small. 2010;**6**(15):1686-1692

[6] Sanchez VC, Jachak A, Hurt RH, Kane AB. Biological interactions of graphene-family nanomaterials: An interdisciplinary review. Chemical Research in Toxicology. 2012;**25**(1):15-34

[7] Hu X, Ouyang S, Mu L, An J, Zhou Q. Effects of graphene oxide and oxidized carbon nanotubes on the cellular division, microstructure, uptake, oxidative stress, and metabolic profiles. Environmental Science and Technology. 2015;**49**(18):10825-10833

[8] Zuberek M, Grzelak A. Nanoparticles-caused oxidative imbalance. In: Saquib Q, Faisal M, Al-Khedhairy AA, Alatar AA, editors. Cellular and Molecular Toxicology of Nanoparticles. Cham: Springer International Publishing; 2018. pp. 85-98

[9] Akhavan O, Ghaderi E, Akhavan A. Size-dependent genotoxicity of graphene nanoplatelets in human stem cells. Biomaterials. 2012;**33**(32):8017-8025

[10] Fahmi T, Branch LD, Nima ZA, Jang DS, Savenka AV, Biris AS, et al. Mechanism of graphene-induced cytotoxicity: Role of endonucleases. Journal of Applied Toxicology. 2017;**37**(11):1325-1332

[11] Malkova A, Svadlakova T, Singh A, Kolackova M, Vankova R, Borsky P, et al. In vitro assessment of the genotoxic potential of pristine graphene platelets. Nanomaterials. 2021;**11**(9):2210

[12] Wang A, Pu K, Dong B, Liu Y, Zhang L, Zhang Z, et al. Role of surface charge and oxidative stress in cytotoxicity and genotoxicity of graphene oxide towards human lung fibroblast cells. Journal of Applied Toxicology. 2013;**33**(10):1156-1164

[13] Zare EN, Zheng X, Makvandi P, Gheybi H, Sartorius R, Yiu CK, et al. Nonspherical metal-based nanoarchitectures: Synthesis and impact of size, shape, and composition on their biological activity. Small. 2021;**17**(17):2007073

[14] Foller T, Daiyan R, Jin X, Leverett J, Kim H, Webster R, et al. Enhanced graphitic domains of unreduced graphene oxide and the interplay of

hydration behaviour and catalytic activity. Materials Today. 2021;**50**:44-54

[15] Lee EA, Kwak S-Y, Yang J-K, Lee Y-S, Kim J-H, Kim HD, et al. Graphene oxide film guided skeletal muscle differentiation. Materials Science and Engineering: C. 2021;**126**:112174

[16] Ranishenka B, Ulashchik E, Tatulchenkov M, Sharko O, Panarin A, Dremova N, et al. Graphene oxide functionalization via epoxide ring opening in bioconjugation compatible conditions. FlatChem. 2021;**27**:100235

[17] Ou L, Song B, Liang H, Liu J, Feng X, Deng B, et al. Toxicity of graphene-family nanoparticles: A general review of the origins and mechanisms. Particle and Fibre Toxicology. 2016;**13**(1):57

[18] Fan T, Yan L, He S, Hong Q, Ai F, He S, et al. Biodistribution, degradability and clearance of 2D materials for their biomedical applications. Chemical Society Reviews. 2022;**51**(18):7732-7751

[19] Khan S, Mansoor S, Rafi Z, Kumari B, Shoaib A, Saeed M, et al. A review on nanotechnology: Properties, applications, and mechanistic insights of cellular uptake mechanisms. Journal of Molecular Liquids. 2022;**348**:118008

[20] Davide L, Marianna D, Federico AR, Carla D. Nanoparticles and potential neurotoxicity: Focus on molecular mechanisms. AIMS Molecular Science. 2018;**5**(1):1-13

[21] Sharma R, Sharma KP, Malviya R. Modulation of shape and size-dependent characteristics of nanoparticles. Current Nanomedicine. 2019;**9**(3):210-215

[22] Kladko DV, Falchevskaya AS, Serov NS, Prilepskii AY. Nanomaterial shape influence on cell behavior. International Journal of Molecular Sciences. 2021;**22**(10):5266

[23] Khalid K, Tan X, Mohd Zaid HF, Tao Y, Lye Chew C, Chu D-T, et al. Advanced in developmental organic and inorganic nanomaterial: A review. Bioengineered. 2020;**11**(1):328-355

[24] Maheshwari R, Raval N, Tekade RK. Surface modification of biomedically essential nanoparticles employing polymer coating. In: Weissig V, Elbayoumi T, editors. Pharmaceutical Nanotechnology: Basic Protocols. New York, NY: Springer New York; 2019. pp. 191-201

[25] Kumar A, Kaladharan K, Tseng F-G. Nanomaterials: Surface functionalization, modification, and applications. In: Santra TS, Mohan L, editors. Nanomaterials and Their Biomedical Applications. Singapore: Springer Singapore; 2021. pp. 405-438

[26] Xie C, Zhang P, Guo Z, Li X, Pang Q, Zheng K, et al. Elucidating the origin of the surface functionalization - dependent bacterial toxicity of graphene nanomaterials: Oxidative damage, physical disruption, and cell autolysis. Science of the Total Environment. 2020;**747**:141546

[27] Hunt NJ. Handbook of surface-functionalized nanomaterials: Safety and legal aspects. In: Handbook of Functionalized Nanomaterials for Industrial Applications. Elsevier; 2020. pp. 945-982

[28] Hussain SM, Braydich-Stolle LK, Schrand AM, Murdock RC, Yu KO, Mattie DM, et al. Toxicity evaluation for safe use of nanomaterials: Recent achievements and technical challenges. Advanced Materials. 2009;**21**(16):1549-1559

[29] Cheah P, Brown P, Qu J, Tian B, Patton DL, Zhao Y. Versatile surface functionalization of water-dispersible

iron oxide nanoparticles with precisely controlled sizes. Langmuir. 2021;**37**(3):1279-1287

[30] Du X, Skachko I, Barker A, Andrei EY. Approaching ballistic transport in suspended graphene. Nature Nanotechnology. 2008;**3**(8):491-495

[31] Francia V, Montizaan D, Salvati A. Interactions at the cell membrane and pathways of internalization of nano-sized materials for nanomedicine. Beilstein Journal of Nanotechnology. 2020;**11**:338-353

[32] Singh RK, Kumar R, Singh DP. Graphene oxide: Strategies for synthesis, reduction and frontier applications. RSC Advances. 2016;**6**(69):64993-65011

[33] Li J, Wang X, Mei K-C, Chang CH, Jiang J, Liu X, et al. Lateral size of graphene oxide determines differential cellular uptake and cell death pathways in Kupffer cells, LSECs, and hepatocytes. Nano Today. 2021;**37**:101061

[34] Wang X, Cui X, Zhao Y, Chen C. Nano-bio interactions: The implication of size-dependent biological effects of nanomaterials. Science China Life Sciences. 2020;**63**(8):1168-1182

[35] Dalzon B, Torres A, Reymond S, Gallet B, Saint-Antonin F, Collin-Faure V, et al. Influences of nanoparticles characteristics on the cellular responses: The example of iron oxide and macrophages. Nanomaterials. 2020;**10**(2):266

[36] Starost K, Njuguna J. The influence of graphene oxide on nanoparticle emissions during drilling of graphene/epoxy carbon-fiber reinforced engineered nanomaterials. Atmosphere. 2020;**11**(6):573

[37] Loh KP, Bao Q, Eda G, Chhowalla M. Graphene oxide as a chemically tunable platform for optical applications. Nature Chemistry. 2010;**2**(12):1015-1024

[38] Li B, Zhang X-Y, Yang J-Z, Zhang Y-J, Li W-X, Fan C-H, et al. Influence of polyethylene glycol coating on biodistribution and toxicity of nanoscale graphene oxide in mice after intravenous injection. International Journal of Nanomedicine. 2014;**9**:4697-4707

[39] Yang K, Gong H, Shi X, Wan J, Zhang Y, Liu Z. In vivo biodistribution and toxicology of functionalized nano-graphene oxide in mice after oral and intraperitoneal administration. Biomaterials. 2013;**34**(11):2787-2795

[40] Sharma H, Mondal S. Functionalized graphene oxide for chemotherapeutic drug delivery and cancer treatment: A promising material in nanomedicine. International Journal of Molecular Sciences. 2020;**21**(17):6280

[41] Tu Z, Donskyi IS, Qiao H, Zhu Z, Unger WES, Hackenberger CPR, et al. Graphene oxide-cyclic R10 peptide nuclear translocation nanoplatforms for the surmounting of multiple-drug resistance. Advanced Functional Materials. 2020;**30**(35):2000933

[42] Lee H, Kim J, Lee J, Park H, Park Y, Jung S, et al. In vivo self-degradable graphene nanomedicine operated by DNAzyme and photo-switch for controlled anticancer therapy. Biomaterials. 2020;**263**:120402

[43] Newman L, Jasim DA, Prestat E, Lozano N, de Lazaro I, Nam Y, et al. Splenic capture and in vivo intracellular biodegradation of biological-grade graphene oxide sheets. ACS Nano. 2020;**14**(8):10168-10186

[44] Dou C, Ding N, Luo F, Hou T, Cao Z, Bai Y, et al. Graphene-based microRNA transfection blocks preosteoclast fusion to increase bone formation and

vascularization. Advanced Science. 2018;**5**(2):1700578

[45] Qi X, Liu H, Guo W, Lin W, Lin B, Jin Y, et al. New opportunities: Second harmonic generation of boron-doped graphene quantum dots for stem cells imaging and ultraprecise tracking in wound healing. Advanced Functional Materials. 2019;**29**(37):1902235

[46] Chen S, Xiong C, Liu H, Wan Q, Hou J, He Q, et al. Mass spectrometry imaging reveals the sub-organ distribution of carbon nanomaterials. Nature Nanotechnology. 2015;**10**(2):176-182

[47] Ip AC-F, Liu B, Huang P-JJ, Liu J. Oxidation level-dependent zwitterionic liposome adsorption and rupture by graphene-based materials and light-induced content release. Small. 2013;**9**(7):1030-1035

[48] Bai H, Jiang W, Kotchey GP, Saidi WA, Bythell BJ, Jarvis JM, et al. Insight into the mechanism of graphene oxide degradation via the photo-Fenton reaction. The Journal of Physical Chemistry C. 2014;**118**(19):10519-10529

[49] Amrollahi-Sharifabadi M, Koohi MK, Zayerzadeh E, Hablolvarid MH, Hassan J, Seifalian AM. In vivo toxicological evaluation of graphene oxide nanoplatelets for clinical application. International Journal of Nanomedicine. 2018;**13**:4757-4769

[50] Jasim DA, Ménard-Moyon C, Bégin D, Bianco A, Kostarelos K. Tissue distribution and urinary excretion of intravenously administered chemically functionalized graphene oxide sheets. Chemical Science. 2015;**6**(7):3952-3964

[51] Yang K, Wan J, Zhang S, Zhang Y, Lee S-T, Liu Z. In vivo pharmacokinetics, long-term biodistribution, and toxicology of PEGylated graphene in mice. ACS Nano. 2011;**5**(1):516-522

[52] Yang K, Zhang S, Zhang G, Sun X, Lee S-T, Liu Z. Graphene in mice: Ultrahigh in vivo tumor uptake and efficient photothermal therapy. Nano Letters. 2010;**10**(9):3318-3323

[53] Jasim DA, Murphy S, Newman L, Mironov A, Prestat E, McCaffrey J, et al. The effects of extensive glomerular filtration of thin graphene oxide sheets on kidney physiology. ACS Nano. 2016;**10**(12):10753-10767

[54] Bussy C, Jasim D, Lozano N, Terry D, Kostarelos K. The current graphene safety landscape—A literature mining exercise. Nanoscale. 2015;**7**(15):6432-6435

[55] Wang W, Gaus K, Tilley RD, Gooding JJ. The impact of nanoparticle shape on cellular internalisation and transport: What do the different analysis methods tell us? Materials Horizons. 2019;**6**(8):1538-1547

[56] Kim SE, Kim MS, Shin YC, Eom SU, Lee JH, Shin D-M, et al. Cell migration according to shape of graphene oxide micropatterns. Micromachines. 2016;**7**(10):186

[57] Vochita G, Oprica L, Gherghel D, Mihai C-T, Boukherroub R, Lobiuc A. Graphene oxide effects in early ontogenetic stages of *Triticum aestivum* L. seedlings. Ecotoxicology and Environmental Safety. 2019;**181**:345-352

[58] Dong Y, Chang Y, Gao H, León Anchustegui VA, Yu Q, Wang H, et al. Characteristic synergistic cytotoxic effects toward cells in graphene oxide dressing with cadmium and copper ions. Toxicology Research. 2019;**8**(6):908-917

[59] Jackman JA, Yoon BK, Mokrzecka N, Kohli GS, Valle-González ER, Zhu X, et al. Graphene oxide mimics biological

signaling cue to rescue starving bacteria. Advanced Functional Materials. 2021;**31**(25):2102328

[60] Ahmad J, Wahab R, Siddiqui MA, Saquib Q, Al-Khedhairy AA. Cytotoxicity and cell death induced by engineered nanostructures (quantum dots and nanoparticles) in human cell lines. JBIC Journal of Biological Inorganic Chemistry. 2020;**25**(2):325-338

[61] Kopac T. Protein corona, understanding the nanoparticle–protein interactions and future perspectives: A critical review. International Journal of Biological Macromolecules. 2021;**169**:290-301

[62] Yuan S, Liu S, Du Y, Wang X, Zhang H, Yuan S. Atomistic insights into heterogeneous reaction of formic acid on mineral oxide particles. Chemosphere. 2022;**287**:132430

[63] Baravkar PN, Sayyed AA, Rahane CS, Chate GP, Wavhale RD, Pratinidhi SA, et al. Nanoparticle properties modulate their effect on the human blood functions. BioNanoScience. 2021;**11**(3):816-824

[64] Gurunathan S, Arsalan Iqbal M, Qasim M, Park CH, Yoo H, Hwang JH, et al. Evaluation of graphene oxide induced cellular toxicity and transcriptome analysis in human embryonic kidney cells. Nanomaterials. 2019;**9**(7):969

[65] Taşdemir S, Morçimen ZG, Doğan AA, Görgün C, Şendemir A. Surface area of graphene governs its neurotoxicity. ACS Biomaterials Science & Engineering. 2023;**9**(6):3297-3305

[66] Tang Z, Zhao L, Yang Z, Liu Z, Gu J, Bai B, et al. Mechanisms of oxidative stress, apoptosis, and autophagy involved in graphene oxide nanomaterial anti-osteosarcoma effect. International Journal of Nanomedicine. 2018;**13**:2907-2919

[67] Seabra AB, Paula AJ, de Lima R, Alves OL, Durán N. Nanotoxicity of graphene and graphene oxide. Chemical Research in Toxicology. 2014;**27**(2):159-168

[68] Mendes RG, Koch B, Bachmatiuk A, Ma X, Sanchez S, Damm C, et al. A size dependent evaluation of the cytotoxicity and uptake of nanographene oxide. Journal of Materials Chemistry B. 2015;**3**(12):2522-2529

[69] Domenech J, Rodríguez-Garraus A, López de Cerain A, Azqueta A, Catalán J. Genotoxicity of graphene-based materials. Nanomaterials. 2022;**12**(11):1795

[70] Gurzęda B, Buchwald T, Nocuń M, Bąkowicz A, Krawczyk P. Graphene material preparation through thermal treatment of graphite oxide electrochemically synthesized in aqueous sulfuric acid. RSC Advances. 2017;**7**(32):19904-19911

[71] Liao Y, Wang W, Li Z, Wang Y, Zhang L, Huang X, et al. Comparative proteomic analysis reveals cytotoxicity induced by graphene oxide exposure in A549 cells. Journal of Applied Toxicology. 2021;**41**(7):1103-1114

[72] Gurunathan S, Kang M-H, Jeyaraj M, Kim J-H. Differential immunomodulatory effect of graphene oxide and vanillin-functionalized graphene oxide nanoparticles in human acute monocytic leukemia cell line (THP-1). International Journal of Molecular Sciences. 2019;**20**(2):247

[73] Zamorina SA, Shardina KY, Timganova VP, Bochkova MS, Nechaev AI, Khramtsov PV, et al. Effect of graphene oxide nanoparticles on differentiation of myeloid suppressor cells. Bulletin of

Experimental Biology and Medicine. 2020;**170**(1):84-87

[74] Ribeiro BFM, Souza MM, Fernandes DS, do Carmo DR, Machado-Santelli GM. Graphene oxide-based nanomaterial interaction with human breast cancer cells. Journal of Biomedical Materials Research Part A. 2020;**108**(4):863-870

[75] Taheriazam A, Abad GGY, Hajimazdarany S, Imani MH, Ziaolhagh S, Zandieh MA, et al. Graphene oxide nanoarchitectures in cancer biology: Nano-modulators of autophagy and apoptosis. Journal of Controlled Release. 2023;**354**:503-522

[76] Bałaban J, Wierzbicki M, Zielińska-Górska M, Sosnowska M, Daniluk K, Jaworski S, et al. Graphene oxide decreases pro-inflammatory proteins production in skeletal muscle cells exposed to SARS-CoV-2 spike protein. Nanotechnology, Science and Applications. 2023;**16**:1-18

[77] Gao Y, Xu A, Shen Q, Xie Y, Liu S, Wang X. Graphene oxide aggravated dextran sulfate sodium-induced colitis through intestinal epithelial cells autophagy dysfunction. The Journal of Toxicological Sciences. 2021;**46**(1):43-55

[78] Liu P, Shao H, Kong Y, Wang D. Effect of graphene oxide exposure on intestinal Wnt signaling in nematode *Caenorhabditis elegans*. Journal of Environmental Sciences. 2020;**88**:200-208

[79] Lalwani G, D'Agati M, Khan AM, Sitharaman B. Toxicology of graphene-based nanomaterials. Advanced Drug Delivery Reviews. 2016;**105**(Pt B):109-144

[80] Akhavan O, Ghaderi E, Hashemi E, Akbari E. Dose-dependent effects of nanoscale graphene oxide on reproduction capability of mammals. Carbon. 2015;**95**:309-317

[81] Lin L, Zhuang X, Huang R, Song S, Wang Z, Wang S, et al. Size-dependent effects of suspended graphene oxide nanoparticles on the cellular fate of mouse neural stem cells. International Journal of Nanomedicine. 2020;**15**:1421-1435

[82] Siqueira PR, Souza JP, Estevão BM, Altei WF, Carmo TLL, Santos FA, et al. Concentration-and time-dependence toxicity of graphene oxide (GO) and reduced graphene oxide (rGO) nanosheets upon zebrafish liver cell line. Aquatic Toxicology. 2022;**248**:106199

[83] Sayadi MH, Fahoul N, Kharkan J, Khairieh M. Investigating the protective effects of Elaeagnus angustifolia fruit extract on hematological parameters and damage of different tissues of male mice exposed to graphene oxide nanoparticles. Nano Select. 2023;**4**(9-10):559-583

[84] Mendonça MCP, Soares ES, de Jesus MB, Ceragioli HJ, Ferreira MS, Catharino RR, et al. Reduced graphene oxide induces transient blood–brain barrier opening: An in vivo study. Journal of Nanobiotechnology. 2015;**13**(1):1-13

[85] Mendonça MCP, Soares ES, de Jesus MB, Ceragioli HJ, Irazusta SP, Batista ÂG, et al. Reduced graphene oxide: Nanotoxicological profile in rats. Journal of Nanobiotechnology. 2016;**14**(1):1-13

[86] Kurantowicz N, Strojny B, Sawosz E, Jaworski S, Kutwin M, Grodzik M, et al. Biodistribution of a high dose of diamond, graphite, and graphene oxide nanoparticles after multiple intraperitoneal injections in rats. Nanoscale Research Letters. 2015;**10**(1):398